SH276.T48 2009

Braving Troubled Waters

AF749099

CHRIST'S COLLEGE CAMBRIDGE
WITHDRAWN
FROM LIBRARY

MARE PUBLICATION SERIES

MARE is an interdisciplinary social-science institute studying the use and management of marine resources. It was established in 2000 by the University of Amsterdam and Wageningen University in the Netherlands.

MARE's mandate is to generate innovative, policy-relevant research on marine and coastal issues that is applicable to both North and South. Its programme is guided by four core themes: fisheries governance, maritime work worlds, integrated coastal zone management (ICZM), and maritime risk.

In addition to the publication series, MARE organises conferences and workshops and publishes a social-science journal called *Maritime Studies (MAST)*. Visit the MARE website at http://www.marecentre.nl.

Series Editors

Svein Jentoft, University of Tromsø, Norway
Maarten Bavinck, University of Amsterdam, the Netherlands

Previously Published

Leontine E. Visser (ed.), *Challenging Coasts. Transdisciplinary Excursions into Integrated Coastal Zone Development,* 2004 (ISBN 90 5356 682 1)

Jeremy Boissevain and Tom Selwyn (eds.), *Contesting the Foreshore. Tourism, Society, and Politics on the Coast,* 2004 (ISBN 90 5356 694 5)

Jan Kooiman, Maarten Bavinck, Svein Jentoft, Roger Pullin, (eds.), *Fish for Life. Interactive Governance for Fisheries,* 2005 (ISBN 90 5356 686 4)

SH276.T48 2009

Braving Troubled Waters

Sea Change in a Dutch Fishing Community

Rob van Ginkel

MARE Publication Series No. 4

Centre for Maritime Research

Amsterdam University Press

Cover illustration: Rob van Ginkel
Cover design: Neon, design and communications, Sabine Mannel, Amsterdam
Lay-out: JAPES, Amsterdam

ISBN 9789089640871
e-ISBN 9789048508136
NUR 741

585034512N
T22300543351

© Amsterdam University Press, Amsterdam 2009

All rights reserved. Without limiting the rights under copyright reserved above, no part of this book may be reproduced, stored in or introduced into a retrieval system, or transmitted, in any form or by any means (electronic, mechanical, photocopying, recording or otherwise) without the written permission of both the copyright owner and the author of the book.

Contents

Preface

'How did you become so interested in doing research among fisherfolk?' This has been a recurring question ever since I began developing an anthropological interest in the occupational world of commercial fishing in the early 1980s. The answer is 'by sheer coincidence'. Before enrolling as an anthropology student at the University of Amsterdam in 1982, I knew little about 'fishing cultures'. For one of the courses I took, I happened to read a fascinating study on folk religion in a North Yorkshire fishing village (Clark 1982). It captured my imagination and I started to read more ethnographic literature on fishing and fishing villages. That got me hooked, to use an appropriate metaphor. I decided that I would eventually conduct research in a fishing community somewhere in Europe. Through a series of contingencies, I ended up doing so in 1986, close to home in the Netherlands. An ongoing conflict in an occupational community of shellfish planters attracted my attention and drew me to the village of Yerseke to conduct fieldwork. While working on my MA thesis, it struck me that maritime anthropologists did not have a journal of their own, which meant that their publications were widely scattered in scholarly journals. Perhaps naïvely, I thought that this void should be filled. With Jojada Verrips, I founded and edited the journal *Maritime Anthropological Studies* (MAST), the first issue of which appeared in 1988. By then, I was firmly committed to the field of fishing cultures. After graduation, I conducted research for my PhD thesis in the fishing villages of the Dutch island of Texel from late 1989 until early 1991, obtaining my degree with a dissertation that was published in 1993. It focuses on the local fishermen's long-term adaptive strategies in view of ongoing debates concerning common pool resource use. Although I subsequently ended up doing mostly non-fisheries-related scholarly work, I also continued to occupy myself with maritime studies and taught undergraduate courses in maritime anthropology. When the occasion arose, I returned to Texel in August 2005 for a second stint of prolonged fieldwork in the local community of fishermen with the aim of examining their occupational culture and practice or what the French dub 'métier'. This notion refers to much more than just a job or an occupation. It conjures up an image of an activity at which one excels, a vocation, an encompassing and existential way of life and making a living.

This ethnography details Texel fisherfolk's engagements with erratic marine living resources, capricious markets and the vicissitudes of political interventions in the fishing industry from the early 18th century until the present day, with an emphasis on post-war developments. The book is empirically grounded, historically specific and theoretically informed. It attempts to situate the occupational community at the interface of local and (supra-)national processes and shows how the latter affect the socio-cultural fabric of the island's fishing villages and prompt particular responses in the fishermen's perceptions and modes of action. Thus, although this is a community study, I will occasionally wander off to events occurring at other levels of integration that impinged upon the local fishing industry. Nevertheless, the book's proponents are Texel's owner-operators, deckhands and others involved in the island's fishing arena past and present. Even though women often play important roles in family firms, fishing per se is a male world. I do devote attention to fishermen's wives, but I am aware that there is a gender bias in this book. In my defence, I can only say that women are conspicuously absent from the official meetings in which fishing and fishing politics are discussed and that they tend to strongly underplay their own contributions to the firms and to the running of households. Despite this lacuna, I am confident that the present book illuminates the building blocks of fishing as an occupation.

Many people have supported my work and helped me in various ways. To the fisherfolk of Texel, I owe much gratitude. They generously shared their knowledge with me and made me feel welcome in their midst. I felt at ease, which made it easy to empathize with them. Many of them supplied me with published and unpublished documents, photographs, videotapes and other materials. Special thanks are due to the skipper-owners who were kind enough to take me aboard their beam trawlers for four fishing trips. These occasions provided invaluable information on work, fishing tactics and relations and the atmosphere on board ship. I was permitted unrestricted access to the archives of the local fishermen's association, the Fishery Cooperative, the local Producer Organization and the Texel co-management group. This is indicative of their leadership's open-mindedness. Additional support came from a number of local institutions, including the municipal archive in Den Burg and the Maritime and Beachcombers Museum in Oudeschild, which gave freely of their information, facilities and coffee. Librarians, archivists and museum staff elsewhere were also helpful. The Netherlands Organization for Scientific Research (NWO) enabled the conducting of research on Texel (project numbers 500-276-202 and 400-04-702), and the Amsterdam School for Social Science Research co-sponsored my fieldwork. NWO also provided a publication grant. I gratefully acknowledge their generous financial support. Two anonymous peer reviewers, and the series editors Maarten Bavinck and Svein Jentoft, made it very clear that I needed to reduce the length of the original manuscript considerably to dispose of excess detail. Although this involved the loss of many a darling paragraph, I must admit that they were right. I am also indebted to Harriet Impey, who cheerfully fashioned my English.

Zeewijk, our residence during my second stint of research, provided a stimulating environment for writing and simultaneously conducting fieldwork. My partner Margreeth and I, and our daughters Emma and Mette, thoroughly enjoyed living on the island and making new friends. We truly felt at home. Owing to a stroke of good luck, we were able to extend our stay on Texel for another year, and, subsequently, we decided to 'go native' completely. This book is a tribute to the island, its inhabitants and particularly its fisherfolk, who for generations have been braving troubled waters.

Introduction

From the late 18th century until fairly recently, Dutch fisherfolk were culture heroes in the national self-image. Braving treacherous seas with frail boats to eke out a frugal living for themselves and their families, fishermen were romantically portrayed in the visual arts and in works of literature and scholarship as embodying 'inner civilization' and national virtues.[1] They were believed to be the epitome of authenticity, uncorrupted by modernity, living austere and pious lives in close-knit communities, preserving in customs and costumes what had disappeared elsewhere in the Netherlands, and maintaining norms and values that simultaneously mirrored and provided a model for the country's 'national character'. In the iconography of how the Dutch presented themselves to the outside world, the archetypical fisherman and fisherwoman – usually depicted in traditional local dress – figured prominently. Highly distinctive and heterogeneous local cultures were thus conflated with nationhood and national identity. The emblematic coastal Arcadia of folkloristic exoticism was given pride of place for purposes of nostalgic cultural nationalism, while in reality fishing communities were rather peripheral – one might even say outlandish – in a highly urbanized country. Through a whim of history, fisherfolk were therefore in the limelight of positive cultural attention. Over the past few decades, however, the mythical image of the pastoral good life has changed rapidly and radically. Fishermen are currently stigmatized as unscrupulous marauders of the sea, knowingly and systematically depleting its resources with a highly industrialized fishing fleet. Their alleged inimical greed is believed to be unequalled, and, according to fisheries critics, had it not been for draconian measures and their strict enforcement, the last fish would already have been captured. The more militant environmental activists even feel that the fishermen's pernicious practices ought to be banned altogether. The shift in imagery speaks volumes about the changed position of fisherfolk in Dutch society.

To be a commercial fisherman in the Netherlands today means to be under the continual scrutiny of national and supranational policymakers, regulators, law enforcers, environmentalists, the media and public opinion. According to fisherfolk, the occupation has lost much of the freedom that arguably characterized it until the mid-1970s. In their own opinion, inshore and offshore fishermen have for centuries been used to deciding how to go about their business: what kinds of boats and equipment to use, when and where to cast their nets, which species of fish, shellfish and crustaceans to pursue, what quantities to catch, where and when to land catches and so on. Currently, among many other things, they have to abide by strictly enforced landing limitations and a myriad of other rules and regulations, make a careful planning of fishing time and keep an extensive logbook administration, whilst being constantly supervised by a satellite vessel-monitoring system. Although the perceived watershed between a time of unmitigated liberty and a time of stifling restrictions and red tape is to some extent a myth, there can be no doubt that the incorporation into the European Union's Common Fisheries Policy has impacted upon the world of Dutch offshore fishing in several major ways. It has thoroughly altered the occupational praxis and culture of Dutch fishermen, and the same goes for fishermen elsewhere in Europe. Along with ecological, enforcement and compliance problems, the external perception of the fishing industry and the public esteem for fisherfolk has changed radically. Yet surprisingly few social science monographs have been devoted to the question of how fishermen have perceived, responded and adjusted to increasingly tight management regimes and changing images of and opinions about the fishing industry, and the ways in which these have transformed the everyday lives and livelihoods of fishermen and their impact upon fisher families and fishing communities. The present book therefore aims to fill a void in exploring, understanding and analyzing these issues in the fishing villages of Texel, a Dutch island facing the Wadden Sea and the North Sea. However, I will not restrict myself to Texel alone, but will on occasion cast my nets wider, and deal with the fisheries and fishermen of the Netherlands as a whole. In addition, I have used a long-term perspective to show that, among many other things, external intervention in the fishing industry occurred early on.

The Dutch fishing industry is relatively small in terms of number of vessels and employees. In 2007, the fishing fleet comprised 345 cutters, 14 large pelagic freezer trawlers and 83 shellfish fishing boats (Taal et al. 2008). Total employment in the fishing and shellfish-farming industry is approximately 2,100 jobs, excluding related sectors such as the processing industry, the fish auctions, the supply sector and the retail trade. Employment in these fisheries-dependent branches amounts to another 15,500 full-time jobs, about half of which are in fish processing and wholesale companies and a third in the retail industry (Task Force 2006; Smit and Taal 2007). The number of jobs in the fishing industry only constitutes a tiny fraction of total employment in the Netherlands. Although these figures are rather modest, the Dutch fish trade occupies an important position in Europe. The value of aggregate fish and shellfish landings in the Netherlands amounted to €476 million in 2007; that of exports of fish and fish products exceeds two billion euros. The most important sector of the fishing industry is the capital-intensive beam trawl fleet, which operates mostly in the southern and south-eastern North Sea to catch sole and plaice and associated flatfish stocks on four- to five-day trips, usually starting at midnight on Sunday or early Monday morning. Together, these species contribute approximately eighty per cent of the total revenues of beam trawling, a fishing technique that is applied by about half of the cutter fleet. The Netherlands holds a significant share of the European Union's total allowable catch for sole (seventy-five per cent) and plaice (thirty-eight per cent). The bigger boats – that is, those with an engine power exceeding 1,500 hp, with a length of forty to forty-five metres, a width of eight to eight-and-a-half metres and a crew of six or more – land over ninety per cent of the total market supply of flatfish (Rijnsdorp, Daan and Dekker 2006:557). The majority of beam trawlers are family-owned and operated. Most flatfish-fishing firms own one vessel; thirty-four own two or more beamers. Important concentrations of beam-trawl fishermen can be found in the towns of Urk, Goedereede, Arnemuiden, Vlissingen, Den Helder and on the island of Texel.

Currently (July 2008), Texel's fishing fleet boasts eleven offshore cutters and sixteen inshore cutters (including several vessels that are on the Fishery Register but not used in commercial fishing). In addition, two boats are foreign-registered but Texel-owned and operated. Although the Texel-registered fleet is rather small, in terms of landing rights it still occupies an important position nationally. The turnover of the Texel fishing industry amounted to €25 million in 2006. About 125 fishermen currently crew the local fishing fleet, while in addition there are a dozen or so co-owners who have terminated their active fishing careers. The local Fishery Co-op employs another dozen people, most firms hire at least one man to mend the nets, and indirectly the fishing industry provides work for many more islanders in ship's maintenance and repair, provision, administration, book-keeping and so on. For two villages in particular, the fisheries are highly important: Oudeschild – where the harbour is situated – and Oosterend – home to most of the offshore family firms. Generally, the islanders take pride in 'their' fishing industry, which used to be much larger than it is at present. Selecting Texel as a research site was linked with my desire to study a community with a long fishing tradition. Moreover, Texel had once had an oyster industry that combined fishing with semi-cultivation. As I had already conducted research elsewhere into the transformation from open-access oyster fishing to oyster farming, opting for Texel facilitated a comparison.

During my first stint of fieldwork (from late 1989 to early 1991), the Dutch offshore fishing industry was in great turmoil. Fishermen often made headline news for overshooting individual quotas and conflicting with law enforcement officers. Fishing opportunities were frequently closed prematurely as the national share of total allowable catches – allocated by the European Community – was exhausted. I collected a great deal of data about contemporary events. However, for my PhD thesis, I failed to cover sufficient ground to include the recent history of the Texel fishing industry and communities. My dissertation pertained to the 1813-1932 era. I was nagged by the thought that I had not really completed the project I had initially envisioned, but I hoped and expected that I would be able to deal with the post-1932 period at some stage in the future. A co-management regime was introduced when I had just become involved in research and teaching unrelated to maritime studies. I attempted to keep informed about the vicissitudes of the occupational world of fishing and I learned of the alleged successes of the new management system. In the social science fisheries literature, co-management had meanwhile turned into a buzzword signifying something of a panacea, which allegedly provided a solution to the failures of top-down modes of managing fisheries. The impression I gained was that in the Netherlands cooperative governance had been a tremendous success. No longer did fishermen make headline news for flouting the rules and regulations and overshooting their individual quotas. Compliance was almost complete, early closures of fishing were history and fishermen still earned good incomes. The Dutch and the European authorities acknowledged their satisfaction with the new system and the fishermen's conduct. Fishermen appreciated the increased stability, flexibility and certainty that group management of individual transferable quotas facilitated. Thus, all seemed quiet on the fisheries front. However, impressions can be deceptive indeed.

When the opportunity arose to continue where I had left off, the profound shift in the fisheries management regime necessitated the updating of my data. Upon my return to Texel in August 2005 to conduct a second round of research, it was immediately evident that the local fishing fleet was considerably smaller than during my first stint of fieldwork. Fifteen Texel registration numbers were no longer on the Fishery Register, while only four new ones had been added. I found two of these newly registered boats – company-owned cockle cutters operated by a father and his two sons from Texel – tied up in the harbour and floating idly on their mooring ropes. Six shrimp-fishing boats, three mid-size stern trawlers and two other vessels that had specialized in herring or round-fish fishing were gone, and a shellfish fishing boat was no longer being used. There were fifteen big-beam trawl cutters compared to eighteen previously. Three more were decommissioned within a few months of my arrival. None of the beamers was equipped to combine flatfish fishing with herring fishing, as had been the case with many boats earlier on. A new type of multipurpose vessel – a smaller version of the beamer adapted to beam trawling and twin rigging in the inshore zone – had made its appearance in the fleet. While several family firms had given up fishing, only one new family firm had meanwhile been founded or, rather, re-established. Generally, the vessels were much older, the composition of the fleet had changed and the engine power of the big beamers had diminished significantly. Apparently, the skipper-owners had been faring less well than I had assumed, and the same applied to the co-venturing crewmembers, whose number had decreased by about seventy-five. Soon I learned that owner-operators faced difficulties in breaking even and that recruitment of deckhands was a serious problem: many were abandoning ship for non-fisheries-related occupations. The situation I encountered only boosted my curiosity as to what had been happening in the local fishing industry since my departure in the spring of 1991 and how the fishermen and fisher families had been experiencing and responding to the recent developments.

The issues that are pivotal to this historical ethnography pertain to fisherfolk's individual and collective adaptations to changing ecological, technological, economic, social and political conditions and legal and regulatory frameworks, their modes of production, and their socio-political organization and worldviews. Briefly, the question regards how fishermen perceive and understand their natural and social environments, how they relate and adapt to them and how they attempt to control them. I will not portray fishermen as one-dimensional human beings who are constantly motivated by, for example, either greed and selfishness or altruism and generosity. Instead, I will attempt to show how not only increasing state interventions and growing social differentiation have influenced their motivations and actions, but also ecological or resource fluctuations and economic cycles of boom and bust in the fishing industry. Throughout the book, I will devote attention to a basic ambivalence of fishermen vis-à-vis each other, dealers and processors, and towards the environment they exploit and the state interventions in their industry. The relationships of power and dependency in the fishing industry have changed with every transformation of the configuration of entry and use rights. However, this has not rendered the fishermen powerless victims of decisions made at higher levels of integration. Every mode of access to resources and every method of allocation creates winners and losers, and those with stakes will seek to defend the status quo while those who expect to suffer may revolt or attempt to dodge the new rules and beat the system. Hence, shifting management regimes often lead to adversarial relations and tensions. As Gary Libecap contends:

> Negotiations in the political arena among competing private interests, politicians, and bureaucrats determine how and when the society will respond to common pool pressures by assigning or adjusting property rights. An examination of the political contracting underlying ownership institutions is necessary to understand how property rights are established and modified and why such a diversity of arrangements exists (1989:28).

What do such regime shifts do to fishermen and fisher families; how do they respond to them; what do they mean to them? Who stands to gain and who stands to lose from specific management measures? This book is about the fisherfolk's understanding of the world as they confront it and as it confronts them. By focusing on *their* actions, *their* routines, *their* stories and *their* views, I will put the fishermen and the fisher families – whose opinions are rarely heard and reflected in policy reports or are muffled in the cacophony of dominant voices – centre stage.

Structures and Contingencies, Strategies and Constraints

I conducted my research with an actor-centred approach, focusing on the strictures fishermen face because of the policy and management regimes in which they are encapsulated, and the ways in which they accept, conform to and comply with or, alternatively, reject, circumvent and counteract such structural constraints or the institutional 'rules of the game'. Choices and actions are moulded in a socio-cultural context, but with certain restrictions fishermen have various options available from which they can choose. Choices are not made by rational, self-interested motives alone; in addition, social demands, cultural and moral conventions and the enactment of routine behaviour play an important role (Jentoft, McCay and Wilson 1998:426). For these reasons, fishermen are not passive recipients of policies or ciphers that can be instilled with rules, standards and norms, bringing about a switch of conduct overnight. Nor do fishermen merely respond to environmental, economic and social change. They also actively act upon their natural and social worlds to create new opportunities and new meanings. The present ethnography is about the fishermen's lived experience of being embedded in the wider society and relating to forces from without. The focus will be on 'how actors "enact," "resist," or "negotiate" the world as given, and in so doing "make" the world' (Ortner 1996:1). Such a perspective demands listening to and observing 'real people doing real things in a given historical moment, past or present, and trying to figure out how what they are doing or have done will or will not reconfigure the world they live in' (ibid.:2). Yet these actors are not autonomous 'agents'. They are embedded structurally in larger social configurations that act upon them as much as they act upon encompassing systems, or at least attempt to do so. Consequently, there is human agency, and we should study 'the impact of the system on practice, and the impact of practice on the system' (Ortner 1984:148).

Arguably, human conduct, including intentional behavioural strategies, can have profound unanticipated, unintended and undesirable consequences. For generations, social scientists have been stressing this point. Indeed, 'the combination of intentional and unintentional actions of different social actors may culminate in significant shifts in environments and ecological dynamics' (Scoones 1999:493). Nevertheless, the question of how 'to relate the unintended consequences of conscious decisions based on the specific ends of competing management units to the patterned outcome and some goals posited for a whole system remains an ill-defined but crucial problem in ecological and economic anthropology' (Rutz 1977:157). It is therefore pertinent to contextualize marine resource exploitation. However, as Bonnie McCay observes, 'It is widely appreciated that context is important to the choices and behavior of people, but the theoretical and empirical underpinning for that observation is woefully lacking' (2002:392). Consequently, the chief objective of this section is to arrive at a sensitizing framework that incorporates a contextual dimension and takes into account complexity, diversity and dynamics. Complexity refers to phenomena that exhibit nonlinear behaviour: that is, positive feedback in which endogenous or exogenous changes to a socio-cultural entity produce amplifying effects (Elliott and Keel 1997:66). Diversity relates to the variety of technological, economic, social and cultural coping responses humans exhibit in exploiting natural resources. Dynamics pertains to the trial-and-error character of adaptation in which human–nature interactions give rise to emergent phenomena that may prove to be resilient in the face of environmental perturbations (T.A. Smith 1997:55), lead to negative or positive externalities or bring about non-equilibrium change (Scoones 1999:482).

We must not incorporate context merely as a background to research data, but focus much more rigorously on it. Context does not necessarily refer to an 'entity' such as an ecosystem or a culture, society or community, which are often defined as relatively autonomous, clearly bounded, stable and balanced for analytical purposes. This insular view obstructs an understanding of the myriad of forces working upon such analytically demarcated but in fact permeable wholes. They are embedded in the surrounding world and a host of remote factors affects them. These include macroeconomic variables such as global fuel prices and commodities costs; interest, exchange and inflation rates; technological developments; food safety, food preferences and availability of alternative foodstuffs. The socio-economic context in which common pool resource users or petty commodity producers operate influences the modes of adaptation available to them. Externally induced changes may be so swift as to allow them insufficient time and opportunity to adjust their socio-economic structures to avoid suboptimal outcomes (Ostrom 1990:21). Due attention to context in the elucidation of actions and consequences may mean dealing with loose, transient and contingent interactions and disarticulating processes from within and from outside predetermined units. The same goes for 'the movements of people, resources, and ideas across whatever boundaries that ecosystems, societies, and cultures are thought to have' (Vayda 1986:310; see also Agrawal 2003:250-254). In a globalizing world, ecological, economic, social and political interdependencies just reinforce the impact of external forces on socio-cultural entities defined as geographically bounded wholes. Consequently, fishing cannot merely be understood in terms of autonomous fishermen, crews, fleets or communities, as I will attempt to make clear. It is thus necessary to show how remote and local interferential factors influence fisheries and fisheries management, and, additionally, how they cause or reinforce resource management problems. Since this procedure involves focusing on the internal–external interface, it is also a restatement of the perennial social science problem of how to relate micro and macro scales. Zooming in on contextual factors may provide precisely the locus where the micro–macro interface can be discerned and studied most advantageously. Such a context-dependent position means 'an open-ended, contingent relation between contexts and interpretations' (Flyvbjerg 2003:43).

Focusing on a particular level of analysis might have important implications. Moran writes that if 'we focus on the impact of state or national forces on local communities, we may find that these wider forces shape the life of local communities in relatively similar ways. However, if we focus on the community, we see "individuals responding actively to actually subvert or alter these external forces, not passively accepting them"' (1990:283). Ideally, we should look at the problem from both angles. However, it is more feasible to use a mode of analysis Andrew Vayda dubs 'progressive contextualization'. It involves a procedure that focuses 'on significant human activities or people–environment interactions by placing them within progressively wider or denser contexts' (1983:265). The researcher can depart from studying specific activities, performed by specific people in specific locales at specific times, and then trace the causes and effects of these activities outwards including the factors impinging on them without *a priori* defining the boundaries of a system. I will therefore be switching perspectives from the individual to the local and from the local to the national and supranational levels, occasionally zooming in on micro events or panning to macro structures and processes. For analytical purposes, it is appropriate to distinguish several levels of exogenous contextual factors, each of which has their own particular impact, although they can also mutually reinforce each other through knock-on or multiplier effects. It is here that micro and macro forces intersect and interact. Researchers may better comprehend and appreciate the often complex origin of the selection of different strategies by focusing on 'the choice sets available to individual users of the resource, the different decision-making arrangements possible and different action strategies; and tracing back the derivation of these choice sets to contextual factors' (Edwards and Steins 1998:367). This requires using a diachronic perspective and retrospective analysis to discern the variety of coping mechanisms to certain types of change in remote variables and to map short-term and long-term processes including feedback responses. Again, responses to external forces and structural pressures are not just a state of mind, but turn fishermen into active agents of change themselves (Butler 2005:253).

Rejecting the assumption of ecological and socio-cultural homogeneity and stability, the present approach focuses on variation and dynamics and looks at how different individual actors and social formations operate in and adjust to their *total* environments through a variety of behaviours, technologies, organizations, structures, worldviews and beliefs (Poggie 1992:51). A useful concept in this connection is adaptation. People adapt to the natural world that surrounds them and of which they themselves are part. The nonhuman environment evolves partly on its own and partly in response to what people do to it. Adaptive strategies and processes result from cybernetics or positive and negative feedback loops. Adaptive strategies involve conscious decision-making. Adaptive processes are feedback loops operating outside of cognitive awareness. Adaptive dynamics are the total of coping strategies *and* processes (J.W. Bennett 1976). Individual and collective adaptive strategies sometimes crosscut each other, giving rise to tensions that may develop into conflicts. Deliberate human adaptations depart to a greater or lesser extent from people's particular views of the world and their place in it. Nevertheless, the effects of people's behaviour upon the natural environment and the constraints that the physical world imposes upon the realization of their goals and aspirations may not be part of the notions that are basic to their actions (Vayda 1986:297). Generally, the properties of people's relationships with nature and with each other 'derive neither from their will nor their consciousness' (Godelier 1986:6). By its very nature, resource utilization is dynamic and adaptations can be either functional or dysfunctional. In the short term, socio-natural regimes may seem to be adaptive, but in the long term, they may turn out to be maladaptive. The ways in which human beings act upon the surrounding world transforms the natural and social environments. In turn, the change influences their social organization, interactions, behaviour and thinking (Wolf 1982:73 74). Nature is thus 'elaborately entangled and fundamentally bound up with social practices and their characteristic modes of cultural representations' (McNaghten and Urry 1998:30).

On the face of it, fishermen across the globe must adjust to similar environments and face corresponding problems, including the vicissitudes in exploiting common pool resources and the economic uncertainties and physical dangers inherent in fishing (Acheson 1981; McGoodwin 1990). Several maritime anthropologists argue that the exploitation of marine ecosystems requires specific economic, social, cultural and psychological adaptations (see, for instance, Andersen and Wadel 1972; Smith 1977; Andersen 1979; Knipe 1984; McGoodwin 1990; Vestergaard 1996). Owing to this fact, geographically disparate fishing communities would seem to share a number of socio-cultural patterns and characteristics. Indeed, in strikingly different settings, one may encounter in fishing communities distinctive identities based on occupational practices, a sharply gendered division of labour, strong kin and family involvement in work, a share system of remuneration, remarkably similar ideas concerning work ethos, an egalitarian ideology, a rhetoric and concepts of independence, individualism, freedom and so on. In early maritime anthropological publications, such feats were often viewed as being ingrained in fisherfolk's adaptations. Some of these studies fit rather well within the cultural ecology tradition in anthropology, including its functionalist or teleological tenets (McCay 1978). What they usually show is that certain modes of behaviour and social organization and specific worldviews prove to be 'ecologically adaptive'. In this sense, the environment is seen to generate particular relationships of work that in turn shape social structure and mark culture (McCay 2001:257).

Other anthropologists reject such ecological-functionalist explanations and point out that despite similarities, there are vast differences among fishermen with different social and cultural backgrounds. Among many other things, this is due to the distinct ways in which they are embedded in and relate to encompassing socio-economic systems and cultures (see, for example, Löfgren 1989; Pálsson 1991; McCay 2001; van Ginkel 2001). Dependence upon marine living resources does indeed influence the fisherman's belief and behaviour, but how it is 'structured into cultural forms depends upon the economic and social structure within which the fisherman is operating' (Löfgren 1977:235). Different modes of production entail different social relationships, rationales and motivations. Factors such as occupational and fleet structure, ownership patterns, degree of indebtedness, boat size, number of crewmembers, division of labour, system of remuneration, degree of specialization, seasonal variation in fish species and species pursued, technology and gear bring along differences in social practices, mental maps, cultural rules, styles, goals and aspirations. Similarly, there is diversity with respect to preferred modes of regulation, institutional arrangements, organizations, management regimes and enforcement. Even spatially and socio-economically proximate fishing communities and the specialist subdivisions within them can exhibit considerable social and cultural differences. It is therefore imperative to devote attention to the social relations of production: that is, differential modes of access to resources, their appropriation, allocation and redistribution, control over the means of production, and the division and organization of labour. Such an approach places actors at the core and avoids the pitfalls of 'systems ecology', which sees man's dealings with nature as an undifferentiated, homeostatic and balanced system where equilibrium is automatically restored. This view should be dismissed as teleology (McCay 1978; Scoones 1999). Human agency is *in* nature and people and environment are mutually constitutive components of the same world (Pálsson 1996). In short, nature acts upon people, and people act upon nature through interactions and social relationships and the mental universe produced, reproduced and transformed in these relations – including their comprehensions and images of nature (Rappaport 1979; J.W. Bennett 1990; Ingold 1992).

This brings us to the intersection of social systems and the fragmentary, heterogeneous, contradictory and ambivalent realm of knowledge, ideas, interpretations, goals, ambitions, values, norms, representations, meanings, beliefs and so on. Human action is always negotiated culturally and effectuated in social relationships (Keesing 1981:169). This is important to understand the shaping and constraining forces of ecological adaptations. In all relations of people with nature and with each other, this mental universe consciously or unconsciously fulfils interpretive, ordering and legitimizing functions (Godelier 1986:11,131ff.). Thus, 'processes of adaptation imply from the outset the development of representations and interpretations of nature shared by the members of a particular society, and the organization of various forms of individual and collective interventions in nature which depend upon these representations and interpretations' (ibid.:6). Actors confront nature through social interactions and social relations and the systems of knowledge, symbols, meanings and values formed and embedded in them, including conflicts, contradictions and ambivalences (J.W. Bennett 1976:40). As Benjamin Orlove states, 'culture and ideology are not ... epiphenomena but ... proximate causes which shape human action. They influence the options among which individuals select and in turn are influenced by the cumulative consequences of such choices' (Orlove 1980:257). Therefore, if worldviews change, human action changes, in turn affecting the environment. These dynamics are infinite; people must continually adapt anew to their natural and social environments (Vayda 1986:297; McEvoy 1988:229). This approach makes it possible to contextually situate the attitudes, actions and conceptions of fishermen, devoting attention to structures and strategies as well as conjunctures and contingencies.

Fishing is an evolving socio-natural regime that must be understood in terms of its wider ecological, economic, social, political and cultural context, 'the actions of other segments of the population, near or distant, that affect any aspect of the fishing industry, fishermen, the waters they fish, or the fish in them' (Durrenberger 1988:196). Like other modes of natural resource exploitation, fishing is a complex, heterogeneous and dynamic system of interaction between humans and the natural environment (Hamilton et al. 1998; Kooiman, van Vliet and Jentoft 1999; van Ginkel 1999a, 1999b; Low et al. 1999). Embedded as they are in encompassing political-economic and cultural structures and processes, many forces act upon or interfere with fisheries and fishing industries – at the same time affecting fisherfolk and fishing communities and the resources they utilize. Consequently, it would be misleading to regard fishermen as autonomous actors in marine resource exploitation (Pálsson 1991). A 'natural model' of fishing that focuses mainly on material contexts, technical activities and ecological relations is inadequate as it fails 'to appreciate the ways in which production systems are differentiated with respect to their social relations' (Pálsson 1989:13). Ample attention must be devoted to the wider ecological and social configurations in which fishermen operate, the unsought and unanticipated consequences of their and other people's conduct, the anticipated as well as the unintended and unforeseen outcomes of fisheries policy and management, and feedback processes that give rise to new coping responses. Such adaptations refer to the modes of adjustment of fishermen to natural and social milieus, or of the natural and social milieus to their lives, needs, wants and goals (J.W. Bennett 1976:246). The intertwined processes of remote influences on resources and resource exploitation – processes that are usually beyond the control of individual actors and communities (Edwards and Steins 1999) – and human adaptations often lead to transformations in socio-natural regimes. These transitions are structural societal changes that occur because of interlinked ecological, technological, demographic, economic, social, political and cultural developments that often mutually reinforce one another. It is therefore pertinent to pay ample attention to macro socio-political forces that operate beyond local communities of resource users but which can and do affect them. Hence, social science research agendas should include interferential factors and feedback loops that affect fisheries, fisheries management, fishing communities and fishermen. I will depart from the idea that there is no straightforward relationship between people and environment in processes of environmental change:

> Environments are dynamically and recursively created in a nonlinear, nondeterministic, and contingent fashion. Social, political, economic, and ecological processes interact dynamically, requiring analysis to be sensitive to the interaction of structural features and human agency across a range of scales from the local to the global (Scoones 1999:492).

However, modernist models of common pool resource management generally devote little attention to external factors impinging upon resource exploitation. Usually, policy and management schemes focus on single species exploited by specific user groups as if they are autonomous systems, instead of being embedded in and thus influenced by wider ecological, economic and socio-political forces. For the sake of such a model's applicability, contextual factors making for complexity, heterogeneity and dynamics are reduced, simplified or ignored altogether. The question of how such factors affect marine resource exploitation or interfere with management objectives and desired outcomes is rarely made explicit. At best, they are considered 'nasty complications'. In their linear view, scientists usually start with the assumption that without human intervention fish stocks are 'ordered, balanced and in dynamic equilibrium' (Smith 1990:5) and any disturbance of this static 'natural balance' must be anthropogenic (Scoones 1999). Consequently, fishermen will inexorably be blamed for overexploitation and it is believed that their behaviour should be checked through more or less draconian management regimes. Many biologists, ecologists, economists, politicians, policymakers, bureaucrats and environmentalists perceive fishermen as a greedy and unruly lot with innate rapacious mentalities, who seek to extract the greatest piece of the common wealth in the shortest time possible. Such a stereotypical and grim view leads to reductionism, obstructing an understanding of what really makes fishermen tick. Fishermen, on the other hand, usually view natural processes as dynamic, unpredictable, complicated, disordered, chaotic and in perpetual flux (Smith 1990:5). Increasingly, ecologists have abandoned the assumption of a natural balance. Instead, they see nature as being in a state of flux, calling for adaptive resource management systems that are responsive to the 'variability, contingency, and openness of ecological systems' (Pickett and Ostfeld 1995:275; see also Klug 2002:702). Natural fluctuations can indeed be quite sharp and fisheries 'are ecologically volatile and vulnerable to any number of external influences' (McEvoy 1988:215). For example, fluctuations of ocean temperatures and salinity, periodic changes in weather patterns, global warming and other environmental shifts have an impact on recruitment and mortality of fish stocks, while diseases and predator–prey interactions can have devastating effects on certain species. Nonetheless, many management models – such as the ones based on the Gordon–Schaefer curve – are 'committed to a simplistic image of marine ecosystems, and a faith in the human capacity to predict and control them' (Holm 1996:178).

Some argue that multi-species fisheries are chaotic systems, with simply too many uncertainties for any kind of predictability and long-term control. This has important implications for policy and management: 'If the dynamics of a fisheries ecosystem are predominantly nonlinear, then *all* fisheries management strategies based on linear cause-and-effect models, single-species assessments, predictions and quotas are profoundly flawed and unlikely to achieve their intended results' (Finlayson 1991:93). They can even have perverse outcomes that are incurred at great expense. Resource management regimes are usually simplified tools to tackle complicated issues. Dealing with ecological, economic and socio-political situations and developments as if they were simple, homogeneous and static provides for easier management tools. However, simple policy is not necessarily good policy, as many policymakers, bureaucrats and scientists seem to believe. Forgetting that simplification has been used for resource management purposes can result in serious complications. For example, models focusing on single-species stock size and fishing effort are inadequate simplifications, but the problem is that 'the realism of multi-species model creates unmanageable complexities' (Holm 1996:184). Nonetheless, simplistic solutions may temporarily alleviate symptoms yet deepen long-term adverse consequences and cause a loss of ecosystem resilience, a problem known as 'the pathology of natural resource management' (Holling and Meffe 1996:330). Against this background, one of the most pressing needs of social science investigations is to determine the ways people understand and relate to their natural and social environments and how they bring about and respond to ecological, economic, political and social change. However, understanding complex and dynamic fisheries systems '*as a whole* presents great challenges, and is rarely if ever attempted' (Hamilton et al. 1998:17, emphasis in original).

Most linear models that assume that a tragedy is inevitable in common pool resource use unless government intervention or privatization is introduced – a view popularized by Garrett Hardin – fail to incorporate contextual factors. Hardin states that without some form of coercion, common pool resources will inevitably be overexploited, a viewpoint he summarized in his famous dictum: 'Freedom in a commons brings ruin to all' (1968:1244). It has led to considerable confusion, as Hardin seems to equate commons with free access. Looking back on his original article after three decades, he concludes that he should have used the adjective 'unmanaged'; in an unmanaged commons, 'overuse of resources reduces carrying capacity [and] ruin is inevitable' (Hardin 1998:683). The homogenizing view of people's rationale and behaviour inherent in 'tragedy of the commons' scenarios grossly underestimates the importance of usufruct, informal rights of access and use, communal management, collective action, socio-cultural heterogeneity and dynamics (McCay and Acheson 1987b). Besides being an oversimplification, the social consequences of departing from such a scenario for policy and management purposes may be enormous and irreversible (see McGuire 1991). The dilemmas inherent in the exploitation of marine resources should be understood 'in terms of the dynamics of conflict and competition between different social groups located in history and culture rather than between the rational economising individual – unspecified – and the group – also unspecified' (McCay and Creed 1989:19). Nevertheless, tensions between the common interest, group interests and private interests do exist. Whether and how these tensions become manifest demands a meticulous contextual inquiry.

The simplicity and range of the 'tragedy' metaphor have contributed to its popularity, particularly among policymakers and state agencies regulating the fisheries (Matthews 1993:239). However, the introduction of new legislation and regulations can disrupt precariously balanced complex adaptive systems based on usufruct and informal distributive rights of entry and use. Extant indigenous management of marine living resources, such as systems of territoriality, is often undermined. Top-down management systems have failed to resolve resource management problems, have led to compliance problems, and have impacted the fishermen's status, cognition, skills, prestige and identity, their interrelationships and the social fabric of fishing communities. It has taken fisheries policymakers, regulators and scientists a long time to realize that there are more than simple cause–effect relations in biology and economy that affect the fishing industry. Although often hesitatingly and firmly based on a means-to-an-end approach, they currently show a willingness to arrive at devolved management systems. In many places, fishermen still feel that their worldviews and practices are misunderstood and misrepresented. They are not necessarily greedy and narrow-minded profit-maximizing automatons, consciously extracting common pool resources to the brink of tragedy. Of course, these remarks should not lead one to think that over-fishing does not occur or that fishermen are continually and necessarily acting wisely from an ecological viewpoint. What I deem important, however, is that social scientists seek to explain why social actors do what they do, how they represent, rationalize and legitimize what they do and what doing so does to them and to their natural and social environments.

The present book's aim is to cast light on the ways in which Texel fishermen – as individual actors and as an occupational community – have actively adjusted to their changing natural and social environments and how, in an interlinked process, their modes of thinking, worldviews and behavioural repertoires have changed. It asks how the articulation of internal and external dynamics has produced specific outcomes for fishing at the local level. This requires devoting attention to how policies emanating from national and European institutions impinge upon local-level social organization, practices, worldviews and motivations, and focusing on the intractable 'real world' and the decision-making processes and interactions of social actors who may not always behave 'rationally'. That is, their actions do not necessarily conform to what policymakers, regulators or economists expect. Socio-cultural considerations as much as economic ones, including the range of choices available to them, guide their quotidian practices. In explaining the 'evolution' of the local fishing industry, it is necessary to take into account a host of important variables to avoid reductionism. There is not a single causal factor that can explain the events and processes that shaped Texel's fishing communities. Nor are developments necessarily one-directional. We can only understand how individuals and social configurations are embedded in particular situations and developments by investigating the gamut of ecological, demographic, social, political and cultural dynamics and structures (McCay 2002:361). What fishermen do is informed and influenced by, inter alia, ecological possibilities and restrictions, demographic variables (including age, composition of the nuclear family, and stage of the family cycle), economic incentives (revenues, remuneration), diversification and specialization options, alternative employment opportunities, strictures and loopholes of fisheries management, vested (property) interests and ease of exit decisions. Social factors such as the social organization of the fishing industry and the fishing firm, the behaviour of fellow fishermen at the local level and beyond, peer pressure and social control, heterogeneous interests and multiple actors, and modes of interaction between owner-operators and between skipper and crew are important. Significant cultural vectors include job satisfaction, past experiences and perceived future prospects, perception of risks and uncertainties, acquired skills, knowledge and images of the seascape and its resources, status considerations, and self and public image and esteem of fishing as an occupation. In a broader sense, religious convictions, community norms, values and morale, trust or distrust, sentiments, and interpretations and meanings attributed to particular circumstances and actions influence strategies and decisions.

On Being There: Notes on Methodology

At the time of the 'repatriation' of their discipline, anthropologists acknowledged that the study of kindred societies and cultures is both compelling and legitimate. They admitted to being ignorant about many social and cultural configurations that at face value appeared to be more or less familiar. This recognition led to reflections on the usefulness of anthropology's theoretical, methodological and conceptual repertoires in settings about which they believed themselves to have an a priori intimate knowledge and comprehensive view, as well as on the pros and cons of field research at home (van Ginkel 1994, 1998). Without going into the details of the epistemological debates on this issue, 'anthropology at home' to me does not seem to be much different from anthropology in remote places, apart perhaps from some practical implications. The crucial point is not where anthropologists hail from, but how they perceive and interpret the reality they confront. Knowledge of the cultural variety of and within human societies will help us to see what is taken for granted as well as what is not so obvious. In doing field research, all ethnographers are 'positioned subjects' (Okely 1992:14), regardless of whether they conduct it at home or abroad. Ultimately, in the generation of knowledge, class, ethnic, religious, gender and other differences between researchers and interlocutors may be more important than place of birth. Besides, given the heterogeneous character of all cultures, it is not easy to find the locus in which any anthropologist would be a genuine insider. In most cases, anthropology at home 'is infested with difference, diversity and division to the same extent as anthropology of other cultures and societies' (Ryang 1997:13). Thus, the island society of Texel and its fishing communities were familiar yet strange to me. I was cognizant of various matters, but ignorant of many others, and I had to get outside and inside the local social and cultural context simultaneously in order to gain a measure of both detachment and involvement. The first such 'familiar yet strange' experience was that I found myself *on* – not *in* – Texel. The Netherlands being a flat country, you are on, and not in, any settlement built on an elevation in the landscape. In this book, I will follow the vernacular.

I conducted field research on Texel from December 1989 to March 1991 and from August 2005 to September 2006. During both stints of fieldwork, I collected data through archival and literature research, interviews and participant observation. I had full access to the archives of the local Fishery Coop, the fishermen's association, the Texel co-management group and the local Producer Organization. I have scrutinized a fairly large number of serial publications. These have been important for establishing the nature of transformations in the national and local fishing industries over the past decades and the kinds of issues and controversies that fishermen have had to relate to. I read through all the issues of the local newspaper, the *Texelse Courant*, that have appeared since its beginning in 1887, and several other regionally or locally important newspapers. Ever since I became interested in fisheries and fishing communities in 1985, I have kept an expanding file of relevant reports in newspapers and weekly or monthly magazines. I also conducted an electronic search for fisheries and Texel-related news in the LexisNexis Academic database, which exhaustively covers national and many regional newspapers from 1995 onwards. In addition, I consulted specialist fisheries publications. Since the Second World War, the Fisheries Department of the Agricultural Economics Research Institute in The Hague has published scores of serial reports, including an annual statistical overview (*Visserij in Cijfers*, 1961-2008), all of which I perused. These sources provide a wealth of information concerning key issues and changes in the Texel and Dutch fishing industries and communities. However, they cannot make up for the experiences and perceptions of real people doing real things. Finding out about the vagaries and vicissitudes of the fishing industry requires talking to fisherfolk.

Hardly ever did I encounter problems arranging interviews, save for practical reasons: it has not always been easy to get hold of crewmembers who work at sea most of the time and cherish their weekends. My interlocutors obviously enjoyed telling me about the past and present of the local fishing industry, shipboard relationships and life in their communities. They take pride in their occupation and do not hesitate to emphasize their fisher identities. During the first stint of fieldwork, I held in-depth, tape-recorded interviews with thirty-six men and nineteen women. I had multiple interviews with eight of them. Interview sessions lasted about two-and-a-half hours on average. Although I used a list of topical questions, I did not straitjacket my interlocutors if they wandered off onto subjects that *they* deemed important. This enabled the discovery of what mattered to them, what attracted them about the occupation and what bothered them. Occasionally, new key issues turned up, which I would then address in subsequent interviews. Photo eliciting was sometimes part of interview sessions and often yielded direct and indirect information on the importance of social and vessel genealogies, particularly in owner-operator families. During the second spell of field research, I formally interviewed eighteen men and seven women (five of whom I interviewed a second time). A few had been among my earlier informants, but nearly all of the interlocutors who had been septuagenarians or octogenarians during my first research venture had meanwhile passed away. I did not record the conversations on tape this time but instead took notes only. Both tape recording and note taking had pros and cons. Recording and transcription meant no details of the conversation were lost. Transcribing tapes was time consuming, and some interlocutors were perhaps more careful about what they said. Taking notes implied less work but also less detail. Moreover, I noticed that a few informants had problems when I jotted down controversial facts. 'You had better not write this down,' they sometimes said. When refraining from doing so, of course, I could add such facts later on. In general, women were more careful with what they said than men, and some women would on occasion tell their husbands not to volunteer so much information. Luckily, the men did not usually heed such advice. Although perhaps avoiding subjects that might compromise themselves or others, I had the distinct impression that they did not withhold their views on particular matters or adjust them in a more favourable direction. On the contrary, in general they seemed frank and open. In addition to formal interviews, during both rounds of fieldwork I talked to a total of about eighty people in less formal ways, for example at the quayside, in the Fishery Cooperative or aboard ship. Such exchanges ranged from brief casual conversations to in-depth discussions.

The fishermen invited me to their informal Monday morning meetings during both the first and the second spell of fieldwork. Here, retired owner-operators, 'shore captains' – that is, owners, mostly in their forties, fifties or sixties, who no longer join fishing trips but are otherwise fully involved in running the family firm – and skipper-owners staying ashore for a week got together. They openly spoke their minds about current fisheries affairs – exchanging news, reminiscing about bygone times and speculating about the future with all its inherent uncertainties. The mood was usually amicable although at times heated discussions developed. After explaining the goal of my research, my presence did not seem to bother the attendants of these gatherings. Usually they numbered about fifteen, with some nearly always being present and others dropping in infrequently or only once or twice. On occasion, they would ask my opinion about certain matters but in general, I just listened in. For me, the meetings provided an illuminating window on the owners' interactions and concerns. In addition to this functional element, I thoroughly enjoyed being there and listening to the protagonists of my research. During the first fieldwork term, Texel fishermen would sometimes teasingly address me as 'Mister assistant-professor'. They said they would tolerate my presence as long as I did not attempt to write anything that would harm their business. In thinking that I could influence it, they grossly overestimated whatever impact my writings could or would have on the local fishing industry. Some shore captains would occasionally question my inquisitive forages. After nine months of fieldwork, one of them asked, 'Don't you know everything by now? You are always asking loads of questions and things such as "do you have this or do you have that for me?" You're spying on us continually and you're becoming a pain in the ass.' He said it in jest and when I riposted that it would not make sense to fabricate data, he smiled, saying, 'No, that's true.' He proved to be a prolific informant. During the second round of research, especially, Texel fisherfolk had come to know me as 'that guy who wrote those books'. Since 'those books' had been received favourably locally, I think they were confident that I would not shake the trust I had gained. As proof of this, I was again welcome to attend their informal meetings when I had returned to the island.

There were several other social occasions when I observed what was going on: at official meetings, funerals, festivals and so on, always slightly reluctant and embarrassed as I felt a bit like a Peeping Tom. I was on a committee of the maritime museum with a retired fisherman and an owner-operator's wife, providing an opportunity to discover how they perceived the history of the local fishing industry and how it should be represented to both local visitors and holidaymakers. I was also frequently present at the quayside on Friday, when the fishing fleet arrives home from a week's fishing trip. Here I could observe the fishermen's work for the upcoming fishing week, hear about how that trip had been and sometimes listen to their stories and tall tales over a cup of coffee in the accommodation on board a cutter. Usually, however, the men wanted to return to the tasks on hand quickly, as they longed for the weekend. The best opportunity to learn about the crewmembers' work world was to join them on a fishing trip. I did so on three occasions during the first stint of fieldwork and once during the second round of research. I noticed that the fishermen appreciated my interest in their way of life and that I gained some esteem by participating in their work and routines aboard. Above all, they were great experiences that unequivocally brought home to me that the everyday life and toils of the fishermen are far removed from the realities of politicians, policymakers, enforcers and academics.

The Structure of the Book

The book's structure is chronological. Each chapter deals with significant trends and developments in the era it covers – whether ecological, technological, economical, sociological, political or cultural in nature. Their emphases therefore vary somewhat. However, there are several recurring themes, of which the primary ones relate to the fishermen's métier, the socio-economic consequences of specific management regimes, and the fishermen's ambivalent views and attitudes concerning collective action and state intervention. In addition, in various chapters I will devote attention to the relations of competition and cooperation among skipper-owners, and the share system of remuneration and its socio-economic dimensions. One of the focuses of this book is on the social dynamic and economic logic of the family firm (including fisherwomen's roles). As David Symes rightly remarks, with respect to fisheries and in contrast to agriculture 'there has been relatively little emphasis on the internal social structures of the "family firm" and its external linkages or on the socially and culturally constrained decision making within these important micro-institutions' (1999:142). As we progress in time, the descriptions will become increasingly dense. That is, the first chapter dealing exclusively with the fisheries spans more than two centuries, while the last chapter covers a time-span of about fifteen years only. This is not due to a paucity of data on Texel's early fishing economy (see van Ginkel 1993), but for the purpose of showing how increasing intervention has intruded on the local fishing communities and the livelihood of fisherfolk, and how they in turn have sought to maintain a grip on their lives.

Chapter 1 presents an introduction to the setting, describing some of the characteristic features of the island and its population. It devotes particular attention to the cultural singularities of Texel as perceived and expressed by the islanders themselves. This auto-image came about in dialectic response to increasing integration into the orbit of the Dutch nation state. The island's two fishing villages, Oosterend and Oudeschild, are depicted briefly. The chapter also includes a concise history of the island, with special reference to its maritime aspects. Situating events and processes in the fishing communities within the wider island history, a bird's eye view of geophysical, demographic, economic, socio-political and religious developments is presented. From archaeological and historical records, it is evident that maritime activities have occupied a prominent place in Texel's economy for centuries. Early on, fishing and shellfish gathering were an integral part of the islanders' subsistence economy, while commercial fishing probably developed as of the 13th century. Due to sand erosion, silting, inundations and other ecological changes, this fishery declined in the second part of the 16th century. At about the same time, Holland emerged as a mercantile world power and the concomitant rise of shipping provided new opportunities. Many Texel fishermen became pilots or seafarers, while other islanders started selling victuals to the Dutch East India Company fleet. In the mid-18th century, nearly half Texel's occupational population directly depended on the sea for a living. Scores of islanders still earned an income as fishermen. By the 19th century, Texel's prominence in maritime traffic was reduced considerably and agriculture gradually surpassed it in significance. This was linked to large-scale land reclamations that more than doubled the island's size. What seafaring subsequently lost in importance, fishing would gain in importance, although times of prosperity and poverty continued to succeed each other. Increasingly, however, tourism turned into the mainstay of the island's economy.

Chapter 2 portrays long-term trends and developments in the local fishing industry from the onset of the 18th century until 1932. It details various forms of Texel fishermen's individual and collective adaptations to ecological and economic change and pays ample attention to the ideational realm. From the early 1700s, Texel fishermen harvested oysters in public waters, replanting young bivalves on plots they claimed in an inlet near their island. For quite some time, the system of fishing and quasi-cultivation proved to be successful. However, myriad problems assailed the oystermen and in the mid-1840s oyster yields diminished dramatically. Several factors contributed to the decline, over-fishing being one of them. There were other causes, too, and the oystermen did not continue their destructive activities until they had caught the very last oyster. They could no longer earn a living from oyster fishing alone. They broadened their economic horizon and shifted to other ventures, including the exploitation of eelgrass, shells and various species of fish, shellfish and crustaceans. Subsequently, oyster fishing turned into a marginal and short-term activity in a varied seasonal cycle in which the islanders utilized a wide range of marine living resources. Diversification, specialization, withdrawal and co-operation were the main strategies used to cope with natural resource problems, while the fishermen also had to adjust to shifting management regimes, especially in the oyster and eelgrass industries. The adaptive flexibility and heterogeneity of the local fishermen made for versatility and resilience. The island's fishing industry – which now included fisherfolk from Oudeschild and other villages – began prospering again, enabled by a growing infrastructure, new means of transportation and the concomitant expansion of markets. The number of fishing boats grew rapidly. However, a prolonged decline of the fishing economy followed its boom times.

Chapter 3 deals with a triple crisis in the local fishing industry, covering the 1932-1960 era. Firstly, the Texel inshore fishermen faced an *ecological* crisis that was caused by the completion of a dam that shut off the Zuider Sea as a North Sea inlet. In addition to closing an important fishing area, the closure also had a devastating impact on the ecology of what is currently known as the Wadden Sea. The crisis was compounded by the global *economic* depression of the 1930s, which hit the island's inshore fishing industry particularly hard. Against all odds, one category – the offshore fishermen – was doing well. The owner-operators modernized with alacrity and with their steel-hulled and motorized cutters, they rose to prosperity, whereas the petty inshore fishermen bore the brunt of the crises. The former mainly hailed from Oosterend, while the latter resided predominantly in Oudeschild. The chapter compares the diverging courses taken by the fishermen of the two village communities with respect to their economic, social and political strategies and ideologies to seek an explanation for differential success. The *political* crisis of the Second World War impinged upon the offshore segment of the Texel fishing fleet as fishing in the North Sea was impossible and the occupying German armed forces impounded the best boats. Inshore fishing flourished temporarily. After the conclusion of the war, the Oosterend skipper-owners succeeded in swiftly picking up the pieces and rising to prominence again. The differences between the owner-operators of the two fishing villages seemed to intensify in the postwar years, in which the position of the local fishing industry's inshore segment deteriorated rapidly. The reasons why one segment thrived while another languished were manifold, the one appearing to amplify the other. In the emic view, however, an ideational element – summarized as 'mentality' – was the key to understanding the divergence. I will attempt to show that the explanation must be sought in the articulation of particular forms of capital that were mainly of an economic and socio-cultural nature.

Chapter 4 chronicles the rapid development of the fishing industry in the 1960s and the first half of the 1970s. The reinvention of the beam trawl – a centuries-old gear type that had fallen into disuse in the early 20th century – led to an astonishing improvement in catching efficiency. Requiring the use of powerful engines, its application initiated a relentless 'horsepower race'. Owner-operators called on external authorities to intervene and limit engine power, but the state did not heed their plea and, on the contrary, subsidized modernization. Although they believed it to be devastating, skipper-owners participated in and contributed to the race, and thus to the exceeding of capacity. The chapter questions this behaviour. The expansive momentum of the fishing industry was also evident in the multiplication of the number of production units. Fissions of family firms and crewmembers that turned independent made for many newcomers in the fishing arena, not only on Texel, but also elsewhere in the Netherlands. The chapter therefore devotes special attention to the family firm and the share system of remuneration and their roles in the rapid upsurge of the local and national fishing industries. Family involvement and co-adventuring made for considerable versatility and resilience, and hence were important for adaptive performance, but the intensification of exploitation in conditions of easy entry to the fishing industry had adverse consequences for the sustainability of fish stocks.

Chapter 5 examines the introduction of quota regimes in the mid-1970s and the impact it had on fishermen and fishing until the early 1990s. Rather than containing the race for fish, the quota management system exacerbated it. Rules and regulations multiplied but were haphazardly enforced, making for uncertainties that contributed to speedy quota uptake until national quotas were exhausted, which was usually the case well before the end of the year. With the introduction of Europe's Common Fisheries Policy in 1983, policing and enforcement increased. A lively trade in quotas – which were individually transferable – made for commodification of landing rights. However, quota busting was still widespread and compliance with the rules was suboptimal. It was a time of considerable turmoil in the Dutch fishing industry. It even led to the downfall of a minister who failed to harness the disobedient fishermen. The management regime had profound economic, social and cultural consequences for owner-operators and crewmembers alike. The chapter attempts to show how Texel fishermen tried to steer their course in an era that was replete with social dilemmas and why they were ambiguous about the regulatory regimes. To understand their active pursuit to achieve a position as 'catch king', it also addresses the socio-cultural issues of rivalry, hierarchy and respect. The troubles in the fishing industry that went along with the top-down efforts to manage it made it abundantly clear that the management scheme needed a fundamental transition.

Chapter 6 outlines developments and events following the introduction of a co-management system in 1993. The state devolved certain responsibilities for quota management to owner-operators who had to cooperate in groups. The system restored peace and quiet in the fishing industry. It enhanced stability, mutual trust and compliance with the regulations. Generally, it was applauded as a model of good fisheries governance by the government, observers and fishermen alike. There were, however, various adverse or unforeseen consequences, including rights hoarding, slipper skippering and quota hopping. Moreover, power balances between owner-operators and co-adventuring crewmembers became increasingly skewed. However, the major threat that endangered the rather well-adapted system was the flow of new rules emanating from the European Union. Although the fishermen still fully support the co-management regime, their support for the European fisheries policy is lukewarm, to put it mildly. In fact, it lacks legitimacy. The chapter addresses the issue of why, from the fishermen's point of view, this is comprehensible. However, the most imminent peril assailing the Texel and Dutch fishing industry is the current economic situation. Overcapacity has been reduced considerably, the fishermen generally work within the rules and they cooperate to arrive at sustainable fisheries, but many are simply muddling through while being deeply pessimistic and wondering how to navigate the current and future seas of trouble.

When examining a time-span of approximately three centuries, it is perhaps easier to see rifts and ruptures, structural evolution and far-reaching transformations than reproduction and socio-cultural continuities and consistencies. Of course, there is both change *and* reproduction and the emphasis on either the one or the other depends largely on the researcher's focus. It is important to remember, then, that in 'the dynamic interplay among external forces, structural ambiguities, and chronic misfit between the appropriate and the possible, the ultimate agent is neither "structure" nor "history," but the individuals acting within and upon them' (Rodgers 1991:44). This book therefore attempts to unravel the threads from which the socio-cultural fabric of fishing and fishing communities is woven. Influenced by internal and external forces, fisherfolk continually make and remake cultural forms and conventions. Their attitudes, stratagems, behaviour and perceptions can only be comprehended in the context of, inter alia, ecological, demographic, economic, social, political, legal and cultural dynamics in a much wider field. Today, anthropologists widely recognize that a community is the *locus* not necessarily the *focus* of fieldwork. We must go beyond it and connect it with structures and processes in the wider society to grasp the economic, political and socio-cultural complexities, contingencies, heterogeneities and dynamics at the local level. This requires looking at the intersection of history, individual and collective experiences and actions, and societal relationships. As outlined above, to understand human-environment interactions, researchers should specify the conditions and institutions that are relevant to people's perceptions, motivations, decisions and actions, and attempt to explain their causes and consequences (see also McCay 2002:393). To do so, we must cast our nets widely. Formulating such programmatic advice is one thing, heeding it is quite another. Nonetheless, I have attempted to take on this formidable task. Drawing upon an extensive body of historical, economic and ethnographic data, I will point to the structures, patterns and interplays of institutional opportunities and constraints and human agency, and the particular articulations of macro and micro events and processes. I will continually shift my perspective to see power in culture and culture in power as people confront the world in which they are embedded and as the world confronts them.

Chapter 1

The Golden Knoll: People, Place and History

The Island and the Islanders

The Dutch island of Texel is the southern- and westernmost of the Wadden Islands, an archipelago extending along the northern coasts of the Netherlands and Germany and the Danish west coast. It is approximately twenty-five kilometres long and on average eight kilometres wide. The Marsdiep Strait separates it from the mainland of the Dutch province of North Holland. Two ferries maintain an hourly connection with the naval port of Den Helder, with a half-hourly schedule during the holiday season. Due to its moderate climate, Texel is a lush, green oval of land with rich and varied vegetation, interspersed with slightly hilly landscapes, dunes, heaths and woods and speckled with seven small villages and a score of hamlets. Until the early 17th century, the island was less than half its current size. Reclamation of marine sediments began in the late Middle Ages, with most land gained during the 19th century. About fifty-five per cent of Texel's surface is farmland, predominantly for dairy farming, flower-bulb growing and arable agriculture. The island's core consists of boulder clay and wind-borne sand deposits from the Pleistocene Age. The highest part is dubbed 'The High Mountain' (*De Hoge Berg*), even though the hummock's top is only about fifteen metres above sea level. Small tracts of land with grazing sheep characterize this area. Sandy beaches and sand dunes defend the island from the North Sea on its western side. On the eastern side, dykes protect it from the Wadden Sea. The sea continually affects the island's morphology, and its encroachment necessitates ceaseless protection. Westerly storms often cause large sections of beaches and dunes to disappear into the sea. To counter this development, dredging vessels frequently suck up sand from the sea and spray it onto the beaches.

The sea is not just a foe, however, as it has been providing many inhabitants with a livelihood in seafaring and fishing for centuries. The Wadden Sea, Europe's largest wetland ecosystem, is an intertidal sea consisting of shoals, sandbars, mudflats and salt marshes intersected by continually shifting channels and gullies. Until 1932, the Dutch part of the Wadden Sea used to be called the Northern Zuider Sea, but the name was changed following the construction of a Closure Dam. A fresh-water lake developed south of the dam. The dam also had far-reaching consequences for the Wadden Sea's ecology (see Chapter 3). The Wadden Sea is rich in nutrients, but due to its high turbulence and tidal movement, it is currently not an attractive habitat for marine living resources: only about a hundred different species of fish, shellfish and crustaceans can be found there. Nevertheless, it produces approximately fifteen per cent of the value of the Dutch fishing fleet's landings. For Texel's inshore fishermen, brown shrimps are an important target species. Several national, international and supranational laws and treaties protect the Wadden area. West of Texel is the North Sea. With a mean depth of ninety metres, it is rather shallow. The southern North Sea is only forty metres deep on average and constitutes an important fishing ground for a variety of species. It is one of the world's most productive seas. The island's offshore fishermen predominantly pursue several species of flatfish, in particular sole and plaice. Today, the local fishing fleet is relatively modest, but it once used to provide a living for many islanders.

Texel is a municipality and boasts approximately 13,450 inhabitants, who fondly refer to their island as 'The Golden Knoll' (*Het Gouden Boltje*). Its largest village is Den Burg, the administrative, economic and service centre, which comprises a population of approximately 6,900. The other villages are Oosterend (1,400 inhabitants), Oudeschild (1,275), De Cocksdorp (1,250), De Koog (1,220), Den Hoorn (965) and De Waal (400). The population figures include inhabitants of nearby hamlets. The remainder of the population lives in the countryside. A dense network of roads and bicycle lanes connect the settlements. The island is currently on the mainland grid for electricity, natural gas and water. Until the mid-1990s, it had its own municipal power plant and fresh-water-making factory. Texel is distinctly rural. Traditionally, agriculture and fisheries have been important sectors of the local economy. Agriculture covers half of Texel's area, with a gross turnover of over €35 million as compared to €25 million in the fisheries (2006). There has never been any large-scale industry on the island. Since the Second World War, tourism has assumed enormous proportions and presently dominates the island economy with a gross turnover of €250 million (2006). Annually, about one million tourists visit the island. During the peak season, the holidaymaker to islander ratio is approximately four to one. Sandy beaches and the island's nature and culture attract many tourists from the Dutch mainland and Germany. The villages on the North Sea coast (De Koog, De Cocksdorp and Den Hoorn) are especially popular and geared to the tourist industry with a host of facilities including campsites, villa parks, playgrounds, an indoor swimming pool, and scores of shops, hotels, bars and restaurants. The island also boasts six museums. Tourism employs more than a quarter of the occupational population directly. The overall impact of and dependency on tourism are much higher, however, with direct and indirect employment amounting to no less than seventy-five per cent. Agriculture accounts for nearly ten per cent of employment, while municipal and state institutions provide another ten per cent.

Thus, Texel predominantly depends on tourism. Holidaymakers began invading the island after 1948, when all Dutch employees acquired the right to a paid holiday. In the summer months, holidaymakers invade the island and island life is geared towards catering for the tourist industry. To some extent, the social fabric of local relationships is temporarily untied. This changes when the tourists have left. In the winter months, the tempo of local life is quite agreeable, although it would be wrong to assume that the islanders go into a winter sleep. They re-establish old ties, get together more often; organizational activities that have been on hold for months are resumed, and several festivals and other social occasions are exclusively celebrated for and by the Texelians (*Texelaars*). As in so many tourist-dependent places, they perceive tourism as both a blessing and a blight. The islanders are ambivalent about holidaymakers, who provide an important source of income to many yet are also perceived as exacting voyeurs, whose presence leads to more traffic, more congestion by cars (and bicycles), soaring property prices, absentee ownership of houses and generally more turmoil. At the same time, largely owing to tourism, Texel is a relatively wealthy society with a well-developed infrastructure of shops and services. On the negative side, due to the dependency on tourism, seasonal unemployment is rather high. Without a doubt, the growth of tourism and the holiday season's extension have had a major impact on the island's society and material and immaterial culture. However, to contain negative impacts the municipality in 1974 adopted a policy of capping the number of tourist beds at 47,000.

It was with the rise of tourism after the Second World War, the immigration to the island of hundreds of mainlanders and the growing impact of outside interference and regulation that an awareness and articulation of local identity gained momentum. Through revitalization and invention or reinvention of traditions, the islanders began recovering their cultural heritage. They apparently realized that adaptation to and ongoing integration within the wider society implied at the same time a 'loss' of their own culture and identity. Once this dawned upon them, they began asserting their right to their own distinctiveness. The islanders currently even indulge in local chauvinism, which arose as a dialectical response to integration processes (see van Ginkel 1995b). The strong sense of localness is also evident in local politics. While in parliamentary elections the islanders more or less follow the national trends, in municipal elections the local political party, *Texels Belang* (Texel's Interest), has invariably won elections since 1966. In the past two decades, the party gained between twenty-five and thirty-seven per cent of the votes. Local interests are indeed central to the party programme and its vision on Texel's future clearly emphasizes this localism.

The islanders, especially those from old Texel lineages, take pride in their island. Material cultural heritage is now meticulously preserved.

With tourism being the local economy's mainstay, many islanders have to make a living by selling 'uniqueness' and 'authenticity': for example 'genuine' Texel products. Many islanders fly the green-and-black island flag and have 'Texel' bumper stickers on their car. Another source of pride is the ferry corporation named TESO, an acronym for *Texels Eigen Stoomboot Onderneming* (Texel's Own Steamship Enterprise). The vast majority of its shareholders are islanders. Local people, ousting a private firm owned by a mainlander who allegedly demanded exorbitant ferry rates and used ancient and untrustworthy steamships, founded it in 1907. Being so proud of it, they continually assess everything related to it rather critically. An institution that is also of considerable cultural significance is the local newspaper: the *Texelse Courant*, established in 1887. It covers the island's news in detail and constitutes an important platform for local discussion. Furthermore, the islanders have an eye for the immaterial elements of Texel culture such as the dialect, local history and folklore. There is a folklore association that organizes various events and a thriving local history association that currently comprises approximately 950 members. Cultural traditions such as the burning of bonfires (*meierblissen*) on 30 April and the celebration of a local festival, *Ouwe Sunderklaas*, on 12 December, are alive and kicking. These calendar fêtes occur outside the tourist season and are inwardly directed (see van Ginkel 1995b, 2007b).

On the face of it, Texel currently seems to be not only a geographic and administrative unity, but also a socio-cultural homogeneous one; a place where all inhabitants reckon themselves to be members of a 'we group' versus a generalized 'they group' consisting of 'other-siders' (*overkanters*), as Texelians call them. The term 'other-siders' evidences a Texel-centric worldview: from the perspective of mainlanders, Texel's location is peripheral and on the other side of the Marsdiep. However, the idea that Texel constitutes a homogeneous socio-cultural unity is a myth carefully maintained for the outside world. Relative to other-siders Texelians regard themselves as a unity, but within the island society a plethora of social and symbolic boundaries are drawn. One important distinction is between 'genuine Texelians', 'Texelians' and 'incomers' (*import*), at least by the first category. The 'incomer' category consists of newcomers who have settled on the island fairly recently. When I interviewed them in 1990, many senior Oudeschilders complained about the fact that because of the influx of people, they hardly knew who was who in the village. They often reminisced about 'the good old times' when all villagers still knew each other inside out and neighbourliness was a matter of course. Although such stories are not devoid of exaggeration and nostalgia, many newcomers have indeed settled in the village. Older islanders especially, but certainly not exclusively, experience this as a loss of community and view the incomers as intruders in their insular world. I heard no such complaints in Oosterend, where the influx of newcomers has been relatively modest. 'Texelians' are usually those who have been born and bred on the island, but whose parents or grandparents were not born there. 'Genuine Texelians' are those who pride themselves in having many generations of ancestors who have lived on the island. The notion of being 'genuinely Texelian' has only been able to develop because there has been considerable immigration to the island. It is a relational concept that presupposes differentiated social knowledge of who can be ascribed to which category. The term refers to being deeply rooted locally. It also has symbolic value, because those who reckon themselves as such take pride in it and feel that they belong to an in-group that gives them the opportunity to distinguish themselves from others. In addition to the differentiation between 'genuine Texelians', 'Texelians' and 'incomers', the members of the first two categories sharply distinguish among the villages' inhabitants. They say that each village has its own character and that the mentality of the inhabitants of the respective villages differs markedly.

Of course, within villages there are several other domains of distinction, for instance based on residence in a neighbourhood, class, occupation, age, gender, and religious affiliation. Villages are small and relatively egalitarian, however, and these distinctions are rather diffuse and not articulated strongly, with – until a few decades ago at least – the exception of religion. There is no recent census data concerning religious denominations. In 1981, the number of islanders without religious affiliation was about thirty per cent. Twenty-eight per cent of the population was Roman Catholic, twenty-seven per cent Dutch Reformed (*Hervormd*), nine per cent Calvinist Reformed (*Gereformeerd*) and five per cent Mennonite. The remaining one per cent belonged to other denominations. In line with general developments, the number of people without religious affiliation has probably grown considerably over the past quarter of a century. There are also marked differences between the villages. Whereas Oosterend is nicknamed 'Jerusalem of the North' for its rather orthodox Calvinist inclination, Oudeschild is locally renowned as a predominantly secular and 'Red' village. Here, support for the Social Democratic movement has traditionally been considerable. It is with the communities of Oudeschild and Oosterend, whose inhabitants refer to their respective villages as 'Skil' and 'Strend', that I will be most concerned in this book.

'Skil' and 'Strend': Two Distinct Fishing Communities

Oudeschild and Oosterend are designated 'scenic' villages, meaning that their picturesque centres must remain intact. The rustic heritage of narrow streets and old gabled cottages with wooden fronts painted in a darkish green is considered worth preserving. Both villages are close-knit communities. Many inhabitants are members of more than one voluntary association – of which there are plenty at the local level. Being well-integrated communities, in both villages social control is rather tight. The construction of space into place is important and despite having a lot in common, the villagers claim that Oosterend and Oudeschild are 'worlds apart'.

In tourist brochures and other representations of the island, the villages are invariably portrayed as fishing communities and this also holds true for the image most islanders – including the respective villagers themselves – have of them. The vast majority of local fishermen reside in Oosterend and Oudeschild. In the Texel telephone directory, many fishermen have included a reference to their occupation (for example, 'North Sea fisherman') whereas this is rather unusual for other occupations. They still take pride in their métier. Even though the occupational community has become rather small over the past decade or so, fishing continues to occupy an important economic and symbolic position at the local level. Scores of inhabitants are fishermen or have been on a crew for a number of years, and most Oosterenders and Oudeschilders have relatives, neighbours, friends or former schoolmates who are working in the island's fishing industry. Emphasizing their fishermen identities, Oudeschild has a fishermen's choir that regularly performs on the island and Oosterend features an annual fish-smoking contest.

With its harbour and Fishery Cooperative, marinas, a Sailors' Chapel, a Maritime and Beachcombers Museum, nautical shops, a sail-loft and fish restaurants, Oudeschild definitely has a maritime touch. Several street names refer to the village's maritime past – for instance Pilot's Crescent (*Loodssingel*), Commodore's Crescent (*Commandeurssingel*) – or bear the names of renowned 17th-century navy commanders (for example, de Ruyter, Tromp, Heemskerck). The village proper currently boasts about 475 houses, with a population of approximately 1200. Approaching the village from the south, we pass a late-16th-century fortress and a pumping station with a sluice, whilst having a clear view of the High Mountain to the left. On the hummock's southern slope are a war cemetery and the village graveyard. The village's southwestern part is called 'the little neighbourhood' (*'t Buurtje*). This peripheral area used to be the village's Catholic section and there is a Roman Catholic church and a vicarage that is presently in use as a B&B. This is one of the village's oldest parts. From here, two parallel roads, intersected by a few alleys, run in a southwest–northeast direction. The easternmost road faces the Wadden Sea dyke, while the other one soon bifurcates in yet another parallel road. Making our way north along the easternmost road, we come across a community centre, a hotel-restaurant annex bar with a terrace, a small supermarket and a hairdresser's. There once used to be many shops and artisan workshops along this stretch of the road, but today most buildings are family homes. This also goes for the buildings to the west, where we meet the Sailors' Chapel dating from 1650 and, a stone-throw away, a secular primary school and kindergarten. The village has been expanding in a northerly direction ever since its establishment in the early 1600s and particularly after the harbour's construction in 1780. Thus, the further north we get, the newer the houses. In a bend of the road, facing the dyke, we meet the Maritime and Beachcombers Museum, which was established in 1980. It focuses on Texel's copious maritime history and heritage and harbours much flotsam and

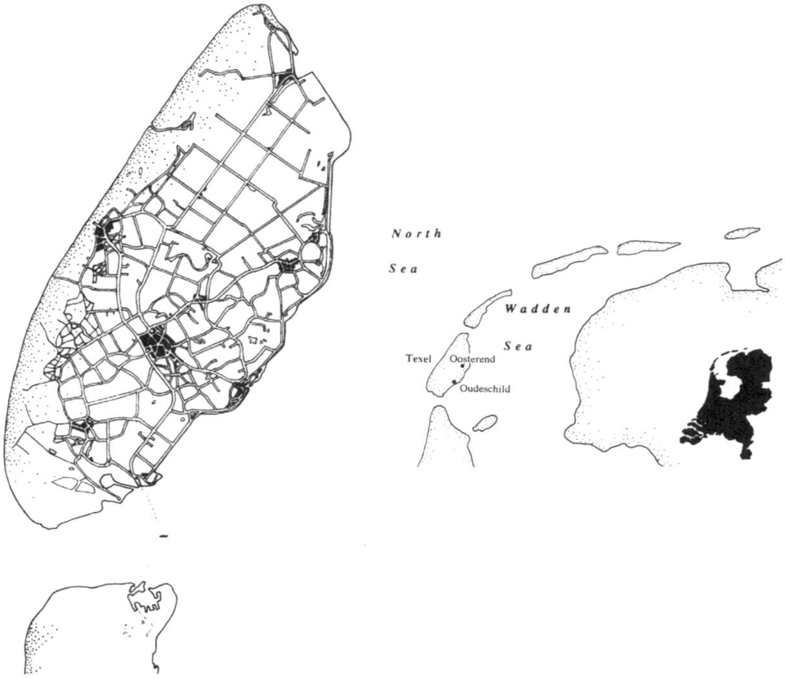

Figure 1. *The island of Texel and the villages of Oudeschild and Oosterend.*

jetsam. There is also a windmill on the premises. Slightly to the north is the road to Den Burg, where we encounter several shops that mostly cater for tourists. It runs in an east–west direction. Going west, we find a new housing project with detached and semi-detached houses on its southern side, and on its northern side an area for houses and small industries that has also been developed recently. Returning east, we hit the dyke that embraces the harbour.

The harbour comprises three sections: the Northern Harbour, the Southern Harbour and the New Harbour. Three boats taking tourists on seal-watching and shrimp-fishing trips and some charter boats for sports fishing operate from the Northern and Southern Harbour quays. The Southern Harbour is used for visiting fishing vessels, commercial barges and the 'brown fleet' of sailing barges now used as charter boats. At the quay are ticket boxes and public lavatories, a few tourist shops, a snack bar, a café, two restaurants and a hotel annex restaurant and bar. To the north is the spacious New Harbour, where most of the Texel-registered fishing boats are moored over the weekend. There is a floating dock for repairs. On the quay is an oblong building: the Fishery Cooperative, with a monument for fishermen who have died at sea. The Co-op contains several compartments for storage of nets, lubricants, chains, cables and other equipment, a workshop and a ship chandlery. The attic is used for net making. The Co-op owns a small oil tanker that takes fuel from three storage tanks. Opposite the Co-op building, across the water, is an industrial site. Northeast of the quay is a marina with a port building and extended facilities, a playground and another restaurant. Four huge wind turbines generate electricity. The quay with the Co-op premises is an ambivalent space, partly embraced by a dyke and symbolically located between the onshore and inshore physical and social domains. Squeezed in between land, sea, and two tourist sites, the quay belongs to the fishermen, yet Texelians and holidaymakers like to observe what is going on there, particularly on Fridays when the fleet has returned from a week's fishing. During the summer weekends, many come to admire the mighty vessels. On summer weekdays, however, old sailing barges converted into charter boats occupy the mooring berths. Oudeschild has clearly benefited from the tourist industry. It is a popular destination for an outing on the island, but the village does not have many tourist beds. Once the tourist season's hustle and bustle is over, it turns into a rather quiet place. The same applies to Oosterend, situated about six kilometres north of Oudeschild's harbour.

Approaching Oosterend from the south, a massive bell tower commands the village's silhouette. It belongs to the island's oldest place of worship, parts of which date back to the 12th century. Originally dedicated to St. Martin, it was converted into a Protestant (Dutch Reformed) church following the late-16th-century Reformation. A neatly kept graveyard, where headstones bear the names of deeply rooted families that have been living here for many generations, surrounds it. The church square is in the village's southwest part, with narrow streets extending in all directions. The plan of the village lends it an atmosphere of intimacy. In the vicinity are a few shops, a small supermarket, a pub, an off-licence, a bank, two restaurants, a butcher's shop and a hairdresser's. Oosterend is the commercial hub for the island's northeastern area. Since the war, the village has expanded predominantly in a northern direction. Here, housing estates and recently built detached and semi-detached houses dominate the scene. Moving further north, we come across two small business sites, a tennis court and, on the northern perimeter, a villa park for holidaymakers. Entering the village from the west, there is a petrol station with a garage. Another garage annex shop is located on Oosterend's eastern side. Situated in a leafy environment, a rather new cemetery sits to the southeast.

In all, the village comprises about 400 houses. Oosterend proper has approximately 1,000 inhabitants. For a community this size, there is a surprising range of local institutions. There is a community centre with a kindergarten. In addition to the 'old church', there are several other churches. A beautiful Mennonite (*Doopsgezinde*) church dates from 1775, but is no longer used for worshipping. A Calvinist Reformed (*Gereformeerde*) church was built in 1897. It became defunct when various Protestant denominations recently merged into the Protestant Church in the

Netherlands (*Protestantse Kerk in Nederland*) and began using the 'old church' for Sunday services. A small ultra-orthodox Calvinist Reformed Community (*Gereformeerde Gemeente*), who deems the Protestant Church much too lenient in terms of religious dogma, also uses the building. A Roman Catholic church dates from 1907. There is a Protestant school in the northeast, while the secular school is in the opposite direction with sports grounds across the street. Although at first glance the villagers would seem divided in religious matters, several voluntary associations serve as vehicles for local social integration. Moreover, every fifth year since 1959, the villagers celebrate a three-day festival in July called Oosterend Present, in which nearly all Oosterenders participate. Its focal point is a historical play: an open-air theatre in which the villagers themselves figure as actors. Begun on the initiative of local entrepreneurs to put the village – which was suffering from an image of devout and dour religious orthodoxy – on the island's map, the fête has meanwhile burgeoned into a spectacular event attracting many onlookers. Its organization and performance tie the villagers close together.

Superficially, little indicates that Oosterend is a fishing community. The Wadden Sea dyke is at a distance of almost two kilometres due east, but there is no harbour. Neither are there any fishing boats. On closer inspection, however, several symbols provide clues to its maritime heritage. For example, the village recently erected two small statues that honour its fishermen and fisherwomen, respectively. They occupy a central place in the village. One statue, unveiled in 2003, depicts three old fishermen talking. The other one, located a short distance away and unveiled two years later, represents two fishermen's wives. In addition, several street names refer to the local importance of fishing past and present, for instance Oyster Street (*Oesterstraat*), Whelk Walk (*Wulkpad*), Eelgrass Street (*Wierstraat*), Cutter Street (*Kotterstraat*), Anchor Street (*Ankerstraat*) and so on. Moreover, many ornaments and paraphernalia are fisheries related, for example weathervanes in the shape of a cutter, or boat names on wooden planks attached to houses. In the local pub, fisheries paraphernalia decorate the walls: old photographs of fishing boats, outdated radio equipment and so on. Paintings and pictures of vessels that are – or used to be – family owned and operated are rather common in the homes of (retired) fishermen. More importantly, in the villagers' self-image, Oosterend is definitely a fishing community.

As mentioned above, in matters of creed Oosterend and Oudeschild differ markedly. The image of the former is that of a pious Protestant community, whereas Oudeschild's is that of a predominantly nonreligious village. Oudeschild's Dutch Reformed congregation merged with Oosterend's in 1950, but when asked how his work was progressing in Oudeschild, the minister allegedly answered: 'I am ploughing on rocks there.' However, in recent decades there have been striking changes in the political and religious ideologies of the two fishing villages. The importance of traditional politico-religious models for living seems to have been on the decline, which was to some extent reflected in the shifting and diminishing support for political and religious institutions. In Oudeschild, forty per cent registered as having no religious affiliation in the 1981 census, while slightly over a quarter were Roman Catholic and another quarter Dutch Reformed. Votes for the Labour Party in parliamentary elections gradually dwindled from sixty-four per cent in 1963 to around forty per cent in 1986 and to slightly less than twenty-eight per cent in 2003. Interestingly, the Christian Democrats – a confessional party including Protestants and Catholics that was established in 1980 – has been on the rise and obtained nearly twenty-nine per cent in 1986 and almost thirty per cent in 2003. In these two years, the Liberal Party got slightly more than twenty per cent and over twenty-three per cent, respectively. Support for the Labour Party thus diminished markedly, whereas other parties, including a confessional one, gained influence. However, this does not mirror a growing religious allegiance of Oudeschilders. Apart from the Catholic minority, most villagers are still nonreligious or nominally Protestant. In Oosterend, the Protestant political parties have continued to gain many ballots. After the Christian Democratic Party's formation, it gained around forty-six per cent of the local votes in the 1986 parliamentary election and almost forty per cent in 2003. Both Labour and the Liberal Party obtained almost eighteen per cent in 1986, the former getting nearly nineteen per cent in 2003 and the latter almost eleven per cent.

In Oosterend, the roaring Sixties affected church attendance, with younger villagers becoming less and less strict with regard to Sunday worshiping. The sharp symbolic boundaries between the religious congregations in general and between latitudinarian and orthodox Protestants in particular would grow more and more diffuse until they finally lost significance for most people, particularly for young church members. The Mennonite church held its last regular Sunday service in 1972. In 1981, a third of the local population was Dutch Reformed, another third Calvinist Reformed. About twenty per cent had no religious affiliation. In his recollection of the 1980s, a Dutch Reformed minister was of the following opinion, 'You either believe or you do not. It does not matter whether you are Catholic or Calvinist Reformed' (*Texelse Courant*, 8 December 1989). Such a view would have been anathema to the villagers only a generation earlier. Church attendance was on the decrease, secularization on the rise. Fishing on Sundays or Christian holidays such as Good Friday or Ascension Day was no longer taboo. Increasingly, church membership was predominantly nominal. Ministers began cooperating more and more, and at national and local levels, Dutch Reformed, Calvinist Reformed and some smaller Protestant congregations participated in a programme called 'On Our Way Together' (*Samen op Weg*) that envisioned a merger. To a degree, the fusion was stimulated by problems of scale that were due to earlier secularization and by a greater emphasis on similarities rather than differences. Meanwhile, most Protestant churches have united as the Protestant Church in the Netherlands. In 2001, this was also the case in Oosterend. The Calvinist Reformed church was closed and today the Protestants use the Dutch Reformed church as their place of worship.

Some churchgoers refused to accept the current of liberalization. Along with the Reformed Church becoming increasingly latitudinarian, in 1967 a small segment of ultra-orthodox believers hived off and established an extremely dogmatic and reformatory congregation of the Calvinist Reformed Community (*Gereformeerde Gemeente*). Currently, it numbers three score members whose core mainly hails from one family and its in-laws. From 2000 until his valediction in 2005, the congregation had its own minister. It even established a school, as it deemed the extant Christian school too lax. Strikingly, the majority of the Oosterenders are rather compassionate about the congregation, volunteering opinions such as 'they still pray hell and doom' or 'they are not allowed much, but, well, as long as they don't bother us it's fine with me'. Such opinions are indicative of the villagers' changed views on religious matters. Latitudinarianism is currently customary, while orthodoxy is regarded with indifference. One might have surmised that the politico-religious convergence between Oosterend and Oudeschild would have led to a blurring of symbolic boundaries between the villages, but this is certainly not the case. Although there is inter-village cooperation, for example in the occupational community of fisherfolk, geographic space still matters in making a community of place. The predominant view on a symbolical level would seem to be: Skil is Skil and Strend is Strend, and never the twain shall meet. There are thus considerable differences between the two villages, even though they share a long trajectory as fishing communities. Maritime ventures have also made a mark on the island's history.

History in a Nutshell: Sundry Facts and Figures

Archaeological excavations on the island have uncovered evidence of occupation reaching back to the Iron Age (800-50 BC), while implements from the Mid-Stone Age (8000-4500 BC) were also found, although they may have belonged to itinerant hunters. Fishing and shellfish gathering for sustenance were undoubtedly an integral part of the subsistence economy of early inhabitants. To protect themselves from the sea, they settled in several settlements on the higher parts of the land: boulder clay emergences that developed in the Pleistocene Ice Age, the High Mountain being the prime location. The lower areas between the emergences gradually filled up with wind-borne sand deposits and marine clay. Small agricultural communities began settling these areas. However, by 250 AD, climate changes of the Post-Roman Transgression caused the sea level to rise, making habitation on Texel increasingly difficult, whereupon people settled on higher ground again. The lower areas were unprotected from the influences of sea and wind, and pastures occasionally flooded. Incursions of the sea eroded large areas of peat to the south of Texel, turning the area into an island. A 9th-century document refers to it as *insula Texla* and also provides proof to Christianization. The diocese of Utrecht owned extensive territory on Texel to which the communities had communal use rights. The villages had wooden churches that were replaced by tuff stone ones in the 11th and 12th centuries. With the rise of urban centres and trade networks in the Netherlands and beyond, commercial North Sea fishing became important on Texel from the 11th century onwards. The island was thus increasingly embedded within the wider world. In the same era, marine incursions led to the formation of the Marsdiep Strait and the Zuider Sea, which would gradually develop into an important fishing area and trade route. At approximately the same time, young dunes began forming. Even today, it is clearly visible that old villages such as Den Hoorn, Den Burg, De Waal and Oosterend are built on elevations in the landscape. This was also the case with De Westen, Texel's most important village in the 13th century, which was mainly inhabited by fishermen, who could reach the North Sea through a channel. Eroding sand dunes – probably caused by deforestation, intensive grazing and the introduction of rabbits for hunting purposes – silted up the channel in the 14th century, whereupon the inhabitants pulled down their houses and moved to Den Hoorn and to what would become the village of De Koog.

It was in the 13th and 14th centuries that Texel's inhabitants began manipulating the landscape to reduce their vulnerability to the sea and extend the island's territory. They started to construct small dykes connecting higher areas, embanked coastal marshes and drained little polders. Initially, the Church was important in coordinating the work. For centuries, the islanders have also planted marram grass to fight large-scale sand drift in the dunes. By the end of the 13th century, Texel was subjected to the Count of Holland. Without the centralized authority of the Church and noblemen, dyke building and maintenance would probably not have been possible. Thus, Christianization, pacification and land reclamation were closely intertwined. However, the Counts' powers were actually limited. They had to extend many privileges and feudalism was weakly developed. For example, in 1415 the island received privileges as a town (*stadsrechten*) with Den Burg as its centre. Among other things, the Counts issued rules and duties concerning proper dyke maintenance, which a bailiff enforced. He was an outsider who leased the office and earned an income from fines, making him extremely unpopular. He had to cooperate with 'native' mayors and aldermen, who were prone to frustrating the bailiff's judicial tasks (van der Kooi 2005). Nonetheless, some progress was made in defending the island against the sea. Land reclamations meant that agriculture gained in importance. Texel sheep cheese and wool became important export products as early as the 15th century. However, reclaimed land was sometimes lost to the sea again.

In the 15th and 16th centuries, impoverishment due to an extremely high tax burden led to a negligence of dyke maintenance. Dyke breaches, inundations, eroding dune sand, epidemics, privateers and pillaging gangs caused many problems. By 1500, Texel's size was approximately sixty-eight square kilometres, but due to major flood disasters declined to fifty square kilometres in 1559. Between 1494 and 1514, the number of houses on the island declined from 750 to 648, diminishing further to 601 by mid-century. In a rough estimate, the island's population fell from around 4,000 to approximately 3,000 inhabitants. Geophysical and political-economic developments thus affected the island's settlement patterns and demography. I have already mentioned the example of De Westen, whose inhabitants moved to Den Hoorn and established De Koog in the 14th century. De Koog developed into an important fishing village with 140 houses in 1514, the same number as Den Hoorn. These two villages alone contained about 500 fishermen who targeted cod, haddock, plaice and herring, sailing as far away as the North Sea's Dogger Bank. However, once again a process of silting up of channels in the early 16th century prevented fishing boats' access to De Koog. This was partly due to the seagoing men neglecting to plant marram grass in the dunes, which led to sand erosion. Along with the process of silting, De Koog's fishing fleet declined rapidly. Still, a mid-16th century map shows various coastal activities, such as beach seining, fishing with small rowing boats, sailing vessels, ships anchored at the Texel Roads and nets drying on the beach. The fish was mostly sold in Dutch Hansa towns. The churches levied taxes on the proceeds of fish sales to maintain their places of worship. However, the situation in De Koog deteriorated to such an extent that the villagers requested to be exempt. The village turned into the poorest community on the island. Heavy storms and floods – which demolished forty-nine houses in 1559 and another seventy houses in 1570 – forced many inhabitants to leave for Den Hoorn. Subsequently, De Koog lost its importance as a fishing community, and to a lesser extent, this also applied to Den Hoorn. Increasingly, the male inhabitants of the latter village turned to piloting and sheep farming.

The devastating floods also led to huge losses in land, crops and cattle, and inundations rendered part of the soil useless. Two more storm surges followed in 1578 and 1590. By the end of the 16th century, the Estates of Holland intervened in dyke maintenance, levying taxes on all islanders and introducing new regulations pertaining to hydraulic engineering. This centralization partly alleviated the difficulties, although dyke breaches would occasionally reoccur. As agriculture gained in importance, privatization of common grazing grounds began as of the late 16th century. To prevent cattle and sheep from wandering off onto someone else's property, the sod banks that would become typical of Texel hedged the pastures. The enclosure, which created a patchwork of tiny fields, was beneficial for arable farming. In addition, it contributed to preventing wind erosion as sheep and cattle could not wander off freely into the dunes and eat vegetation that was important for stabilizing the sand. In the wake of geopolitical developments and the rise of Holland as a mercantile nation, the island started to prosper. In a sense, Texel rose as a hub of international traffic and trade. Ships departing from the Zuider Sea for trade in the Baltic region often cast anchor at the Texel Roads situated off the island's east coast. There, they took shelter from the strong westerly breeze and awaited favourable winds. It was therefore also strategically important, and for purposes of defence, a redoubt was constructed on its southeastern coast in the early 1570s. It was a time of considerable upheaval. Spain had declared war on the Low Countries in 1568, and the Reformation and its iconoclast movement made Protestantism the official religion. Until this time, commodities from the Far East were imported via Spain and Portugal. However, when Spanish troops conquered the port of Antwerp in 1585, the Dutch developed their own Far East trade with the assistance of wealthy Flemish merchants who had fled their country. In 1602, the Dutch East India Company (*Verenigde Oostindische Compagnie*) was established and soon turned into a successful enterprise, shipping merchandise such as herbs, spices, cotton, silk and china to Europe. Many of its vessels departed from the Texel Roads, linking the island with the sinews of maritime commerce.

The island's economy benefited, and a prolonged period of unprecedented stability and prosperity ensued. The village of Schilt (currently Oudeschild) owes its existence to seafaring and the East India Company. It began as a small group of dwellings for the Admiralty, but by 1630 there were already more than sixty houses. In the course of the 17th century, Schilt rapidly developed into a lively small town with the hustle and bustle of arriving and departing sailors. A well in its proximity supplied the merchant fleet with fresh water for its lengthy overseas voyages. As the water's iron content was high, it could be preserved for months and was in high demand. With occasionally more than a hundred vessels weighing anchor, Oudeschild attracted many artisans and traders, including shipwrights, chandlers, sail makers and rope makers. A Protestant Sailors' Chapel was erected in 1650. Another new settlement, Nieuweschild (New Schild), situated about six kilometres to the northeast, came into existence a short time later, whereupon Schilt was renamed Oudeschild (Old Schild). Pilots, who mostly lived in the villages of Oudeschild, Nieuweschild and particularly Den Hoorn, guided the incoming and outgoing vessels on their way to and from the North Sea and the Zuider Sea. As early as 1615, piloting had tight regulations pertaining to tasks and duties, and fixed tariffs. In slack times, pilots often combined their work with fishing. Fishing and whaling also provided a livelihood for scores of islanders, with Oosterend and its satellite hamlet of Oost meanwhile becoming an important farming-fishing community. The island's coat of arms symbolized its dependence on the sea: a golden shield with two red lions facing each other, standing on the stock of an upturned black anchor. The island prospered and its population increased from 4,663 inhabitants in 1622 to 5,420 only ten years later. The booming maritime economy stimulated population growth, but agriculture was also flourishing. The island's territory had been declining due to inundations for a long time, but now land was gained from the sea again. This was mainly because Texel was connected to a small dune island to the north. Between this islet of Eierland (Egg Land) and Texel extended an area of wetlands with large sandbars and creeks. In 1629-1630, the islanders built a sand dyke across the sandbanks' highest part. They also planted marram grass and deployed reed mats to trap sand. Gradually, more and more sediment built up there. In 1633, the island's area comprised sixty-two square kilometres of land. In the course of time, an ever more complex and centralized social organization arose to defend the land against the sea.

During the entire 17th century and a considerable part of the 18th century, Texel benefited from the upsurge of mercantile seafaring. By 1700, about thirty Texel captains skippered merchant ships and many more islanders – some contend up to 400 – crewed on them. By this time, Oudeschild boasted eleven inns 'where kings and princes might be lodged' and many more to cater for sailors. The population continued to expand and so did the number of houses. In the mid-18th century, approximately 6,000 islanders inhabited more than 1,200 houses, nearly a third of them in Den Burg. It was and would continue to be the island's administrative and economic centre. A 1742 census indicated that nearly half the island's occupational population earned a living in maritime ventures as captains, helmsmen, sailors, pilots, whalers, barge skippers or fishermen. Less than a quarter of the occupational population were farmers. The wealthiest seamen generally lived in the island's capital, while the 145 pilots – whose number would rise to 266 less than four decades later – lived predominantly in Oudeschild, Den Hoorn and Nieuweschild. Oudeschild had 132 men who earned a living in a maritime occupation, while Oosterend – including Oost and Nieuweschild – had 136. Traditionally, the inhabitants of Oosterend and environs earned a living in agriculture, which many combined with fishing. Along with the emergence of oyster fishing in the early 18th century, maritime pursuits grew in importance. Most of the 136 mariners were therefore probably fishermen. The census mentioned 151 Den Hoorn men in maritime occupations, mainly pilots-cum-fishermen. However, Den Hoorn began facing the problem that a bay near the village, where they moored their boats, was silting up, whereupon the local pilots started to move to Oudeschild.

Following repeated requests, in 1780 a harbour, including a shipyard and a slipway, was constructed in Oudeschild. Previously, ships anchored at the Texel Roads while smaller vessels moored near the dyke or in inlets. Storms had occasionally caused heavy damage as craft broke from their anchors and sank, often taking a heavy toll of human lives. Although the harbour satisfied the need for a safe haven, the decline of the East India Company in the second half of the 18th century and its subsequent bankruptcy in 1799, adverse political and economic times and shifts in maritime traffic implied that the village's heyday was over. In addition, neighbouring Den Helder – situated on the mainland – rose as a naval port. It obtained a harbour in 1782 that seriously competed with Oudeschild's. More devastating yet was the Batavian and French Era (1795-1813), when warfare and the British blockade of commercial ports, privateers and the trade barrier of Napoleon's Continental System virtually blocked overseas trade. The island's economy suffered profoundly, as did the Dutch economy as a whole. To add to the misery, taxes increased, prices rose and the French began impounding commodities, sapping the island's resources. It also standardized and centralized the administration, the law and the judicial system, taxlevies and so on. With no mercantile traffic at sea, piloting became redundant and the pilots rapidly impoverished. They earned a meagre income of five guilders per month and most consigned with the navy, where they earned four times that amount. By the turn of the century, only thirty-three pilot vessels were left of a once considerably larger fleet. In addition, there were forty-two barges, thirty-one oyster boats and thirty small fishing craft. Oudeschilders, who previously had not occupied themselves with fishing, owned nine of the smaller vessels. The number of sailors dropped to less than a hundred. The French were expelled from the island in 1813, but Texel's misery was not over yet. Piloting suffered when the North Holland Canal was opened in 1825. Vessels destined for Amsterdam no longer needed to navigate the shallow Zuider Sea and the Texel Roads swiftly lost its function. By the end of the decade, the number of Texel pilot boats had dwindled to fifteen. Moreover, a state reorganization of the pilot service in 1835 demanded that all pilots should be stationed on the mainland. The villages of Den Hoorn, Oudeschild and Nieuweschild thereupon rapidly declined. Nearly all of Nieuweschild's fifty houses were torn down. Many residents of Oudeschild and Den Hoorn moved away, scores of buildings were broken down and those who hung on in Den Hoorn predominantly turned to agriculture. Along with the decline of maritime trade and piloting, the wealthiest inhabitants of Oudeschild – including many of those who directly or indirectly earned a living as artisans or suppliers of the merchant fleet – migrated to the mainland. Oudeschild rapidly became impoverished and many inhabitants depended on poor relief from the church and the municipality. The remaining male villagers mainly opted for inshore fishing and eelgrass harvesting, while Oosterend and Oost reinforced their positions as fishing communities. Thus, over the centuries we can observe a relocation of fishing villages from the North Sea coast (De Westen, De Koog and Den Hoorn) to the Wadden Sea coast (Oosterend and neighbouring hamlets, Oudeschild). With about 300 fishermen, fishing turned into a dominant maritime pursuit, while in turn agriculture surpassed maritime ventures in importance. Despite the economic malaise, as of 1829 the island's population increased again and would continue to expand for several decades, not least because of land gains (see figure 1). The demographic fluctuations were mainly due to migration.

New land reclamations were the main cause for agriculture to begin to dominate the island economy. In 1835 it became possible to embank and drain the coastal marshes east of the sand dyke that had been built two centuries earlier and connected Eierland with Texel. At almost thirty-two

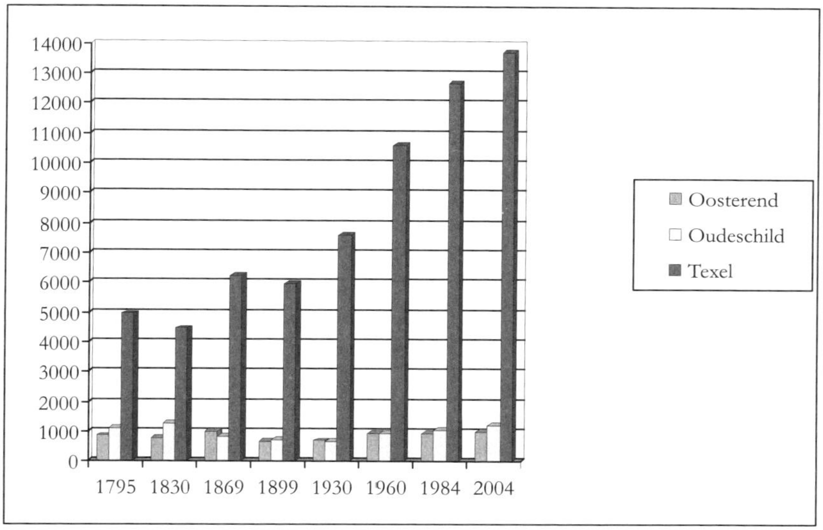

Figure 2. *Demographic development of Texel and the villages of Oosterend and Oudeschild.*

square kilometres, it still is the island's largest polder. The affluent Flemish ship-owner and trader Nicolas Joseph de Cock initiated the reclamation, which attracted scores of new settlers from the mainland. Texel's seventh village, for obvious reasons dubbed De Cocksdorp (De Cock's Village), was built at its northern extremity a year after the reclamation. Although to a large extent populated by farmhands, it would also gain some importance as a fishing village. By 1841, it had Dutch Reformed and Catholic churches, a school, and twenty dwellings accommodating 120 inhabitants. Being relative newcomers from non-Texel lineages, the villagers occupied a special position on the island. The vast new territory was turned into farmland and could be rationally exploited according to the latest knowledge and technology. As early as the mid-19th century, a degree of agricultural mechanization came about there. Initially, farming proved to be difficult since the soil was rather poor and required intensive fertilizing and drainage. However, three more polders were reclaimed in the 19th century. Along with some minor land reclamations, these hydraulic works added another fifteen square kilometres to Texel's land mass. There have been no such major coastal engineering projects since.

For several decades, agriculture thrived. Towards the end of the century, however, a severe agricultural crisis made many decide to leave the island, among them several of the settlers of the new polders. Relatively cheap corn and mutton from overseas were dumped on European markets and the effects were felt on Texel. Hundreds of islanders migrated to the United States. While farmers suffered, the fishermen were still doing well. Their numbers rose from 390 in 1875 to 534 in 1891. The fishing fleet's growth necessitated enlargement of the Oudeschild harbour in 1890, but subsequently the fisheries headed towards a prolonged depression. In the early 20th century, agriculture began prospering again and many fishermen chose to become farmhands. Push-and-pull factors in various economic sectors affected the fishing industry. By the early 20th century, tourism was on the rise. The establishment of a new ferry company in 1907 facilitated this development. From the mid-19th century, a daily steamboat service had been sailing between Oudeschild and Den Helder, but out of discontent with its owner's policy, the islanders inaugurated TESO, which gradually expanded the daily schedule of crossings. The company also operated buses on the island. When cars and motorcycles made their appearance, the ferry company adapted to the development by launching a new vessel in 1926 that could transport vehicles. Roads were improved, the island got electricity, telegraph and telephone, public lighting, sewerage, banks, more schools and general practitioners, and so on. In short, the infrastructure improved considerably as did education and health care, while contact with the mainland intensified. It became much easier for mainlanders to pay a visit to the island. Meanwhile, a local tourist information office opened and De Koog had several seaside hotels. The former fishing village gradually developed into a thriving seaside resort completely geared to catering for holidaymakers. Later on, De Cocksdorp underwent a similar development and became a tourist resort too.

Despite the upsurge of the tourist industry, agriculture still held pride of place in the local economy. For various reasons, fishing was not doing well (see Chapter 2). When the global economic crisis began affecting the island, it was compounded for the fishermen by the profound consequences of the construction of a Closure Dam (*Afsluitdijk*) that closed off the Zuider Sea from the North Sea in 1932. The dam caused an ecological disaster and threw many skippers and deckhands out of employment, while alternative jobs were scarce due to the depression. By the 1930s, agriculture also suffered an economic decline and could not absorb extra hands. Scores and scores of Texelians ended up on the dole or in work relief projects. By the time the economic depression began to wane, war had broken out. In May 1940, German troops invaded the Netherlands, including Texel. The fishermen suffered: their boats were impounded, fuel was soon in short supply, and several able-bodied men were forced to work in the *Arbeitseinsatz*. Nonetheless, few lives were lost and the islanders seemed to be escaping atrocious hostilities. However, in the spring of 1945 this changed dramatically. In February 1944, an infantry battalion consisting of approximately 800 Georgians who had been enlisted by the Germans was stationed on the island. By this time, the German army had suffered heavy losses in Western Europe and needed to deploy ancillary troops. The Georgians, who had served in the Red Army, were in a difficult situation. They could choose between becoming prisoners-of-war or lending themselves as the Germans' auxiliaries. They opted for collaboration, but when it became abundantly clear that the German Army's defeat was imminent and that the Georgians would be assigned to the front, they decided to rise up against the Germans. They feared that as 'defectors', they would probably not receive a warm welcome upon return to the Soviet Union, while the allied forces would be likely to treat them as if they were German soldiers. In early April 1945, they killed nearly 400 of their German overseers in a surprise action, but German officers succeeded in calling in reinforcements from the mainland. Heavy fighting ensued. It took five weeks to suppress the revolt, turning Texel into one of the Second World War's last combat zones. This so-called 'Russians' War' (*Russenoorlog*) took the lives of 565 Georgians and an unknown number of Germans (estimates of their death toll vary between 420 and 800). The civilians on the island, who were caught in the crossfire, suffered 117 casualties. Damages were enormous. Texel was liberated on 20 May 1945 – a fortnight after the German occupying force had officially surrendered to the allied forces.

As elsewhere in the Netherlands, the post-war era was primarily devoted to economic reconstruction. Heavily damaged buildings and scores of farmsteads needed repairs or had to be rebuilt. The majority of money and effort went into making things work again. In Oudeschild and Oosterend housing left much to be desired, with buildings being quite small, damp and lacking plumbing and piped water. This would change only slowly, as economic reconstruction had priority. Rapid modernization of agriculture ensued in the 1950s. Land re-allotment schemes and rationalized production changed the landscape made up of small plots into a pattern of larger ones, tractors replaced horses, and ongoing mechanization in general made many farmhands redundant. Along with these developments, agriculture stopped functioning as a seasonal escape route for petty fishermen. Increasingly, the Texel fishing industry was divided into a growing section of modern cutters and a rapidly declining segment of old wooden-hulled sailing vessels with auxiliary engines. By 1960, employment in the agricultural sector had diminished to almost thirty-eight per cent, while fishing accounted for slightly less than four per cent. The island's growing tourist industry provided new jobs – albeit with high seasonal unemployment in the winter. Until 1964, the ferry's home base was Oudeschild, but with the increasing number of cars that needed to be transported – a consequence of growing wealth, the extension of leisure and holidaymaking and democratization of car ownership – relocation was deemed necessary to enable the operation of a roll-on-roll-off vessel. Two score Oudeschilders worked for the ferry company. Fearing that many villagers would become unemployed and their village depopulated, the locals put up a formidable fight to maintain the ferry terminal, to no avail, however. It was relocated to the hamlet of Horntje, and soon a second ferry was launched. Contrary to expectations, however, Oudeschild began to thrive following the upsurge of tourism and aquatic sports and the rapid expansion of the North Sea fishing fleet from 1960 onwards. Particularly after the harbour's enlargement in 1973, which provided more quay space and a marina, Oudeschild turned into a popular port of call for yachts. With the establishment of a local museum in 1980, the village boasted another tourist attraction. By then the harbour had a floating dry-dock for vessel maintenance and repair, owned and operated by two brothers whose siblings were skipper-owners of fishing boats. Oosterenders still owned most of the fishing vessels, and the fishing industry made their village prosper. In the next chapters, we shall see how this wealth came about.

Chapter 2

Trimming the Sails to the Wind

Under conditions of common pool resource use, fishermen rarely exploit one single species or a single ecological niche. It is precisely this aspect of the utilization of commons that is often neglected by theorists who regard enforcement of rules by external agencies or privatization as solutions to tragedies of the commons. If not tied to a single resource – for example when one 'owns' this resource as a tenant or as a person with specific entitlements – fishermen tend to take optimal advantage of the variety of marine ecosystems, choosing to use specific niches as they deem fit. Switching and diversification may not necessarily be rational choices, but they are certainly reasonable strategies in view of declining stock abundance and environmental uncertainty. Under common pool conditions, such flexibility may be greatly enhanced through institutions that emphasize the right to include rather than the right to exclude users from the streams of common or communal benefits (Runge 1986:625). Returning to the question of the unplanned, unintended and unforeseen consequences of adaptive behaviour outlined in the introduction of this book, we should distinguish between on the one hand the intentional strategies of actors, what they lead to, how these results produce outcomes that are partly undesirable and how actors respond to them, and on the other hand types of behaviour that are unintentional and in the long run may have either favourable or unfavourable consequences. Put differently, we should be aware of the simple fact that human adaptations are dynamic and involve continual feedback loops: 'Humans can assert control over environmental deterioration by conscious discipline or planning, ritual regulations, social interaction, or cultural precedents, and/or by very occasional automatic feedback mechanisms operating outside of cognitive awareness' (J.W. Bennett 1976:148). However, the sum total of conscious adaptations can turn into a long-term process that slips from human awareness and may ultimately prove to be maladaptive. People manipulate their natural and social environments, but the properties of their relations with nature and with one another are often contradictory and derive from neither intention nor awareness (Godelier 1986:6).

In this connection, a common but ill-understood adaptive pattern of fishermen is important; they tend to turn away from declining resources (Townsend and Wilson 1987:323). In capitalist societies fishermen have always had to acquire new markets as part of their coping responses to ecological variations, natural resource and price fluctuations, and to adjust to state interventions in the fishing industry, the supply of alternative foodstuffs, changing consumer preferences, fluctuating exchange rates and overt and covert trade barriers. Combined with local demographic and other factors, these have generated a variety of adaptations among fisherfolk (Löfgren 1979:85 86). Social and political processes constrain or enhance the ability of actors to grapple with problems they face adequately, whether these problems are of an ecological, economic, political, or social nature. Adaptive responses to resource deterioration, trade-cycles, rules and regulations, the vicissitudes of a market economy, and social constraints often include diversification, intensification, specialization or withdrawal. Diversification refers to spreading risks and increasing alternative modes of exploitation and employment. Intensification – or expansion – refers to a growing commitment to invest in one or another mode of resource utilization. Specialization means restricting one's activities to a particular niche in the ecological or economic system. Withdrawal implies leaving the specific mode of resource exploitation altogether to find alternative sources of income, such as wage labour or social welfare payments (McCay 1978; Pettersen 1996). Another mode of coping with the uncertainties of resources and markets is cooperation, which may take on various guises from informal and occasional to institutionalized and structural. How such adaptations operate in practice is fully comprehensible only by taking into account contextual factors in a diachronic perspective and by devoting attention to the economic, social and cultural embedment of human behaviour. It is therefore my explicit aim to describe and analyze the adaptive strategies of Texel fishermen in the 18th, 19th, and early 20th centuries within this wider context.[1] More specifically, I will focus on diversification, intensification, specialization, withdrawal and cooperation as modes of adapting to the uncertainties of natural resource availability, market fluctuations and state interference. These forms of adaptation can be considered 'ecological regimes': that is, systems of regulation and sanctioning to which people submit collectively and individually in their attempts to adapt to and control their natural and – by extension – their economic and social environments.

Usually, changes in ecological regimes come about because of resource decline or economic perturbations. Often, a pattern of real or mimicked ownership is instituted or altered to counter resource and/or price deteriorations. Underlying such an adaptive strategy is a process of political manoeuvring and bargaining, involving various competing interest groups. Although the participants in the game of formal and informal contracting – the efforts of individuals or groups 'to assign or to modify property rights' (Libecap 1989:4) – may agree on a given strategy at some point in time, the outcome is never certain. As Klug rightly maintains:

> In order to understand how and why particular property regimes are applied to particular resources at particular times, it is important to understand the historical development and shifts within these property regimes,

> as well as their 'social construction' through the community of users, managers, and policymakers ... who shape them in the context of both domestic and international legal frameworks (Klug 2002:698).

The relationships of power and dependency in the fishing industry change with every transformation of the system of access and use rights: 'Different bundles of property rights, whether they are de facto or de jure, affect the incentives individuals face, the types of actions they take, and the outcomes they achieve' (Schlager and Ostrom 1992:256). Specific forms of labour organization and (re)-distribution of (the revenues of) goods produced accompany specific territorial forms of property. It is thus pertinent to identify the actors and social formations that are involved in political contracting, their relative bargaining positions and their power bases.

Political decisions concerning access to and use of the marine domain are bound to have intended and unanticipated social and environmental consequences (Hamilton et al. 1998:17). They also set in motion political processes in which individual actors and configurations of actors try to secure their respective interests. New regulations affect different categories of fishermen in different ways. Moreover, such transformations can bring about new cooperative or competitive forms of resource use. There will be political bargaining among competing interest groups and contracting for property rights: 'All things equal, those interest groups with greater wealth, size, and homogeneity will have more resources to influence politicians regarding the assignment of property rights' (Libecap 1989:17). Economic and socio-cultural heterogeneity and a concomitant lack of trust among fishermen will render collective action more difficult, exacerbating the problem of free riding (Ruttan 2006:843). If actors succeed in convincing the state that entry to and appropriation and (re)-allocation of renewable marine resources should be restricted to a greater extent than the existing situation, then a number of fishermen are likely to be barred from extractive resource use. It follows that '[d]istributional conflicts inherent in any new property rights arrangement, even one that offers important efficiency implications, can block or critically constrain the institutions that can be adopted' (Libecap 1989:121). Stakeholders with more leverage are prone to win in the process of political bargaining. Those who lose under a new resource management system will try to alter it, sometimes by changing, ignoring or transgressing the rules of the game, while those who benefit will attempt to maintain it and consolidate their interests. By definition, each shift in property allocations has distributional implications that are potentially divisive and may bring about multiple-use problems and conflicts. In what follows, I will explore the reasons why new institutional arrangements and ecological regimes in the fishing industry came about, how specific configurations of rights affected allocation and distribution, how fishermen responded, the ways in which they attempted to maintain control over what they perceived as 'their' domain, and the means they used in this process.

Oystermen and Nature's Nemesis

Oyster fishing began to flourish in the early decades of the 18th century and gained a prominent place in the island economy. The inhabitants of the village of Oosterend and the neighbouring hamlets of Oost and Nieuweschild used about fifty flat-bottomed boats to harvest the bivalves south of the islands of Texel, Vlieland and Terschelling and around the island of Wieringen, situated approximately sixteen kilometres southeast of Texel. The Texelians caught or gathered considerable amounts of European flat oysters (*Ostrea edulis*), using dredge nets or small rakes when the receding tides left the flats exposed. Fishermen hailing from the other islands also pursued oysters, as did some of the inhabitants of Zoutkamp, a town situated on the mainland coast. The Texel oystermen marketed their catch in the country's capital of Amsterdam and in the German port of Hamburg. Frequently, they arrived with 'shiploads of oysters' (le Long 1727:581) and their revenues were so considerable that the Estates of Holland intended to levy a tax on the oystermen. In 1727, they held an inquiry, from which it appears that the oystermen believed that the oyster banks 'sprouted from nature through God's blessing and were not planted by man' (quoted in Dijt 1961:100). They further stated that the oyster banks 'are not owned by anyone, nor has anyone ever had a privilege to them. Anyone from East or West who wished to do so, could fish these banks freely and unhindered' (ibid.). British and French sailors, whose vessels were anchored at the Texel Roads, indeed took the opportunity to catch oysters without having to pay any kind of tax or tribute. Thus, in principle, entry to these waters was open to all and so was the exploitation of the resources in them. It was only by catching or gathering the oysters – in other words through their labour – that the fishermen could appropriate the bivalves. Once harvested, they were considered their property.

Texel oystermen resented the situation of *res nullius*. In 1754, they submitted a petition to the Estates of Holland, asking to terminate the free right of foreigners to harvest oysters (van der Vlis 1977:196). Their plea was ignored. However, their intimate knowledge of the marine domain enabled them to use tactics of secrecy to prevent outsiders from using particularly lucrative locations. Whenever they had spotted a productive oyster bank, they attempted to monopolize this niche. Although access was legally unrestricted, systems of territoriality and usufruct existed that mitigated unbridled exploitation. Texel fishermen claimed special use rights to a shallow cove indenting the island's northeast. Here, they replanted mature and immature oysters they had harvested in the public domain. Each oysterman staked out a plot, demarcated with branches on the corners, where the oysters were tended and cared for until they could be marketed (Paludanus 1776:252). The islanders considered these waters *res communes* and they carried on a form of oyster cultivation. Even if the fishery was free in a formal sense, informal regulations arranged for access and use of these plots by Texelians and for the exclusion of outsiders: each plot was 'habitually respected as someone's property' (Verslag zeevisserijen. 1860:36). The area was guarded on the oystermen's common account. Their aim of claiming access and use of certain locations for themselves was to exclude outsiders, not so much to protect or conserve the resource.

However, with its forty-one square kilometres of shallow, relatively warm sea water shielded from the westerly winds and its firm seafloor, the cove where these nursery beds were located provided excellent conditions for oyster reproduction and growth. Moreover, the technological means available to fishermen were rather simple; gear efficiency was concomitantly low; the vessels' radius of action was but small; the oystermen often could not sail due to storms and ice-drift; they did not market oysters between April and October; and they refrained from sailing on Sundays. In addition, certain practices and arrangements prevented unbridled fishing of oysters. For example, without the pressure of any external authority, Schiermonnikoog fishermen limited their fishing season and the size of marketable oysters. They returned immature oysters to the sea. Although a striking example of self-management, the arrival of newcomers who did not abide by these informal agreements led to violations while Texel fishermen did not feel obliged by the arrangements of their Schiermonnikoog compatriots who, unlike Texelians, never replanted oysters. This meant that the fisherfolk of these islands had opposed interests, 'and the interests of Schiermonnikoog fishermen compel them to also begin fishing earlier than has been arranged' (Paludanus 1776:251). Nonetheless, the communal use and management of the nursery beds near Texel probably advanced oyster reproduction. Each vessel had to land approximately 100,000 oysters annually to provide a living for all its crewmembers (ibid.:257). Since sixty oyster boats sailed from Texel, it is safe to assume that the local fishermen shipped approximately six million oysters per year. In the autumn, they began shipping the bivalves to the markets, of which Hamburg was still an important destination. The oyster traders were among Oosterend's most affluent villagers.

Both catches and revenues fluctuated sharply. For instance, between 1794 and 1798 the Texelians sold between 1.4 and 2.4 million oysters annually to Amsterdam merchants and approximately 1.5 to 5 million to Hamburg dealers. However, these amounts declined during a number of consecutive years and in 1805 they marketed only a total of about 150,000 oysters in both cities. Harsh winters and ice-drift brought about serious damage to the oyster stocks and several cold summers hampered the oysters' reproduction. Fluctuations in demand and supply led to price fluctuations, and trade barriers during the Batavian and the French Era (1795-1813) impeded oyster shipments. Shipments to Hamburg (whose market used to be their 'gold mine') practically came to a standstill. Taxes rocketed and fishermen were severely restricted in their operations. Declining oyster shipments after 1799 were a consequence of the introduction of permits for oystermen and the French blockade of the British merchant fleet – Napoleon's Continental System – thwarting overseas trade from the Netherlands to other countries. Oyster sales in Amsterdam, which had become poverty-stricken under the French occupation, diminished not only because of the citizens' decreasing purchasing power, but also because of a tax levied on fish and shellfish by the city government. When the French were finally expelled from the island and the Netherlands, the Texel oystermen found themselves in dire straits. Their deplorable state had serious ramifications in the wider community: 'The oyster fisherman has bought on tick from the baker and the shopkeeper, but is incapable of paying his debt, so that the baker and the shopkeeper also suffer because they lose their credit.'[2]

The tide did finally seem to turn. From 1815 until 1840, the oyster banks were very productive and the oystermen shipped from one to eight million oysters annually (Anonymous 1852:363). In 1839, they estimated their aggregate annual revenues at 60,000 guilders. Largely, the complex system of capture-and-culture fisheries worked out well. Nevertheless, after 1840, catches declined year after year and the oyster banks became less and less productive until they were nearly exhausted. As early as 1841, the governor of the province of North-Holland asked the mayor of Texel whether it would be expedient to limit the oyster fishery. However, the mayor deemed any limitation 'detrimental to both the society and the fishery itself'.[3] Nor were the Texel oystermen in favour of intervention. To keep their trade going, they began to import oysters from France and England and additionally, they fished oysters in Zeeland waters to replant on their plots. Initially, they did not worry seriously about diminishing catches. They had encountered these before and the 'experience of previous years has shown that adversity is not lasting but has always yielded to better times'.[4] However, their misplaced confidence would soon vanish. Catches near the island declined from several million oysters to a few hundred thousand. The image of the sea as a cornucopia with an inexhaustible supply of oysters faded rapidly when the crisis persisted in the following years and eventually turned out to be an irreversible tragedy.

What caused this tragedy? It seems to have been brought about by a number of factors, one reinforcing the other. Firstly, land reclamation in 1835 had meant that more than three-quarters of the cove situated on Texel's northeast side was lost as a location for replanting and tending oysters (see Chapter 1). Texel fishermen could henceforth only re-lay their oysters in what was left of the cove, which, to make things worse, silted up. Thus, the nursery grounds vanished and the natural milieu for the reproduction of oysters deteriorated. This ecological deterioration is important, since the increasing scarcity implied that the level of exploitation of the oyster stocks rose relatively, because initially catching efforts did not decrease. Secondly, natural circumstances also contributed to the tragedy. Oysters are very sensitive to changes in the ecosystem. Even slight fluctuations in water temperature, salinity, seabed morphology and food supply (phytoplankton) can cause considerable mortality. Severe winters caused marked oyster mortality, and cold summers had a negative impact on reproduction. Moreover, storms and changing currents also affected the oyster stocks. It is further conceivable that the imported oysters did not adapt well to local conditions and that along with them micro-organisms were introduced that were harmful to oysters. Thirdly, developments in infrastructure and transportation were important and the peaceful and prosperous times stimulated demand. Oysters could be distributed to remote markets with the advent of steam navigation and railroads. Prior to the steam era, catches were attuned to the demand in markets that could be reached by sailing vessels within a few days. However, the advent of steam-powered vessels and railways implied an enormous expansion of the distribution network. 'Never before,' the fishermen remarked in 1839, 'has it been possible to ship Texel oysters with sufficient speed and certainty at such distances.'[5] Oyster runs to Hamburg increased in number and Texel oysters were even landed in Russia's St. Petersburg. Fourthly, catching efforts and techniques changed in connection with what seemed an insatiable consumer demand. The local oyster-fishing fleet expanded between 1836 and 1846: from sixty to eighty boats (a crew of three manned each boat).[6] In the same era, the catching technology also changed. Dredge nets were improved and used more widely and more often, virtually replacing small hand-operated rakes. Each vessel would use three such dredges, towing them across the seafloor. Although the Texel oystermen still practised a kind of quasi-cultivation, the exploitation of public waters with a growing number of vessels and more efficient gear possibly undermined the oyster banks' carrying capacity. Aggregate catches increased initially, but per boat catches declined – a fair indication of excessive fishing. Moreover, increased catches at the same time implied lower prices and falling incomes. To keep their earnings at an acceptable level, the Texelians were forced to harvest even more oysters.

State officials began to be concerned and pondered on measures to protect the oyster banks. They were of the opinion that 'to permit that some annihilate this prosperous and important industry to find a scanty means of subsistence is not in compliance with the well-understood interests and duties of the State' (Anonymous 1854:142). However, fearing that regulating one fishery would disadvantage another, the state refrained from intervening. In the 1850s, it had appointed a committee to investigate the state of the fisheries, and this committee proposed to leave all fisheries unregulated. Previously, measures to protect the offshore fishing fleet's interests had been in place. Tariffs and taxes, and prohibitions on landing certain species, had restricted possibilities in the inshore fishing industry. It was therefore decided to liberalize the Fisheries Act in 1857. The Texel oystermen themselves were opposed to any kind of regulation because oyster prices, which had risen again due to the scarcity, initially kept their income up to the desired level. It was precisely because of these high oyster prices that the fishermen marketed all the oysters – mature and immature – that they were able to harvest. However, higher prices could not make up for lower catches, and eventually the income of Texel oystermen fell. Although the fishermen's behaviour was damaging to themselves as a collective, it was perfectly rational for each individual to catch as many oysters as he could. The mechanism of subtractability applied: almost all of the gain would go to each individual fisherman, whereas the costs (over-fishing and ultimately exhaustion of the oyster banks) were passed on to the collective of users. The fishermen were also caught in a zero-sum game: if a fisherman were to throw immature oysters back into the sea, another would probably catch and market them.

From the mid-1840s onwards, many oystermen began harvesting eelgrass in the summer and autumn, and shells (the raw material for the production of lime) in the spring and winter to make up for their declining incomes (see below). Consequently, in the 1850s and 1860s only thirty to forty vessels were active in the oyster fishery for a short season (mostly October and November). Probably because of these adaptive strategies, the oyster stocks recovered somewhat, and from 1858 until 1862 the annual catch averaged approximately two million oysters. Due to this slight recovery, some fishermen were re-attracted to oyster fishing, which again led to declining catches. In 1864 it was stated that the oyster banks were once more 'fished dead'.[7] A few years later, what was left of the cove where they used to plant the bivalves was reclaimed, so that the fishermen completely lost the communal underwater grounds in which to deposit immature oysters. Henceforth, the fishermen were compelled to deliver their catch to the oyster dealers immediately. They also did so in the summer – the time of reproduction (spat-fall). The fishermen were aware of the harmful consequences of their behaviour, but they maintained that they were not in a position to act differently. It was hard to earn money with other fisheries during the summer season and not all fishermen could – or desired to – be active in the eelgrass industry. For this reason, 'the poor fishermen, to eke out a living – a living that is getting increasingly meagre – slight and destroy their future capital' (Verslag zeevisserijen 1880:67). The fishermen's impoverishment thus contributed to excessive exploitation.

The oyster tragedy was not a strictly local phenomenon. In many coastal waters of Europe and the United States, stocks were depleted by the mid-19th century (Berghahn and Ruth 2005). Being immobile and thriving in shallow waters, oysters could be harvested rather easily and were particularly prone to overexploitation. In addition, the supply of oysters initially seemed inexhaustible, an image that faded swiftly when the crisis persisted. With the rising consumer demand and the concomitant intensified commercial exploitation of the species, the ingredients for covetous antics would seem to have been present. This view resonates in a state report deeming oystermen 'not in the least the sort of persons to be able to act with judgment in the exploitation of the fisheries; their short-sightedness, their greed, and also their lack of funds, make them care only for the moment and not for the future' (Verslag zeevisserijen 1863:24). In the Netherlands, the stereotype of the selfish fisherman who passes on the costs of his behaviour to nature and the collective of users was conventional wisdom early on. In the Texel case, however, the decline of oyster fishing cannot be attributed to the fishermen's conduct alone. Coastal engineering, ecological changes, development of infrastructure and transportation were also important. The report rightly points to the fishermen's lack of funds. To make a living, they simply had to catch as great a share as possible from the declining stocks. This mode of behaviour had little to do with an innate rapacious mentality, but everything to do with the fact that the fishermen's economic existence was endangered. Debts to shopkeepers and suppliers had to be discharged and the costs of living had to be met. The fishermen's short-term interests did indeed prevail, not because they were purblind and greedy per se, but because other options were as yet lacking. As we shall see, however, many turned away from oyster fishing. In addition, efforts were made to counter the oyster crisis by changing the mode of exploitation.

Copying the example of French oyster planters, attempts were made to farm oysters on leased plots near the islands of Texel and Wieringen. The Board of Sea Fisheries (*Collegie voor de Zeevisscherijen*) took the initiative in 1859, and later three private individuals hailing from Amsterdam followed suit. This form of oyster farming differed from the Texel system of quasi-cultivation in that the lessees tried to catch oyster spat using 'collectors' (usually shells) to which the spat could attach and grow, whereas previously the Texelians had only gathered or fished young bivalves and replanted them on plots they had staked out. A few Texel oyster traders perceived advantages in oyster farming, but they never were very successful. The oyster farming experiments failed signally for several reasons. Severe winters, storms and deteriorated ecological conditions caused poor results, and on top of that, oysters were frequently stolen from the plots because there was insufficient policing. Most Texel fishermen deemed privatization unattractive and despite their frequently destitute situation, they remained opposed to government intervention. When the Board of Sea Fisheries asked their opinion about oyster farming, they replied: 'The Texel oystermen are very satisfied with the destiny that is afforded them by nature, and they also think that nobody is powerful enough to lay down the law for nature in this respect.'[8] In their Protestant worldview, they perceived nature as a God-given entity in which earthlings should not intervene except through investing labour. The Oosterend fishermen embraced the Protestant ethic of working hard and living frugally – or what Max Weber (1969 [1920]) dubbed *innerweltliche Askese* (this-worldly asceticism). In a cascade of petitions and letters, they depicted themselves as belonging to the 'humble but industrious lower classes', as diligent folk who were always looking for ways to make a living. They perceived nature as an entity that was there to be exploited. Whosoever would be successful in his earthly work would also be rewarded in the hereafter. This disposition implied that the fishermen could hardly be considered stewards of the marine commons. However, as we shall see below, they used several adaptive strategies that – albeit unintentionally – prevented complete exhaustion of a single species.

By 1880, the Board of Sea Fisheries insisted on government intervention in the inshore fisheries, to wit regulations concerning seasons and methods of fishing, and the enforcement of these measures. It regarded policing necessary,

> because due to practices existing since time immemorial, the Zuider Sea fisherman has become used to the idea that everything he finds in the sea is his property. Therefore, it will demand a profound effort to dissuade him from this view, and this task will surely be accomplished only slowly (Verslag zeevisserijen 1880:67).

A legal season was introduced in 1884; oyster fishing was prohibited from April until October. Gradually, it dawned upon Texel fishermen – who began to complain louder and louder about the oyster stocks' deterioration – that the government should indeed regulate the fishery. However, their attitude towards state intervention continued to be ambivalent. They sometimes urged the state to open the legal season earlier, while at other times they wanted tighter surveillance to enforce the rules. In hindsight, the state regretted its earlier policy of non-intervention in the fisheries:

> The government, which supposed wisdom where it was in fact lacking, left the care for the prosperity of this fishery in the fishermen's hands. The disastrous consequences of this policy were imminent. It is easy to fish in most gullies, and thus fishing could be continued until the oysters were completely, or almost completely, extinct (Verslag zeevisserijen 1893:114-115).

The measures taken in the early 1880s proved to be too little, too late. The state was obliged to look assiduously for other solutions.

The idea that the productivity of tenure-based oyster farming could exceed that of unrestricted common pool resource exploitation gradually gained acceptance. The state fishery advisor, biologist P.P.C. Hoek, maintained that owners would take better care of their resources than commoners (1878:390 391). In the southern province of Zeeland, the introduction of oyster farming as of 1870 was – at least initially – hugely successful (see van Ginkel 1989). Some of the Texel oyster traders who had previously attempted to farm oysters near the island were among the pioneers. Although their efforts had not been particularly fruitful, they had remained convinced of the privatized system's advantages. They had therefore migrated to Zeeland, where ecological conditions were much better. The Zeeland successes led to renewed efforts to stimulate oyster farming near Texel. In 1884, the state expropriated oyster grounds to the island's south and introduced a lease by public auction. Texel fishermen vehemently opposed the enclosure of the commons for fear of being ousted from business. They favoured what they termed 'free labour'. The Board of Sea Fisheries supported them, in part at least. It was not so much opposed to the lease system as such, but to the system of auctioning the rights to the highest bidders. It preferred a system in which the fishermen themselves would gain control over the plots as lessees without having to pay huge rents. The Board deemed it unacceptable that only a few wealthy leaseholders would gain access and that the less affluent fishermen would be excluded from entry. It also feared opposition from the fisherfolk. The Secretary of the Treasury, who had jurisdiction over the public waters, ignored the Board's advice, probably because the system it proposed would not line the state's coffers. The state perceived the lease system as in its 'rational' economic interest, granting many opportunities to the forces of capital to capture the commons. The fears of fishermen and the Board of Sea Fisheries proved to be right. Nearly all the lessees were successful and wealthy Zeeland oyster planters and shippers. They could easily afford to outbid the islanders. Obviously, the political process of defining and enforcing property rights was socially divisive because of its distributional implications (Libecap 1989:4; van Ginkel 1989; McCay 1998). Capitalist entrepreneurs from without became the captors of the locations that had until the lease system's introduction been a communal good.

Nevertheless, oyster farming near Texel never really took off. Within two years after the introduction of the lease by public auction, most Zeeland lessees had given up their attempts to farm oysters in the Zuider Sea. The sums of money offered for the lease of a plot fell dramatically. What seems to have caused this failure is an irreversible ecological deterioration due to changing currents and water temperatures, reclamations and silting up of some locations that had also hampered oyster fishing. In addition, the leaseholders did not give sufficient care to the plots, and policing and supervision were inadequate, so that poaching and theft were rife. The severe winter of 1890-91, which also caused serious problems in Zeeland, dealt a deathblow to oyster cultivation. This gave rise to fishermen's attempts to reclaim the commons. In 1893, for example, Texelians sent a petition to the government in which they stated that the lease 'is very damaging to the free pursuit of various fisheries, and does not appear to be profitable to the lessees' (*Texelsche Courant*, 26 March 1893). Under mounting pressure from Texel fishermen and their compatriots from neighbouring islands and communities, the privatized plots were returned to the public domain two years later. Consequently, several fishermen were able to earn a meagre living for a few weeks after the opening of the legal oyster season. Oystering never regained its prominent place in the local economy. Once constituting the 'gold from the water', the bivalves increasingly turned into a by-catch in whelk fishing, particularly following an oyster disease in the early 1920s. A dwindling number of fishermen continued to dredge oysters, until after the building of a dam in the Zuider Sea in the early 1930s they disappeared completely (see Chapter 3). By then, most Texelians had long since converted to other fisheries.

Eelgrass Mowers, Monopoly and Machinations

Initially, harvesting eelgrass (*Zostera marina*) in the summer and autumn seemed to offer a supplement to the dwindling importance of oyster fishing in the 1840s. Combining both meant that the Texel fishermen could augment their declining incomes from oyster dredging. In contrast to oyster fishing, which constituted an example of common pool resource use in which the state only intervened through input measures and a temporary enclosure after an irreversible deterioration part of the eelgrass industry was regulated and privatized early on. This intervention had nothing to do with overexploitation (as in the case of oyster fishing), but was closely linked to an attempt to arrive at the control of market forces. Like the exploitation and appropriation of natural resources, production for the market brought about a number of uncertainties that could potentially be checked through limited access and limited production schemes.

For centuries, eelgrass had been used in the construction and maintenance of dykes, as roofing material, and to stop leaks in vessels. It abounded in the shallows around the island of Wieringen and south of the Wadden Islands, where tidal currents were weak. Texelians were involved in its exploitation to only a small extent. Demand dwindled in the second half of the 18th century, when stones and clay replaced eelgrass as building materials for dykes, but in the early part of the 19th century it regained some economic importance as packing and insulating material, as stuffing for mattresses and chair seats, and as fertilizer. Eelgrass began growing in April and 'ripened' in August or September, which meant that it broke loose from its roots and stem and started to float. Texelians and their southern neighbours from Wieringen gathered the washed-up or floating eelgrass using rakes, hooks and nets. As long as it was not harvested with its roots, this salt-water vegetation would grow again the next season. An improvement was the introduction of eelgrass mowing in the 1830s, which enabled harvesting a better quality. From June to September, the islanders used scythes to cut off the eelgrass when the receding tide left the flats near to exposed. The eelgrass was brought ashore, put in fresh ditch-water for some time to get rid of the salt and then spread out on the side of the ditches to dry. When dry, the eelgrass was transported to sheds, where it was pressed into bales and prepared for marketing.

It was the mayor of Wieringen, Jacobus van Hengel, who played an important role in this development. He also contracted some skippers to transport the eelgrass to Amsterdam and Rotterdam, where dealers bought the dried material for fixed prices. Increased competition soon led to a flooding of the market and a concomitant fall in prices. In years of abundant harvests, the market was soon saturated. The excess stocks could be stored and preserved for several years, but this meant that the traders needed little if any new supplies in the next year(s). In the early 1840s, about 80,000 kilos of dried eelgrass were annually shipped from Wieringen, while Texel's market share was probably less, though attempts were made to expand marketing opportunities. To keep the increasingly impoverished Wieringers at work and away from the poor-relief funds, van Hengel successfully requested the government to grant the municipality a concession on the eelgrass fields. As of January 1845, the mayor privately leased the exploitation rights of a number of eelgrass fields. He aimed to gain control over production to adapt it to the inelastic demand, expand the market, guarantee the quality, and maintain a fixed price of six guilders per bale of a hundred kilo of dried eelgrass. He licensed Wieringers to mow eelgrass, stipulating that they pay a small charge per bale so that he could settle the lease fee. He further demanded that the islanders sold the eelgrass to the skippers who had a monopoly on the trade of eelgrass mown on the leased beds. Policing aimed to prevent unlicensed mowers from cutting eelgrass on the leased grounds.

When later that year Texelians were mowing eelgrass near their island, some Wieringers summoned them to cease doing so, since they believed that only they held licenses to do so. The Texelians thought that they had 'equal rights' and demanded similar permits from the provincial governor. The governor turned to the mayor of Texel, Pieter Keijser, for advice, and the latter wrote to him saying that he did not support his fellow islanders and that he acknowledged the concession given to his Wieringen counterpart. At the same time, he requested the governor not to renew the lease term, since he was an advocate of free exploitation. However, the Wieringers incorrectly regarded themselves concessionaries of *all* eelgrass grounds. Access to several locations was still open, and this proved to be the Wieringen consortium's weak point, since it did not in fact have a monopoly after all. Moreover, the trade in gathered eelgrass was still free, and Texel fishermen tried to lease some eelgrass fields. They had noticed that the Wieringen system of exploitation and trade was advantageous. Though their attempt failed, they could still mow eelgrass in the public domain, and they voluntarily arranged a deal with two Texel traders who were prepared to also pay them six guilders per bale and give them cash advances if the fishermen delivered exclusively to them. These two traders then consulted their Wieringen competitors to prevent the market from being flooded; they concentrated on different areas of distribution. Under this management arrangement, the poor fishermen-cum-eelgrass mowers were integrated into a capitalistic mode of production and lost a great deal of their independence.

When the first lease term was about to expire, van Hengel requested a renewal, this time supported by Keijser, who had meanwhile become an advocate of privatization. Keijser thought that price fixing could continue, as he deemed it necessary for the Texel fishermen to earn a living. He feared that if prices fell, given the prevailing crisis in oyster fishing, they would abandon fishing altogether, and many would have to apply for poor-relief funds. Both mayors agreed that the eelgrass beds near Texel had to be leased to Texelians and those in the environs of Wieringen to Wieringers to preclude 'envious people' from becoming involved in the eelgrass trade. The state and the provincial administration heeded their advice and as of 1850, the privately leased rights to mow eelgrass were assigned to four Wieringen skippers and three Texel oystermen – Gerrit Vlaming, Hendrik Timmer and Pieter Wuis – for six years. The lessees had to give concessions to those wishing to mow eelgrass on the leased grounds in exchange for a small remuneration, earmarked to settle the lease sum. The Texel eelgrass mowers followed the Wieringen model and entered into a voluntary contract with the lessees. They had to deliver all the eelgrass they mowed and dried to the lessees, who would pay them a fixed price and buy all the eelgrass before the beginning of the new season.

Unexpectedly, the mowers supplied such large quantities that the dealers could not entirely sell their stocks. This was also a consequence of the fixed prices, which – given the small demand – made the eelgrass too expensive on the market. Therefore, the traders only bought dried eelgrass in small batches, leaving the mowers with the lion's share of the harvest. Although they occasionally received small advances from the lessees, mostly in the form of groceries, cash earnings lagged behind. They badly needed the money, but their contract with the lessees prevented them from selling to others at lower prices. When the new season's eelgrass could be harvested, the mowers still had a large supply from the previous year. Feeling that the dealers had not kept their word of buying all the eelgrass, they finally decided to breach the contract and sell their eelgrass at lower prices to traders who were not involved in the Wieringen–Texel lease arrangement. This behaviour not only outraged the Texel lessees, but also went against the grain of the Wieringers, who apprehended that they would be unable to sell their eelgrass. As a consequence, van Hengel decided to introduce free trade for a few weeks. His plan was to saturate the market, resulting in a fifty per cent drop of prices in 1851. He wanted to make it perfectly clear that market regulation was necessary, since unbridled competition would inexorably lead to overproduction. He was convinced that his Texel neighbours would now understand the importance of concerted action. They did indeed, since the proceeds of the 1851 harvest were negligible.

This state of affairs dealt a serious blow to the Texel eelgrass industry, which was further undermined by mutual competition, conflicts and jealousy. The three Texel lessees attempted to prevent persons who were not supplying eelgrass to them from exploiting 'their' eelgrass beds. However, some Texelians entered the leased beds without permission, and sold the eelgrass to other merchants. They were caught and fined, but this infuriated scores of islanders. Although the state had leased the eelgrass beds to the trio, it was understood that *all* Texel fishermen would have the right of access and exploitation. The three lessees requested several times to be exempted from the obligation to permit entry to those who asked access for a small charge, while several mowers demanded the state to undo the lease contract with the lessees. Keijser advised the state to ignore both requests. He merely pointed out that the contract was meant to ensure the Texel eelgrass mowers' rights so that they would not suffer from any possible arbitrariness of and exclusion by the lessees. Although the mowers remained dissatisfied, most of them renewed their agreement with the lessees and accepted a new price of five guilders per bale.

Some mowers did not want to commit themselves, opted for free trade, and supplied the eelgrass to merchants who paid slightly more. This lack of unity was furthered by the fishermen's ambivalent attitude: one moment, they were strongly in favour of a monopolistic organization; another moment, they vigorously supported free trade. In this regard, mayor Keijser spoke of 'the incomprehensible contradictions and sudden changes in the way of thinking' of the fishermen, who would 'pray and beg in the deepest of plaintive tones to repeal the lease contract' one day, and then argue for its renewal the next.[9] This opportunism was inextricably connected with the trade cycles in the eelgrass industry. If demand was high, the mowers only perceived advantages in fixed prices, whereas if trade was slow and stocks piled up, they preferred lower prices so that they would at least have some income. Thus, short-term interests prevailed. The mowers had little choice. They belonged to the 'poor class' and in the summer had few alternatives to eelgrass harvesting. In 1854, for example, 140 Oosterend fishermen with seventy vessels and eighteen Oudeschild men with nine boats participated in the mowing, transportation and drying of eelgrass. The eelgrass industry provided an important source of income for the fishermen in their precarious métier. If they experienced their agreements with the lessees as too restrictive, they opted for freedom so that they could at least earn some money. At times, this worked well, but when remunerations tumbled, partly because of their behaviour, they were quick to turn to the lessees again.

Nevertheless, the lessees became fed up with the mowers' volatile behaviour. They repeatedly tried to evade the rule of having to permit access to whoever chose to mow eelgrass. Two Texel merchants, Dalmeijer and Voigt, protested the lessees' wheeling and dealing. In July 1855, they informed the local administration that the lessees misused their position by demanding that the mowers should exclusively deliver to them. Dalmeijer and Voigt wished to buy more eelgrass, but could not because of the mowers' contract with the lessees. They asked the local administration and the Minister of Finance to repeal the contract, since in their opinion it had been wrested from the mowers. Their request stemmed from interested motives. Upon its rejection, they offered a higher lease sum but demanded expunction of the rule to permit access to the mowers for a small charge. Other merchants who were not lessees argued in favour of a lease by public bidding. They maintained that a private lease hindered fair competition, impaired state income, and led to a truck system and other usurious practices. These renewed frictions were linked to the lease term's expiration. Local and state administrations consulted each other as to the future arrangement. Texel's mayor stated that the Texel eelgrass mowers were dissatisfied with the monopolistic system, but even more so with free trade. Therefore, they had aligned again with the lessees. Wieringen's mayor was a proponent of the status quo, and the state deemed a return to free exploitation and trade undesirable because it would be prejudicial to the mowers' interests. They perceived them as poor fishermen who in order to survive needed to sell their harvest for almost any price. Competition would be unbridled and to the fishermen's detriment. Therefore, it was decided in 1856 to renew the lease contract with the same Wieringen and Texel lessees for another six-year term. Again, this led to myriad petitions for and especially against the lease system. Nonetheless, the efforts of Wieringen's mayor brought about a more or less regulated production, expansion of markets and stability of prices.

Despite recurring discontent, the eelgrass industry was an important link in Texel's fisheries cycle. The fishermen badly needed the money earned in eelgrass harvesting to tide over the winter season when storms and ice-drift often impeded sailing. Moreover, the eelgrass served as a security for buying goods and groceries on tick. In this era, the lessees imposed quotas on the mowers' eelgrass landings. Each mower was allowed to land approximately 400 kilos per month. With the extant prices, this meant that during the eelgrass-harvesting season the mowers earned twenty-four guilders per month. The gathering of floating and washed-up eelgrass also continued to be important. Scores of Texelians earned a small income exploiting it, even though its quality was poorer and its price less than half that of mown eelgrass. Access was free and huge harvests could compensate for low prices. Since the eelgrass began to wash up after the mowing season, many fishermen participated in both modes of exploitation. By the end of the 1850s, they could earn a maximum of 300 guilders per person with eelgrass mowing and gathering. They nevertheless remained discontented and whenever fishing was more lucrative, many refrained altogether from the arduous work of the eelgrass harvest.

Repeatedly, mowers and merchants submitted petitions to alter the monopolistic system of organization. Shortly before the lease term expired, the state consulted provincial and local administrations about the lease system's renewal. It was decided to introduce a lease by public bidding. This would line the state's coffers and give interested parties an opportunity to become lessees. The first lease by public bidding was held in 1861. The eelgrass fields had been divided into three large plots, so that only a few persons or associations could become lessees. Wieringen's mayor and his partners succeeded in leasing the two largest plots, while the wealthy Oosterend oyster trader Gerrit Vlaming rented the remaining plot. The total annual lease sums amounted to 4,000 guilders. Soon the disadvantages for the mowers became clear. Almost immediately, they received fifty cents less per bale than previously. The lessees needed higher margins to pay the lease fee. However, the mowers were allowed to supply more, so that their income was levelled. Nevertheless, they had become more dependent upon the lessees, who now decided who would gain or be denied access to the eelgrass grounds. The lessees still needed the mowers' labour, so that they were interdependent, but the power balance was skewed. The lessees' dominance implied a further corrosion of the fishermen's independence. The small entrepreneurs were forced into a temporary role as wage labourers and the capitalistic relations of production brought about class antagonisms.

Again, the Texel fishermen – whose financial position deteriorated – desired a return to free access. In 1863 they argued that the public lease was incompatible with their interests, because they could not mow as much eelgrass as they liked, some were excluded from entry, and the prices had fallen. They also criticized the fact that the eelgrass was not shipped immediately and that they did not receive payment upon delivery. They suspected that the lessees wanted to encourage a truck system. Some leaseholders were shopkeepers and the mowers could buy their groceries in their stores on credit, albeit more expensively than in others. For these reasons, the mowers requested nullification of the lease contract. The opposition to the public lease was primarily a resistance against the altered power relations, which gave the fishermen the feeling that they were enslaved. They formed an interest coalition that had, however, little other power means than the relatively weak weapon of petitioning. Despite the torrent of requests to return to freedom in eelgrass mowing, the Minister of Finance decided to continue public leasing. With growing competition to become a leaseholder and concomitant outbidding, the lease fees rose. The Texel and Wieringen municipal councils had anticipated this and filed a joint protest. They stated that they preferred free access in eelgrass mowing to a public lease, taking for granted intensified competition and lower prices. The mowers would 'most certainly experience more benefits than under continued public leasing with its inflation of the lease fee. Under no condition will they then be forced to accept the lowest wages from a lessee who has to pay a high lease fee or find himself excluded'.[10] Their request was to no avail.

The Texel entrepreneur Pieter Koning and two outside business associates leased two plots. Oosterend oyster and eelgrass trader Gerrit Vlaming and a Wieringer leased the other plot. The new lessees had to count in the risen fees, but the market price of eelgrass could not be raised, since French eelgrass and alternative products such as wood-wool, moss and kapok had saturated the market. Moreover, during the last year of its lease, the Wieringen combination had built up huge stocks and tried to compete with the new lessees, a strategy that would also be used later on. The result was that in 1868 almost 900,000 kilos of eelgrass were marketed, while the extra costs of the lease fee were passed on to the mowers. They only received four instead of five or six guilders per bale. In the late 1860s, Texel fishermen rarely earned more than ten guilders per week with the mowing of eelgrass, while the truck system was maintained; they had to buy Koning's goods and groceries.

Following continual complaints, a government committee that had to investigate the lease system's pros and cons was inaugurated in 1869. It concluded: 'The problem with both the private and the public lease is that it delivers the eelgrass mower to the lessees' arbitrariness. The lessee determines how many bales the mower can harvest and fixes the wage. The eelgrass mower is powerless and cannot resist the decisions of his principal' (Anonymous 1870:227). In the mowers' interest, the committee deemed it unwise to continue the public lease system. Should the government decide to continue with it despite their opinion, it was advisable to lease smaller plots so that competition would be enhanced. Another option was to let the municipalities regulate and supervise the eelgrass fields' exploitation to prevent overproduction. For a small charge, they could hand out permits to the mowers who would be free to deliver to a dealer of their choice. The committee considered this important because in that way 'the local eelgrass mowers will find work and will not have to be supported by poor-relief funds' (ibid.:229). The state opted for the eelgrass fields' division into smaller plots but continued the public lease system for the higher state revenues.

The next public lease of fourteen plots in May 1871 yielded a total lease fee of 30,000 guilders. The prices for the mowers were maintained at the level of four guilders per bale. For some time, the industry prospered, harvests were good and at least the merchants benefited. This was partly due to an expansion of the market through improved communication and transportation. However, in 1873 the harvest was small and the mowing wages weighed heavily upon the lessees' exploitation bill, while trade prices had decreased. For this reason, the lessees – who had gained the rights of exploitation for the duration of twelve years – could not pay the lease fee and terminated the lease contract (which was possible after three years). In 1874 and 1877 new public leases followed, resulting in an annual lease fee of 18,610 guilders. A Wieringen man leased most of the plots. He raised the mowing wages but often deducted money for alleged poor quality of the eelgrass. Therefore, he often paid less than his predecessors. There were also other ways to pass on the trade's risks to the mowers. Koning, who still leased some small plots, invited Texel eelgrass mowers to mention the wage they were willing to work for. Although it would be to their advantage to bargain collectively, they were hopelessly divided and Koning benefited. By the end of the 1870s, the eelgrass industry had lost its attraction for many fishermen. Scores of Oosterenders turned to North Sea fishing in the summer. It was one of the few alternatives in this season. However, Oudeschilders were financially less strong and could not afford to buy seaworthy boats. To eke out a living, they had to continue working in the eelgrass industry.

In 1882, a new public lease was imminent. Koning sent a petition to the Minister of Finance, pointing out that a coalition of lessees had been formed that would keep wages low, while cashing considerable profits. He thought it more correct that these advantages should be passed on to the mowers, which was possible if the Minister granted permits to the fishermen for a small charge. The fishermen themselves also demanded 'free labour and free trade'.[11] Texel's municipal council supported Koning and the eelgrass workers and deemed it pernicious that one lessee should have a monopoly. Again, the request was ignored. At the next public lease, the bids almost doubled; one lessee even paid a fee of 31,000 guilders. However, he soon had to terminate the contract, and subsequent leases yielded considerably smaller lease fees. The demand for eelgrass had decreased dramatically and competition was fierce. Repeatedly, the costs were passed on to the mowers, who took the brunt of the industry's crisis. Not only did they have to accept lower prices for the eelgrass they supplied, they also had to pay for transport, the rent of ditches, and the storing and pressing of eelgrass. Although the amounts of money for these services were fairly small, they could not really be missed. Small wonder, then, that the enthusiasm for participation in eelgrass mowing continued to dwindle in the second part of the 1880s. In 1886, for instance, only thirty Texel boats (mostly from Oudeschild) were involved in the work. The mowers landed so much eelgrass that the sheds were still stocked in the next two seasons and the lessees did not need new supplies. The price of eelgrass declined so sharply that the exploitation costs were hardly covered. The Texel fishermen, especially those whose vessels were unfit to be used for North Sea fishing, again experienced the lease system's disadvantages and begged the appropriate state institutions to end the lease system. As of the first private lease they and their Wieringen colleagues had regularly complained about the low wages and wage cuts, the small advances upon delivery, and the late final account, their dependency on the lessee and the truck system. The fishermen, who despite their ambivalent attitude in this respect valued economic liberty, often experienced the interdependency between themselves and the lessees as a burden. Nonetheless, they continued to be opportunists.

Following the crisis in the eelgrass industry, the municipal council of Wieringen began to urge the Minister of Finance to withdraw public leasing. Its aim was to regain control of eelgrass exploitation, so that it could once again act as buyer and trader. The mayor wrote to the Minister: 'Public leasing is an immoral system based on extortion, swindle and theft vis-à-vis the mowers' (quoted in Hoekstra 1939:VI-17/18). These complaints and the diminishing lease fees led to the Minister's proposal in 1895 that municipalities should lease the eelgrass beds privately. Texel refrained from doing so, but Wieringen grasped the opportunity with both hands, leased the eelgrass beds and established the *Gemeentelijke Zeegras-Exploitatie* (Municipal Eelgrass Exploitation), a not-for-profit enterprise. Henceforth, the Wieringen administration distributed the concessions to mow eelgrass. It also determined the maximum quantity that could be harvested. Partly because of limited production, the season was shortened from June to mid-July. The Minister demanded that the mowers earn a fixed price per bale, part of which had to be paid in cash upon delivery and the remainder within a set term. A municipal inspector supervised the quality, and if it did not meet the standards, the price could be decreased.

The mowers were de facto on the payroll of Wieringen municipality. It supplied some large dealers who paid predetermined prices. These dealers often bought large quantities in years when prices were low and speculated that they would benefit if market prices rose. However, if they thought the prices set by the Municipal Eelgrass Exploitation were too high, they bought only a little, leaving Wieringen stuck with huge stocks.

It was not to the Texelians' liking that Wieringen had gained almost complete control of the eelgrass industry. They insisted on a lease by public bidding and did so not because they were in favour of such a lease system, but because it appeared to be the only way to improve access to the industry. Texel's mayor met with the Oudeschild eelgrass workers, who told him that Wieringen preferred to hand out mowing concessions to its own fishermen. The mayor judged the situation no better than before. Changing his mind, he tried to lease eelgrass fields in Texel's vicinity. The Minister was susceptible to his arguments and publicly leased some extra eelgrass plots. Still, the municipality of Wieringen, eager to maintain its monopoly, turned out to be the highest bidder and acquired the exploitation rights. Henceforth, it even succeeded in leasing these beds privately, so that it could set the amounts that could be mown in order to prevent overproduction. This went against the grain of Texel's eelgrass mowers, who had to ask for permits through the local agent of Wieringen's municipal eelgrass enterprise, Albert Dros, a dealer in washed-up eelgrass, shells and cockles. Dros was an important Texel merchant and like so many others before him, he also used the truck system. Fewer and fewer Texelians applied for permits, the reason also being that the Wieringen enterprise only paid a small amount upon delivery, settling the final accounts the following summer. Many could not wait that long for their money and gave up eelgrass mowing. In most years between 1895 and 1915, no more than twenty Texelians worked in the eelgrass industry.

The Texel fishermen's struggle to ameliorate their destitute situation continued. Occasionally, they obtained minor successes as when they were allowed to mow more bales of eelgrass and the state reduced the lease fees, but wages remained low and final accounts came irregularly and usually were long overdue. Whenever there was a surplus stock, they could supply less and when all costs were subtracted, their net incomes were very small indeed. They repeatedly requested the local and state governments to come up with solutions, to no avail. When in 1912 Wieringen again leased the eelgrass fields for a six-year term, all efforts to improve the situation seemed wasted. The next year, the Texel and Wieringen mowers applied a new strategy: they refused to work for the wages offered, called a strike and under this threat the Wieringen eelgrass enterprise raised the wages from 4.50 to 5.50 guilders and in 1918 even to seven guilders per bale. The eelgrass workers' bargaining position had slightly improved through the establishment of associations and because some fishermen had become representatives in the municipal councils of Texel and Wieringen. Especially in Oudeschild, where the Labour Party had recruited a large following, the inhabitants attempted to amend their situation through political and economic organization. This increased leverage was of little use when the fishermen had to cope with the consequences of geopolitical forces. During and after the Great War, the demand for eelgrass declined and the industry languished. The most important markets, Belgium and France, which used to buy three-quarters of the eelgrass harvest, could not be supplied due to the hostilities. Following the war, trade limitations and unfavourable exchange rates deepened the recession.

It was only in the mid-1920s that the market recovered. Subsequently, the eelgrass mowers received a share of the profits Wieringen municipality gained with the eelgrass trade. On Texel, eelgrass gathering had become even more important than eelgrass mowing. The amount gathered exceeded the amount mown by approximately three to four times, but net returns were low because prices were bad and many intermediaries were involved in the trade of washed-up eelgrass. Nonetheless, entry was free and gathering eelgrass enabled scores of islanders to weather the winter season. By the end of the 1920s, the islanders earned an annual sum of between 68,000 and 70,000 guilders on aggregate with the gathering of eelgrass and another 10,000 to 11,000 guilders with the mowing of the sea-plant. Besides gatherers and mowers, many others were involved in the trade, transportation, drying and pressing of eelgrass. The exploitation of eelgrass constituted an important link in the annual fisheries cycle. It also provided an income during the summer for fishermen with insufficient capital to buy a seaworthy vessel. In the 1930 season, most reapers who owned a small vessel earned fifty to eighty guilders during the eelgrass campaign; un-propertied fishermen received slightly less. The mowing wages were increased to eight guilders per bale, but soon serious problems assailed the eelgrass industry, as harvests diminished sharply. According to biologists, a mysterious disease – the wasting disease that struck the entire Atlantic coastal area – caused this decline. Yet fishermen maintained that the main cause of the decline was the Closure Dam, the construction of which was completed in 1932. Currents and salinity changed and within a few years the eelgrass vanished not only from the Zuider Sea, but also from the waters north of the dam, currently known as the Wadden Sea. Whatever the exact cause, the events ended the industry and left many dependent on eelgrass harvesting in dire poverty (see Chapter 3).

Casting the Nets Far and Wide

The various management regimes in eelgrass harvesting led to recurring resistance of eelgrass mowers against their concomitant restrictions and the distribution of exploitation rights to private persons or corporate organizations. These regimes were not related to ecological problems but to economic ones. The sustainable use of eelgrass would probably not have been endangered in an unlimited entry situation, for despite growing exploitation the quantity of eelgrass did not diminish, on the contrary, it increased. In 1870, the eelgrass fields in the northern Zuider Sea measured over sixty-five square kilometres, while half a century later this area had more than doubled. Yet complete freedom would probably have resulted in the fishermen not showing self-restraint in the commodity's appropriation. In a situation of unbridled competition, the market would have been flooded, and the eelgrass workers' remuneration would have dropped drastically. It is clear that the regulations were primarily intended to control the distributive channels through restrictions of production. In the oyster industry, by way of contrast, the privatization of plots was closely connected to overexploitation. Despite the enclosure of some underwater grounds for eelgrass exploitation and oyster cultivation, access to a large proportion of waters near Texel remained free, while North Sea fishing was hardly restricted at all. The fishermen thus had many opportunities to switch target species and exploit other ecological niches.

As we have seen in Chapter 1, offshore fishing has been an important venture on the island for centuries. Initially concentrated on the island's western – North Sea – coast, by the early 19th century the fishing communities were located on the eastern coast. With the demise of piloting, many poor Oudeschild men switched to inshore fishing. Their fishing fleet comprised only nine vessels by 1800, but it would gradually expand. Several inhabitants of the new village of De Cocksdorp, established in 1836, also turned to fishing. The former initially targeted herring and later also several other species; the latter fished cockleshells that were processed into lime in mainland kilns. In 1839, the island's fishing fleet boasted 111 boats; a decade later there were 130. By this time, most fishermen combined harvesting oysters, eelgrass and shells to earn a living, occasionally targeting other species depending on such factors as weather conditions, availability of resources, equipment and market prices. For many, fishing constituted an escape route in adverse economic times even though external authorities did not stimulate this development. For a long time, the state had been favouring the large-scale offshore herring and cod fisheries, taking protective measures to ensure that the fishing companies (*rederijen*) had a monopoly on providing markets. However, the rules and regulations turned into a straitjacket and obstructed fair competition by the petty fishermen. The near-shore and inshore fisheries were pretty much ignored by state institutions. In 1854, a government committee was installed to look into the state of affairs in the fisheries. It concluded that the restrictions led to unjust relationships. Three years later, a new Fisheries Order was launched that withdrew the protective measures with regard to the large-scale fishing industry and opted for a laissez-faire policy concerning the petty fishing trade. The inshore fishermen would benefit considerably from this change of affairs, as there were very few obstacles for anyone wishing to enter the fishing industry.

The cultural ideal was self-employment and independence, to be achieved through the ethos of entrepreneurialism, thrift and diligence.

Many fishermen therefore wanted to own their own vessel. A considerable investment of between 1,500 and 4,000 guilders was required for a new boat, so the fishermen often had to rely on creditors to achieve the aim of becoming an owner, at best giving them limited freedom. Most creditors were suppliers of fisheries goods and services, dealers or processors. For this reason, a truck system could easily develop, leading to higher fishermen expenditures. Even if fishermen redeemed their debts, they usually felt morally obliged to continue the relationship with their former creditors. Some suppliers used their position to acquire boats that were operated by a hired crew. For example, an Oudeschild sail-maker owned six vessels. Creditors ran limited risks, as they were usually suppliers of goods and services and knew their customers would return to them. Heavily indebted as they were, the fishermen generally had to match high production with low consumption. Even second-hand vessels could often only be acquired through loans. Being free of debt was symbolized by gilding the masthead, a sign of pride most fishermen could not afford themselves.

Those in the best position to break free of disadvantageous ties with moneylenders and middlemen were the fishermen operating in family firms. Young boys not even in their teens joined their fathers aboard, usually got paid nothing or only some spending money, received on-the-job training and gradually learned the ropes of the occupation, in particular to grasp the clues provided by the seascape for navigation and fishing proper. Having a son could thus save on money that otherwise had to be spent on paying a non-related deckhand. Pooling their resources, agnatic kinsmen were able to maintain flexibility and, with a bit of luck and ardent labour, accumulate capital. Often, a skipper's wife or his daughters would also contribute money they earned to the firm. In many cases, fishermen's spouses for example ran a small shop, did the laundry for well-to-do families, sewed clothes or worked as cleaning ladies. Young girls would usually also contribute to a household's revenues by running errands or working as a maidservant. As family firms were units of production and units of consumption, economizing on household expenditures was an oft-used strategy to weather adverse times. Ideally, each son of a skipper-owner would have his own boat, while aspiring deckhands often attempted to become independent. Using the share system of remuneration, owners and crewmembers alike bore the risks of production and benefited from capital investments, making them co-adventurers with overlapping interests in keeping costs low and proceeds high. The system became widespread in the last quarter of the 19th century. It was based on a division of proceeds – after subtracting costs – between the owner and the deckhands. A boy would begin as an apprentice 'third man', gaining a smaller percentage than a full-blown deckhand did. Crewmembers were hired from Christmas until Christmas the next year. The social distance between skipper and crew was mitigated by the fact that they were often related and shared the risks of the trade. The emphasis was on cooperation and equality, and the skipper was a *primus inter pares* (Löfgren 1972:99). Moreover, when the fishing economy boomed, many deckhands succeeded in setting up their own businesses.

The fishing fleet continued to expand – not just on Texel, but on the entire Zuider Sea coast. Because of infrastructure and market developments, there was a growing demand for fish and fish products and more opportunities for Texel fishermen to land a variety of species. Previously, the island's relatively remote location and the perishable nature of all fish species proved to be an obstacle to market what was available. It is no coincidence that the fishermen focused on shells and dried eelgrass, which did not go off, and on oysters, which could be kept alive fairly easily during market runs. As of the 1870s, Texel catches could be transported further away by rail and by a regular steamboat connection between Harlingen and Hull. They began supplying markets at home and in Belgium, France and England. According to various reports in the local newspaper, the island fishermen could sell whatever they caught or harvested at sea. Since access to a large proportion of waters near Texel remained free, the island fishermen had ample opportunities to switch target species. Actually, many began pursuing species they had previously disregarded either because it was difficult to find market outlets, or because of the rapid deterioration of fish. With faster and more regular transport and new conservation methods, these problems were overcome. For example, the boiling and salting of shrimp and cockles was introduced on the island in the mid-1870s. The preserved seafood was shipped mainly to England and, after the turn of the century, to France.

In addition to oysters, eelgrass and shells, Texel fishermen began catching periwinkles, cockles, whelks and mussels, starfish, shrimp, flounder, sole and plaice, rays, anchovy, garfish, eels and several other species of fish. Depending on the season and the economic situation, Texel fishermen sailed the Zuider Sea and/or the North Sea, taking optimal advantage of the variety of ecosystems in these marine environments and utilizing the niches and species they deemed most profitable. A boom in coastal fisheries characterized the period 1870-1895. Deckhands began earning considerably more than farmhands. In Oudeschild, where until mid-century there had only been a small fishing fleet, scores of inhabitants entered the fishing industry and the Oosterend fleet expanded and modernized. At the peak of the fishing fleet's expansion in 1895, the island counted approximately 175 boats, crewed by over 500 skippers and deckhands. Along with this development went a degree of functional differentiation. The owner of the local shipyard, two sail-makers, a few tanners and blacksmiths, and several fish processors and traders benefited from this expansion. Together, they hired scores of labourers and the fishing industry thus provided work for many hands directly and indirectly. Near the harbour were two warehouses for the storage of ice that was cut from canals in the winter and several warehouses for the storage of dried eelgrass. Elsewhere on the island, there were dozens of eelgrass storage sheds. By the mid-1890s, Oudeschild boasted six plants where shrimps were boiled and salted and there was another one in De Cocksdorp that preserved cockles in a similar way. The fishermen could still buy commodities and services on tick with artisans and shopkeepers – some of whom were also fish dealers and processors – on the understanding that they would be loyal customers in future. Occasionally, the processors would introduce quotas to balance supply with demand and keep prices at levels that made exploitation worthwhile. However, self-restraint was usually short-lived as fishermen from other communities also operated on the market for shrimps and cockles, making control of supply and prices impossible as long as supra-local agreements were lacking.

Some fishermen were part-timers, including those who beach-seined for herring in the spring. A score used small flats (*vletten*), operating hoop nets (fykes). A number of Texel fishermen specialized in mussel fishing or temporarily caught mussels as a part of their annual fisheries cycle, switching to other species when these fetched comparatively better prices. Mussel fishing took flight between 1873 and 1890, when Texel fishermen landed between 3.5 and 13 million kilograms of mussels annually. They exported mussels to England, where they were used as bait in the offshore fisheries. Although some Texelians attempted to establish market outlets for consumption mussels in Belgium, they were unsuccessful as Zeelanders already operated there. After 1890, the demand for 'bait mussels' diminished sharply. Most Texelians gave up mussel fishing for the time being. In addition, prices of shells fell when mechanized harvesting with steam-powered vessels was introduced in the 1880s. The approximately sixty-five fishermen who were involved in shell fishing could not compete with mass production. This turning of the tides forced fishermen to diversify even more, but most of the fishermen from De Cocksdorp, who had been focusing on shell fishing for a long time, either opted for a job ashore or emigrated (see below).

Texel fishermen behaved as veritable opportunists. They worked as farmhands when fishery activities slackened, as was usually the case in the summer. Several fishermen owned some plots of land, but farmers never worked for fishermen. In general, occupational pluralism was particularly practised in the ranks of the poorer fishermen, who lived mainly in Oudeschild, De Cocksdorp and the hamlet of Oost. The Oosterenders mostly concentrated on plying the dangerous North Sea, but they, too, occasionally picked up other pursuits, especially during the winter when inclement weather forbade offshore fishing. From figure 1 it is clear that only flatfish fishing was feasible throughout the year, but that a rather small number of Texel fishermen continued during the winter season. The reason for this was that few vessels were fit to sail the North Sea in rough weather. A considerable portion of this segment of the Texel fishing fleet converted to shrimp fishing from November until March. North Sea flatfish fishing yielded particularly good results during the times of boom. Weekly gross revenues of 150 to 200 guilders were not exceptional. Scores of fishermen combined shrimp fishing with eelgrass harvesting. Others switched between pursuing several species over the annual cycle of fishing. From April until October, several fishermen had to find temporary employment outside the fisheries, mainly as farmhands. Seasons for some species were rather short, for example in the herring, anchovy, eel, garfish, and ray fisheries (see Appendix A for images of the main species Texel fishermen landed). Herring and anchovy migrated to the Zuider Sea in early and late spring, respectively. In some years, catches were so abundant that they made for an excellent grossing that surpassed the proceeds of all other fish the fishermen landed during the remainder of the year. This was the case in 1890, for example, when the gross revenues during the brief anchovy season amounted to between 1,000 and 1,700 guilders per crew. However, rich catches did not necessarily imply good proceeds. Looking back on his fishing career, a 73-year-old fisherman told a news reporter in 1939: 'I can hardly think of any other occupation that is so precarious. If you had a fine catch and returned home early in a good mood, the mood would change when you learned that others also had many fish. For a large supply implied such low prices that you would have made even more money with half the amount of fish landed' (*Texelsche Courant*, 30 September 1939).

A short-term orientation made for flexibility with respect to deciding which niches and species could be exploited best. Decisions on fishery cycles and gear switching were largely made on the basis of resource availability and accessibility, fish prices, fishing knowledge and expertise, vessel size, personal preferences, and alternative sources of income. For most, these strategies implied success. By the end of the 19th century, Texel fishermen utilized a wide spectrum of marine resources, and oysters were caught during a short season only. More and more, the fishermen alternated their catches. Thus, after a short spell of intensification, the Texelians opted for diversification, a strategy fishermen elsewhere also applied if the catches of any one species declined (McCay 1978:409ff.). The unintended and unforeseen consequence of the fishermen's adaptive strategies was that the pressure on resources was dispersed, alleviating pressure on any single species. Therefore, the carrying capacity of these species was not undermined, which could have been the case if all the fishermen had concentrated their efforts on one or only a few resources. There were brief spells when many people focused their endeavours on a single resource because prices were high, the catches of other species proved disappointing, or inclement weather prohibited sailing the open sea. Some migratory species – in particular herring and anchovy – were available during a brief season and did not always show up, and if they did, many fishermen from around the Zuider Sea coasts competed for a share of the catch. This would sometimes lead to crowding, especially when initially prices were high. As the number of competitors increased, prices usually dropped and if better proceeds could be obtained in other fisheries, many fishermen would switch to other species. Pluri-activity had obvious merits as an effective insurance against resource decline and market fluctuations, although

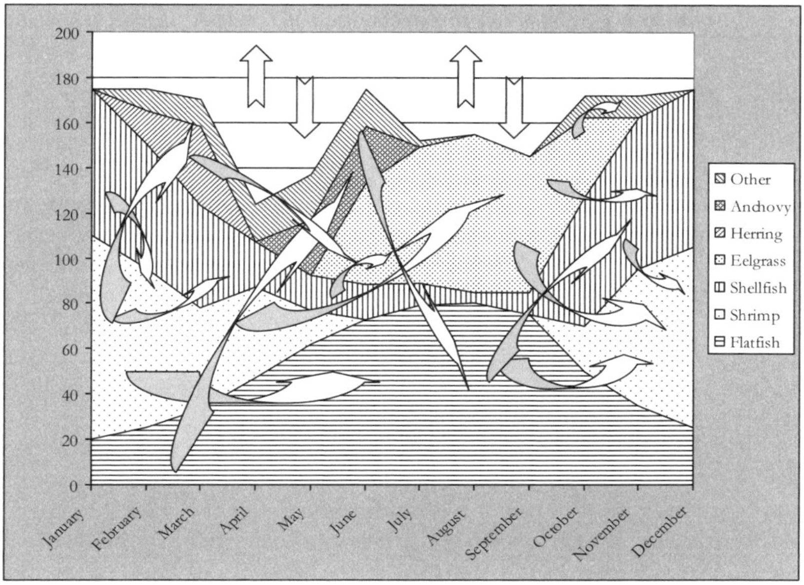

Figure 1. *Seasonal cycle of the Texel fisheries, ca. 1890. The vertical axis represents the number of vessels in specific fisheries. The arrows indicate species switching and seasonal exit or entry. From April until October several boats were idle as fishermen preferred taking on seasonal land-based work. Flatfish include sole, plaice, flounder, turbot, brills and dab. Shellfish include oysters, mussels, cockles, periwinkles, whelks and (from April until October) shells. The category 'Other' includes rays, garfish, eel and starfish.*

it required greater capital outlays. Their strategy of diversification required the fishermen to have various sets of gear and to acquire the necessary knowledge and skills to pursue specific species. In addition, it was a prerequisite that fishing grounds were freely accessible.

It was the fact that marine resources were held in common that enhanced the fishermen's versatility and shock-absorbing capacity. However, multiple use of the inshore zone could lead to gear conflicts between various groups of fishermen using the same locations. For example, applying mobile gear such as trawls or dredge nets was incompatible with using stationary gear such as hoop nets, set nets or fish traps. A limiting factor on gear and niche switching were the claims some fishermen made on exclusive entry and use of certain locations, especially if they used fixed gear. Initially this was done on a first-come, first-serve basis, but as of 1907 the Fisheries Inspection handed out permits. In addition, leased beds for shellfish cultivation were inaccessible for non-lessees. Disputes over space were usually zero-sum games in which there were winners and losers. As we have seen, Texel fishermen generally opposed the enclosure of the commons. Even in circumstances where natural resources are not the *de jure* property of anyone, there may be rules regarding access to and use and control of these resources. These claims can turn into regimes of appropriation and control. Yet rules and practice are not always in accordance with each other, exclusion of outsiders may constitute a problem, and under certain conditions the pressure on natural resources may increase strongly. Repeatedly, the ambivalent attitude of Texel fishermen towards territoriality was a problem, as there were competing rights and claims to legitimate use. Although they claimed the exclusive use of certain fishing grounds, they rarely acknowledged the rights of others who claimed fishing grounds elsewhere. This phenomenon, widespread all across the globe, is known as the 'tragedy of incursion' (McCay and Acheson 1987a). More importantly, territoriality should not be seen as a form of resource management. Rather, it is based on the motive of privileged space. Exclusion of outsiders is highly problematic; whereas some individuals or groupings abide by customary practices, others may not. In this respect, the fishermen did indeed consider the marine domain a commons they could freely exploit. Although some fishermen were sometimes able to exercise informal use rights, entry into the fishing industry was relatively easy. This freedom also had a darker side, in that competition was fierce and on occasion, the fishermen themselves called for measures restricting their liberty.

Reluctant Competition, Ambiguous Cooperation

Partly on the insistence of fishermen, and partly because of the onset of biological investigations of fisheries, the state increasingly intervened in the fisheries as of the late 19th century. Through the establishment of organizations, fishermen attempted to achieve political participation, gain leverage and increase their bargaining power. They frequently resorted to informal contracting and institutionalized solidarity to mitigate open-access conditions. However, diverging interests of specific categories of fishermen led to disagreements on the kinds of measures to be taken and, generally speaking, their attitude towards state intervention was ambivalent. With a keen eye for the fishing industry's sociological dimensions, the fisheries biologist Hoek pointed out 'how much the social position of fishermen differs internally and how difficult it is to bring about cooperation in such diverging social units' (1902:115). A rift existed between small-scale fishermen and capitalist boat owners, lessees of fishing grounds and adherents of *mare liberum*, and between producers and dealers, because each of these categories cherished its own wishes and posed its own demands. These remarks regarding the diverging interests in the Dutch fishing industry are apt and also apply to the Texel situation. If they perceived advantages, Texel fishermen favoured regulation and urged the state to intervene, but as soon as they saw disadvantages, they vehemently opposed it and demanded the withdrawal of regulatory regimes. Sometimes, they asked for shorter seasons to protect immature fish and shellfish, but when the fisheries were struck by crisis, they often opted for an extension of the season. There was also ambivalence concerning gear, mesh and fish-size regulations. Fishermen would demand tight enforcement at one moment and at the next call for more lenient restrictions. In addition, the fishermen protested against the part-enclosure of the marine commons when the state leased privatized oyster beds and eelgrass plots to the highest bidders at public auctions. For instance, a Texel Fishermen's Union was established in 1887 whose chief purpose was to seek abolishment of privatization. The Texelians rhetorically contended that the seas around their island had always been 'free', but the fact that they had claimed and gained communal territorial use rights in oyster fishing and eelgrass harvesting in an earlier stage contradicts this viewpoint.

In the late 19th century, there were several attempts to arrive at a national organization of fishermen, usually as a federation of local associations. However, because of the lack of concerted action 'the fishermen's organizations attained hardly anything' (Bossaers 1987:210). The efforts failed repeatedly, as local branches of a national fishermen's association were converted into local organizations that steered their own courses. It was much easier to identify with community members than with outsiders. Even Oudeschild and Oosterend fishermen predominantly clung to their own organizations (see below). If cooperation at the island level was difficult, this applied even more to solidarity with colleagues from outside, which was usually short-lived. The switching of allegiance in part explains the frequent rise and fall of voluntary associations on Texel. Between 1870 and 1932, there were twenty-odd fishermen's organizations on the island, equally divided between both fishing communities. They ranged from general interest organizations and mutual insurance funds, local branches of national or sectional unions, cooperatives and trade associations to single issue action groups. These associations, funds, unions and cooperatives commonly had a multi-purpose character and besides short-term objectives, they developed long-term strategies. They were a means of adaptation to changing circumstances (Orbach 1980:55). Among many other things, they sought state support for infrastructural and other improvements, more bargaining power and leverage vis-à-vis state institutions, fish traders and fish processors, and self-help to support fisher families in case of mishaps and damage.

Between 1894 and 1904 alone, thirty Texel fishermen lost their life at sea. Such tragedies were aggravated if the mortal remains could not be recovered or identified. Apart from burial being important for religious and ritual reasons, a wife could not claim an inheritance or remarry if her spouse was still missing. The increasing number of accidents and deaths, which were partly linked to the growing number of fishermen sailing the North Sea and the share system that encouraged taking risks, led to the foundation of a widows' and orphans' fund to make the victims less dependent on charity. The fund, *Texel's Belang* (Texel's Interest), was established in 1904 and for some time united all of Texel's fishermen but membership fell after financial troubles and rumours about fraud, leading to internal strife and dissension. Following the introduction of a state insurance system, the organization stopped functioning in the early 1920s. At this stage, *Texel's Belang* was the only voluntary association that catered to the entire occupational community of fisherfolk on the island. All other organizations functioned at the village level.

In general, the membership of local fishermen's organizations was large. Few stood aloof as there was strong social pressure to join and the boards kept fees modest in order to encourage membership. Nonetheless, sustained collective action proved to be difficult. During economic recessions, many could not even afford the small fees and cancelled their membership. The associations therefore occasionally faced a free-rider problem. Even at the local level, dissension arose easily, sometimes over minor details concerning policy or personnel. In some cases, factional strife disrupted a local association, often leading to fission. Two organizations in one village then competed for resources and members, but since the occupational communities were only small, eventually one of them would perish or they would reunite. Their rapid rise and decline might give the impression that cooperation came about swiftly and spontaneously, dissolved quickly, and was rekindled easily. To a certain extent, this impression is accurate, though in many cases an existing association, union or cooperative was continued under a new name, but with the same board, the same members and more or less the same goals as its predecessor. Often little changed in practice. At the local level, organizations brought about the social integration of the occupational community and contributed to reinforcing the fishermen's bargaining position. Through their associations, Texel fishermen attempted to wield power and to defend their economic, political and social interests with local, provincial and national governments; to gain leverage in their dealings with traders and processors, on whom they depended for the marketing of their catch; and, more generally, to prevent marginalization and to ameliorate their socio-economic situation.

As has been noted above, the quarter of a century after 1870 brought relative prosperity to the fishing industry. Subsequently, however, several problems confronted the Texel and Zuider Sea fishermen and fishing communities. Many voted with their feet and an exodus of Texel fishermen ensued. Between 1880 and 1913, no less than 152 fishermen with 494 relatives left the island and settled on the mainland – many of them taking on non-fishery related jobs – or emigrated (Schreur 1953:24-27). Seventy-eight fisher families were from Oudeschild, forty-six from Oosterend (including many from the hamlet of Oost), twenty-seven from De Cocksdorp and one from Den Hoorn. Seventy fishermen continued their old occupation elsewhere, mostly in neighbouring Den Helder. Twenty-six fisher families migrated to the USA; seventeen of them were from De Cocksdorp, five from Oudeschild and four from Oosterend and environs. The majority of those who migrated to the mainland began working as unskilled labourers in industrial areas. In the Netherlands, the belated industrialization provided many job opportunities. Mainland industries could absorb labour that had become redundant in agriculture and fisheries or had not yet entered it. For example, thirty-seven fisher families, mostly from Oudeschild, migrated to Amsterdam and environs, where the adult men found employment as dockworkers. There, they became acquainted with socialist ideas that they took with them upon their eventual return to the island.

In the impoverished village, neighbourliness and solidarity reigned supreme. In addition to informally assisting each other, the villagers founded a number of local associations aiming at mutual support in times of sickness, death and other adversities. Strikingly, the Dutch Reformed church – to which more than seventy per cent of the local populace nominally belonged – did not play a significant role. After an early 1860s conflict in the church that involved the minister and the church authorities, most villagers began avoiding Sunday services. They opted for self-help. A significant minority of Oudeschilders were Roman Catholic (23.3 per cent in 1849). Less than three per cent were Mennonite. Oudeschild was a close-knit and socially cohesive community, where social control was tight. Although the fishermen were self-employed petty commodity producers, they confronted capitalist relations of production and they considered themselves to be labourers rather than independent entrepreneurs. The Social Democratic ideology could explain their marginalized position and provided a model for a way out of poverty. As most Oudeschilders had previously turned away from the church, they perceived socialism as a vehicle to betterment and emancipation. Oudeschild subsequently turned into the island's 'Red Village'.

In contradistinction, churches played a pivotal role in Oosterend community life. In the mid-19th century, almost three-quarters of the inhabitants were Dutch Reformed. There were Mennonite (eighteen per cent) and Roman Catholic (eight per cent) minorities. Sunday church attendance was high, but the Protestant locals did not refrain from pub-crawling, sprees and dancing and in general were rather latitudinarian. This changed with the arrival of orthodox ministers. Revivalism began to be a force. In 1850, conventiclers attracted more and more followers and later a Calvinist Reformed (*Gereformeerde*) church would be established. The orthodox Protestants were dubbed 'the fine' (*de fijnen*), while the heterodox Protestants, Catholics and non-believers were called 'the coarse' (*de groven*). The religious differences would set the tone for local relationships for decades. They not only determined in which church one worshipped on Sunday, but the village's entire communal life was deeply divided according to religious denomination as nearly all organizations were based on religious affiliation, while there was both a secular and a Protestant school. Mixed marriages were taboo, and shopping was done with shopkeepers of the same creed. The differences even had a spatial dimension as 'the fine' predominantly lived in one neighbourhood and 'the coarse' in another. Generally, the orthodox proved to be more tightly integrated than the latitudinarians. Many fisher families were orthodox and seemed less inclined to migrate than the Oudeschilders.

For those skippers who stayed behind, it was increasingly difficult to recruit deckhands. The mean grossing of boats fell from 1,650 guilders in 1905 to 750 guilders in 1914. By this time, nearly all skippers worked with the share system of remuneration. It varied according to the type of fishery. For example, in shrimp fishing a deckhand would get a third of the net proceeds, in North Sea fishing twenty-five to thirty per cent, where an apprentice 'third man' would get a fixed amount of money. The weekly incomes were usually less than seven guilders per fisherman. Few men wanted to join a fishing crew. This forced several skippers to also find work on shore permanently or temporarily, for example as farmhands. The remaining fishermen tried to cope with the uncertainties by pursuing new markets and making mutual arrangements regarding production and minimum prices. Along with the expansion of markets and the introduction of fish conservation methods, the chains between producers and consumers grew longer. This left the fishermen in a position at the production-market chain's outer end, making them vulnerable to recessions and sharp price fluctuations. As Sinclair states: 'the large number of petty producers for a given market makes it difficult to adjust production in the aggregate, with the result that particular commodities tend to be overproduced relative to demand and prices are driven down' (1985:20).

Whereas the skipper-deckhand social relationship was usually rather egalitarian, relations between fishermen and merchants were often asymmetric and antagonistic. Given the rapid deterioration of fish, crustaceans and shellfish, the fishermen had little 'withholding power' (Löfgren 1977:228). They depended on intermediaries for marketing their catch and they were price-takers not price-makers.[12] For instance, dealers in the nearby port of Den Helder, where many Texelians landed a part of their catches, had developed usurious practices. By collectively protesting this state of affairs, the fishermen succeeded in countering the dealers' behaviour. We also have seen that a truck system with extremely unfavourable conditions developed between Texel fishermen and large-scale shell, cockle and eelgrass dealers. They often had to dispose of their catch in exchange for commodities in the dealers' shops while they usually had standing debts that firmly tied them to their creditors. The fishermen's bargaining position was rather weak. In the eelgrass industry, this even led to incipient class divisions. At the local level, however, several traders were petty entrepreneurs and depended on the fishermen for the supply of fish and shellfish. Both categories were involved in a 'moral economy': 'a consistent traditional view of social norms and obligations, of the proper economic functions of several parties within the community' (E.P. Thompson 1971:79). The fishermen attempted to strike deals with individual traders concerning production and prices. This moral economy often amounted to a rhetorical ideology: we *should* all cooperate with each other here. In fact, cooperation was usually short-lived, since fishermen and traders were opportunistic and ambiguous with respect to price and production restrictions. Eventually, fishermen forgot or violated informal agreements, turning cooperating individuals into fiercely individualistic competitors. For example, by the turn of the century the fishermen increasingly avoided supplying shrimp and cockles to processing plants. Instead, they boiled, salted and marketed the crustaceans and bivalves themselves. The processors eventually had to close down their plants; at the outbreak of the Great War not a single one was left on the island. It was also difficult to maintain voluntary arrangements, because Texelians were of course not the only fishermen supplying the market. Therefore, even when engaging in collective action or entering contractual relations with dealers, they were unable to control prices. Although in general dealers or processors had stakes in maintaining an attitude of competition among fishermen and hindering organized collective action, the Texel fish traders were not in a position to prevent the fishermen from organizing themselves. The latter began establishing trade cooperatives in order to restore the balance of power vis-à-vis fish traders and to obtain higher incomes.

It was a mainlander, J. Zwier Visser, who took the first initiative to establish a trade cooperative on Texel. In 1899, he toured Dutch fishing communities in an attempt to make them cooperate in a national federation of fishermen's associations. He also visited Oudeschild and Oosterend, and among other things emphasized the importance of circumventing as many middlemen as possible in the marketing of catches. These intermediaries not only creamed off their proceeds, they also attempted to maintain a monopsony on the island. When outside traders visited the island to buy fish and shellfish, the local traders outbid them. Thus, fair competition was impossible, but once the outsiders had left, the Texel traders returned to offering low prices. The petty coastal fishermen depended on the local dealers, so that they had few alternatives to land their catches. Furthermore, the price-fixing that was practised elsewhere also left much to be desired. Zwier Visser criticized these practices and told the Texel fishermen that it was possible to bypass the middlemen by marketing the catch directly to large dealers through trade cooperatives, which would have the advantage of higher returns and the expansion of markets. Trade cooperatives were then established in both Oudeschild and Oosterend, and immediately joined by thirty and twenty-one members, respectively. Following a difficult start, the cooperatives became fairly successful, but Zwier Visser's aim of founding a national federation was crosscut by rumours of mismanagement and fraud. Nonetheless, both local trade cooperatives continued their work under new names. The son of an Oudeschild teacher, W.A. Muller, was appointed as manager of both organizations. Thus, he could coordinate their activities and though *de jure* they remained separate institutions, de facto they cooperated. Under Muller's leadership, the cooperatives succeeded in obtaining higher prices for their members.

They even expanded their activities, found new markets and purchased fishery equipment, baskets, salt and other goods at considerable discounts. They also bought storage facilities and leased oyster plots from the state on favourable conditions. To ensure high prices, the cooperatives sometimes introduced quotas, especially for shrimp.

Until the onset of the Great War, the cooperatives worked quite well and the fishermen clearly benefited from their activities. However, as of 1914, fish and shellfish prices rose phenomenally, although the market for eelgrass collapsed. In Western Europe, Dutch fishermen faced less foreign competition, as the Netherlands remained neutral, whereas fishermen in neighbouring countries could not fish due to the hostilities. Mussel fishing, in particular, yielded excellent results. In addition, the Texel fishermen targeted various other species in the winter. The Zuider Sea herring and anchovy fisheries were profitable in the spring. In the summer, Texel fishermen who had large flat-bottomed boats otter trawled for flatfish and other species in the North Sea. With almost unlimited marketing potential and rocketing prices, the fishermen wanted to get rid of the cooperatives' restrictions. Most members withdrew and the cooperatives dissolved. Thus, in times of economic boom, most fishermen preferred sailing an individualistic course. They were ambivalent opportunists, but this very same opportunism also constituted the basis for their return to cooperation when the need arose. During the First World War, the local fleet was no longer on the decline. As can be seen from the table below, Oosterend fishermen were predominantly oriented on North and Zuider Sea fishing, while Oost and Oudeschild fishermen mostly exploited the inshore domain.

	Oosterend	Oost	Oudeschild	Cocksdorp	Horntje	Total
North/Zuider Sea boats	18	3	2	1		24
Inshore vessels	8	13	37	6		64
Rowing boats	4	9	14	5	5	37
Total number of boats	30	25	53	12	5	125
Crew	63	46	99	21	5	234

Table 1. *Texel fishing fleet in 1918 according to type of fishery and village (adapted from Schreur 1953:14).*

The upsurge of fishing during the Great War and the closing down of trade cooperatives necessitated another way of marketing the catch. Following repeated requests, in 1916 a small municipal fish auction was established at the Oudeschild harbour quay. Soon, seven local fish dealers competed to acquire part of the landings, mostly consisting of shrimp. The North Sea fishermen preferred landing their catches in Den Helder or as far away as IJmuiden and Scheveningen where they obtained higher prices. Others sent their mussels and shrimp to the market through intermediaries. Annual gross proceeds increased considerably. In 1917, they amounted to between 2,000 and 4,000 guilders. With such amounts of money, recruitment of deckhands was no problem whatsoever. The boom was, however, short-lived. As soon as peace was restored in Europe, the Dutch and Texel fishermen faced foreign competition again. Markets fell away and fish prices declined sharply. Moreover, devaluation of French, Belgian and German currencies, inflation, diminished purchasing power abroad, high shipping tariffs, rising import taxes in France, trade barriers in Germany, increasing costs of fishing equipment and soaring taxes at home exacerbated the difficulties. On the island, the fisheries depression returned in all its devastating dimensions. Along with it, the ambiguity concerning mutual agreements between fishermen and traders also surfaced again. A deal was struck to auction all landings in Oudeschild, but much to the dealers' chagrin, more and more inshore fishermen began dodging the self-imposed rules. One of the local fish traders, Abraham Boon, said the fishermen had a 'moral obligation' to abide by them and that 'an iron whip' would have 'to move them towards their ill-comprehended self-interest'.[13] Mandatory auctioning on the island turned less problematic after that, in part because the local authorities henceforth imposed and monitored the rules. The system only applied to the fleet's small-scale segment. The North Sea fishermen landed their catches elsewhere. In general, the petty Texel fishermen grew more and more vulnerable to adversities. Many fishermen became dependent on unemployment relief work or on poor relief.

This did not apply to all Texel fishermen. Several Oosterend skippers, who had been quite successful when fish markets thrived, continuously invested in vessels and equipment. For them, accumulation of capital for reinvestment was important, whereas most fishermen from Oudeschild and Oost were content to satisfy their needs and remain self-employed. The differences between the Oosterend and the Oudeschild fishermen were already observed in a 1891 fisheries report: 'There is a marked distinction in character between the inhabitants of these two villages. Added to this disparity should be the jealousy that emanates from the fact that the fishermen of Oosterend have gradually surpassed those of Oudeschild concerning the size and seaworthiness of their vessels as well as the perseverance and profitability of their business' (Verslag zeevisserijen 1891:102). As stated above, Oudeschilders perceived themselves as labourers rather than independent entrepreneurs. Through self-help and solidarity, the Oudeschilders tried to cope with the problems they encountered. The Social Democratic movement rapidly gained influence in the community. After universal suffrage was introduced in the Netherlands in 1917, the Social Democratic Labour Party would get the majority of votes in the village (forty-four per cent in 1922, rising to sixty-six per cent in 1929). Some of the local socialist leaders worked in the fishing industry and obtained seats in the municipal council. Oudeschilders were predominantly latitudinarian Protestants with only nominal ties to the local church. Between 1885 and 1925, the position of minister was vacant most of the time and in the 1920s, church attendance was negligible. Allegedly, visiting ministers even had to wipe the cobwebs off the pulpit and the Bible.

Things were quite different in the village of Oosterend. Orthodox Protestants had been gaining more and more influence. The Calvinists, who had congregated in a wooden church since the mid-1850s, built a new church in 1897 as it saw its membership increase, while under the influence of a new minister the Dutch Reformed simultaneously adopted a much more orthodox doctrine. The Oosterenders nonetheless took minor dogmatic differences to be fundamental religious oppositions and the village remained deeply divided in matters of creed. They also believed that socio-economic differentiation was God-given and they regarded the socialist goal of emancipation as being opposed to Christian principles. The support for Protestant political parties of various persuasions was large. In elections, they obtained seventy-two per cent of the votes in 1922 and sixty-four per cent in 1929. The religious divisiveness notwithstanding, at the local level there was a modus vivendi and the entire village joined forces in times of despair. Oosterend fishermen were not as indigent as their northern neighbours, exploited other niches and considered themselves as capitalist entrepreneurs. They were deeply religious and regular churchgoers. They were inclined to take more risks, as they believed that the Lord would not take their lives prematurely. The Oudeschild and Oosterend fishermen's different religious orientations, worldviews and economic attitudes to some extent hampered inter-village cooperation in voluntary organizations (see also Chapter 3).

An old cause of disagreement also contributed to this. For a brief period, the Oosterend fishermen had had their own harbour. It had been built in 1843, but its construction was such that it began silting up soon. With the demise of oyster fishing and a lack of revenues for maintenance, the harbour fell into disuse by 1852. The Oosterenders immediately began demanding the right to a new harbour and more forcefully so by the late 1880s when their fleet had expanded, arguing that Oudeschild's harbour was too remote to keep an eye on their vessels. It took Oosterend and Oost fishermen a ninety-minute walk to the harbour, sometimes in the dead of night as sailing depended on the tide. Since travelling back and forth was so time-consuming, they only returned for the weekend. This partly explains why Oosterenders turned into week-trippers as opposed to Oudeschilders, who were mainly day-trippers. When Oosterenders requested that a new harbour near their home village be constructed, the Oudeschild fishermen wanted theirs to be deepened and enlarged. This clash of interests had an impact on the social relationships between the two communities, particularly after the Oudeschilders won. As we shall see in subsequent chapters, for decades to come the fishermen from the two villages would stick to their own organizations.

With the money they had earned during the First World War, several Oosterend skipper-owners began motorizing their fleet. For example, in

1918 a skipper had a new flat-bottom built that was equipped with a 35 h.p. engine, requiring the formidable sum of 13,400 guilders. Such investments demanded a particular attitude of deferred gratification and consumptive restraint, while they often called on the entire family to contribute. With the exception of occasionally participating in eelgrass gathering, Texel fisherwomen and girls did not work in the fishing industry proper. Nevertheless, their contribution to the household budget often enabled weathering lean times or investing in the boat and equipment. A woman, born in 1908, recalled: 'My parents had to be thrifty, mother too. She was so happy if she received clothes from the family.' She continued: 'She made oilskins herself, as it saved money.' Living extremely economically enabled her parents to invest in the firm. It was a common pattern. Women would do the laundry, make and repair clothes, take care of the children and in many cases do odd jobs or run a small shop to earn extra money. Boys began crewing upon finishing primary school, while girls would do household chores and earn some money, for example as a maidservant. The latter usually saved for a trousseau, but also chipped in their bit to cover expenses. The fishing fleet's modernization led to an expansion of the catching potential of some production units, but at the same time, the island's fishing fleet was much smaller than before. In 1931, eighty-one firms operated eighty-three vessels: four cutters, fifty-six sailing boats – most of these equipped with an auxiliary engine – and twenty-three flats. In less than four decades, the number of vessels was halved. Like gear switching, turning away from the fisheries also implied a relaxation of pressure on the marine resources. The Oosterenders specialized in North Sea flatfish, and only occasionally switched to inshore shrimp fishing in the winter when shrimp prices were high or storms made it dangerous to sail the North Sea. An important element in their success was the fact that their crews were usually composed of agnatic kin. Family firms were able to continue fishing because skipper-owners working with close kinsmen did not have to pay full shares (see also Chapter 4).

Generally, the motorization of Oudeschild vessels occurred much later than in Oosterend. Most of the Oudeschild fishermen began specializing in shrimp fishing, which usually gave 'dependable' outcomes – at any rate relatively dependable compared with other fisheries. They still combined this activity with other maritime pursuits or occupational switching if necessary, and between June and November catching mature shrimp was impossible because they disappeared from the nearby waters. As of the late 1920s, small fry were reduced down into poultry feed in a local fishmeal plant. Four brothers who ran a duck farm for down and eggs established and owned it. They used the dried shrimp as fodder. Initially, Texelians protested against catching immature shrimp and demanded a prohibition, but as it was to no avail they also began landing small fry. It helped them stay in business, even though most of these fishermen could barely make ends meet. The petty fishermen were dealt severe blows during the recessions. With their feeble vessels, they could not sail the North Sea and they tried to eke out a meagre living from the crowded inshore niches. This occasionally led to overexploitation, for example in whelk fishing. Sometimes the fishermen refrained from sailing altogether because the wear and tear of their equipment would cost more than they could earn. They usually curtailed consumption and cut expenses by neglecting maintenance of their vessel. Some would not replace half-rotten planks and had their boats covered with tin sheets: the vessel's 'shroud'. This made for economic survival in the short run, but undermined future chances of recovery.

It became ever more difficult for petty skipper-owners to recruit crewmembers, especially if they could not count on their own male kin. Given the share-system of remuneration, it was financially unattractive for crew to join a skipper-owner who did not earn much. Thus, in an era of economic crisis many crewmembers left the fishery. This led to a shortage of workers, forcing some skipper-owners to temporarily leave their vessel in the harbour and find a land-based job, especially in the summer. Others sold their boat and acquired a smaller one that they could operate alone or with one deckhand instead of two crewmembers. Although some petty fishermen continued fishing part of the year, the fact that they had to eat into their capital sapped their resilience. A teacher who had been working on the island for some time noted:

> The Texel fishermen will attempt to help themselves as long as possible. It is precisely because of this fact that needs in many families are greater than the outsider presumes. Whomsoever can do without will definitely not call for support. However, when asked for it, one can be sure that it is absolutely necessary.[14]

Fewer and fewer fishermen's sons went into their fathers' line of work and more and more fishermen – crewmembers and skipper-owners alike – quit fishing altogether. They either found jobs on Texel (commonly as farmhands) or on the mainland (mostly as factory workers). Consequently, the number of Texel fishermen steadily decreased from 234 in 1918 to approximately 176 in 1930. As we shall see in the next chapter, several developments would contribute to a further decline of a specific segment of the local fishing industry.

Out with the Tide

'If the sea can be fished dead, let us at least obtain our share of the funeral meal' (Anonymous 1854:143). With this phrase, an official report stereotyped the Dutch fishermen's mentality and behaviour as early as the mid-19th century. Underlying this idea was the perception that the sea was no-one's property and could therefore be freely exploited without any restraint. Later, Garrett Hardin (1968) popularized this view in his 'tragedy of the commons' proposition. He postulates that each 'rational being' exploiting commons seeks to maximize his gain in the shortest time possible. This 'rational being' puts his short-term self-interest above the long-term collective interest. The inherent logic is that he receives the benefits of extra effort and only pays a fraction of the costs, subtracting from the potential benefits of others. The disadvantages of his behaviour are passed off on the ecosystem, the user group as a whole and society, a problem known as detrimental externalities. Because each user of the common good thinks and acts this way, the inexorable result is ecological deterioration and, finally, destruction of the natural environment. Therefore, 'rational' individual behaviour has irrational ecological and social consequences. Tragedy, says Hardin, can only be avoided through intervention by an external authority (usually the state) or by privatization: 'The social arrangements that produce responsibility are arrangements that create coercion' (ibid.:1247).

On the face of it, the tragic story of *Ostrea edulis* would seem to provide a fairly straightforward illustration of Hardin's theorem. However, as I have argued, the reasons why overexploitation came about cannot be explained by merely citing the deterministic tragedy model. Hardin assumes a direct and unmediated relationship between individual behaviour and the ecosystem, thus obscuring a myriad of factors relevant to people's use of natural resources. Hardin and his adherents equate commons with open access; assume that individual actors behave as autonomous beings, are selfish and uncommunicative, and act as if devoid of social norms and values. They also stress individualism and competitiveness over community spirit and cooperation; neglect institutional contexts; use a priori reasoning without empirical substantiation; and are ethnocentric because they view state intervention or privatization as the only solutions to the problem of overexploitation of natural resources.[15] However, fishermen do not operate in a historical, economic, social and cultural vacuum. Their beliefs, norms and values, relationships of cooperation and conflict and the institutions they have developed should be taken into account to understand the shaping and constraining forces of ecological adaptation, as these adaptations operate through systems of cultural meanings and social relationships.

Importantly, Texel fishermen considered some fishing grounds to be *res communes* (communal property). There were de facto rules and rights as to who could exploit the resources held in common, even though the fishermen were rather ambivalent about territoriality. For a long time, the system of quasi-cultivation of oysters worked rather well. It was only under particular conditions that overexploitation emerged. Exogenous forces, ecological and technological developments, and fluctuations in a market economy were important. The Texel example demonstrates that in situations of ecological deterioration (not necessarily caused by fishing), indebtedness and impoverishment, fishermen face dilemmas that – if they have no or few alternatives – may force them to opt for survival in the short run with all the detrimental ecological and social consequences that it entails. For the individual, it may be a reasonable choice to intensify exploitation, particularly if he depends entirely on a single resource (Ostrom 1992:297). However, this is rarely the case under common pool conditions and usually the pressure on the resource is alleviated as a result of various adaptive strategies. We have seen that Texel oystermen began combining oyster fishing with other modes of marine exploitation. Moreover, the state intervened and enclosed a part of the commons. According to neoclassical economists, the introduction of property rights would theoretically: reduce overexploitation of resources; increase the rewards for conservation and protection of resources; improve the use of labour and capital resources (thus increasing efficiency) and lead to internalizing the costs of resource use (thus diminishing transaction and enforcement costs). It would also prevent the disappearance of economic rent; maximize resource rent; bring about more stable prices; give the producer a relatively greater proportion of the benefit of his activity than under commons conditions, and provide higher net incomes to lessee fishermen.[16] In practice though, the lease of fishing grounds to individuals and corporations did not bring about an improvement for the oyster industry. Oyster fishermen and planters often opted for collective action to solve resource management problems. These strategies often failed because of free riding and evasion of the rules and regulations. Therefore, the fishermen and planters asked the state to wield its authority and enforce the rules and regulations. Once again their attitude was ambivalent: when they perceived advantages in state involvement, they were in favour of intervention; when they felt their freedom to act was stifled too much, they would try to circumvent the rules or ask the external authorities to withdraw specific measures.

The same applied to the eelgrass industry. The state leased the rights on the exploitation of eelgrass to individuals or municipalities that in turn permitted access to Texelians (and Wieringers), while excluding outsiders. Depending on their stakes, the islanders repeatedly attempted to change the extant management regime or, alternatively, to hinder transformations. People benefiting from the continuation of a regime sought to maintain it, while others urged for changes to share in the benefits. This led to varying coalitions and attempts to influence decision-making processes. The opportunism and ambivalence of Texel eelgrass mowers-cum-fishermen in the political game must be understood in this context and against the background of economic fluctuations: 'In contracting over proposed property rights, the bargaining stands taken by the various parties depend upon how they view their welfare under the new arrangements relative to the status quo' (Libecap 1989:11). Each of the ecological regimes had consequences for the fishermen and influenced their behaviour. If the mowers perceived advantages in the allocation of temporary property rights to private individuals or institutions, they argued in favour of its introduction; but if they thought that it would be disadvantageous and felt estranged from the 'free' sea, they would equally passionately demand its abrogation or alteration. Especially when they had experienced becoming increasingly dependent on the lessees after the introduction of the lease by public bidding, they tried to redress the power balance by collective actions. The fishermen, whose values included independence and freedom, organized in anti-enclosure movements, a development that also occurred elsewhere following the enclosure of the marine commons. This culture of resistance must be understood against the background of the starkly deteriorating socio-economic position of these petty commodity producers. Many lacked the capital to diversify or change their activities and therefore depended on eelgrass exploitation to make a living. They were encapsulated in capitalist relations of production. The lessees and traders paid wages and absorbed the shocks of fluctuating market prices by lowering labour costs. In addition, they exploited the mowers through a truck system. The eelgrass mowers developed a germinating class consciousness. The Texelians stood up for their right to an existence and gained results despite their weak bargaining position. In the end, the state was sensitive to their arguments and asked the municipalities involved to take over the regulation and supervision of eelgrass mowing again. The municipality of Wieringen did so. Although the eelgrass mowers henceforth worked under slightly better conditions, they remained dependent and dissatisfied. The gathering of washed-up or floating eelgrass gave many an opportunity to supplement their meagre incomes. So did the exploitation of various other marine living resources.

As economists Townsend and Wilson maintain, there is a 'normal tendency of fishermen to switch away from declining stocks' (1987:323). They usually apply the same strategy once it has become unrewarding to catch a particular species. Such switching behaviour is 'the most important adaptive strategy used by fishermen' (Acheson 1988:49). It may bring about a dispersion of pressure on marine resources, a consequence neither intended nor foreseen. This is exactly what Texel fishermen have been doing for quite a long time. They turned to the exploitation of various ecological niches in an attempt to cope with resource deterioration, market fluctuations, state intervention and limitations of access such as those of the oyster and eelgrass industries. They generally pursued different species with different kinds of gear over the annual cycle. They were continually trimming their sails to the wind. The common pool nature of resources enables this kind of flexibility: 'Common property is created by the guarantee to each individual that he will not be excluded from the use or benefit of something; private property is created by the guarantee that an individual can exclude others from the use or benefit of something. Both kinds of property, being guarantees to individual persons, are individual rights' (Macpherson 1975:107). Of course, the availability of multiple niches and resources is a precondition for resilience and efficient adaptation under conditions of common pool exploitation. Marine commons offer more opportunities in this respect than, say, a small communal tract of land that is suitable only for cattle grazing.

However, one should not consider these adaptations as cybernetic processes automatically leading to homeostasis. Nor should one mistake them for evidence of control over nature or signs of ecological wisdom. These adaptations themselves are often quite diverse, partly owing to socio-cultural factors. Unless we consider these integral dynamics – that is, the interdependencies and interactions of factors and actors in a system of resource use and the processes and transformations they bring about – we cannot fully comprehend the complexity, heterogeneity and dynamics of renewable resource utilization. As long as diversification provided an opportunity to survive economically, Texel fishermen opted for this adaptive strategy. Depending on their financial opportunities, social and cognitive capital, preferences and expertise, and such incentives as resource availability and market prices, they switched target species and gear or turned to occupational pluralism. This led to a diversified annual fisheries cycle, with harvesting efforts distributed across a host of species of fish, shellfish, crustaceans and other marine resources by the 1890s. Diversification – which was feasible under common pool regimes and stimulated by an expanding market – meant that overexploitation of any one single species became less likely. We should of course bear in mind that a precondition is that fish are plentiful and catches are relatively small. The existence of alternative fishing opportunities also meant that distribution of income from exploiting marine resources was spread across any number of islanders who wished to participate in the fisheries. In addition, informal territoriality reduced the number of fishermen able to exploit a specific niche. However, the attitude towards territoriality was ambivalent and incursions were rife. On the one hand, Texel fishermen tried to monopolize access to common resources to which they claimed customary rights while, on the other, they attempted to encroach on other people's territory.

The same ambiguity characterized cooperation. Despite their rhetoric of individualism and independence, Texel fishermen had a penchant for cooperation. Given the overlapping interests of all participants in the local fishing industry, there was a strong impetus to organize. Their voluntary associations were the vehicles through which they defended their interests in economic, political and social arenas, but cooperation was particularistic. It was usually limited to the local level because of socio-economic differentiation within the occupational community and the concomitant specific interests and problems of its component parts. Socio-cultural divergence within the island society, particularly between the Oosterend and Oudeschild fishing communities, also proved an obstacle to supra-local institutionalized cooperation (see also Chapter 3). Splits in local associations, free riding, dwindling membership during recessions and dissolving trade cooperatives during booms undermined sustained collective action. Cooperation – whether formalized or not – could bring about factional conflict and such particularistic conflict in turn often gave rise to cooperation again. Informally, the fishermen would mutually defend their common interests and repeatedly reunited after schisms. The shared experience of particular uncertainties and risks forged entrepreneurially inclined individualists into a close moral 'contractual community' (Taylor 1983:xv). Generally, petty independent entrepreneurs are well aware of their vulnerability under conditions of change due to, for example, economic fluctuations or state intervention: 'The sense of precariousness, of contingency, leads to the awareness of life as "struggle" and to ambiguity in their relationships to others in the major classes' (Bechhofer and Elliott 1981:184). This ambivalence was indeed an important aspect of the fishermen's relationships with each other, with external authorities, and with traders and processors ultimately linking them with consumers.

The market expansion meant that the chains between producers and consumers grew longer, making fishermen more vulnerable to national and international competition and economic cycles beyond their control. In the mid-1890s, Texel fishermen were hit hard by a depression that would last for two decades. To avoid its effects, many invested in specialized gear. Oudeschilders predominantly turned to shrimp fishing, while Oosterenders increasingly focused on the North Sea to catch flatfish. According to the 'economics of flexibility' hypothesis, specialization usually follows once intensification and diversification have become inadequate coping responses. This proposition states that:

> minimal, less costly, and more reversible responses to environmental perturbation are predicted to occur first. If an environmental problem worsens or is not adequately met by the initial responses, 'deeper,' more costly, and less reversible responses take over, restoring flexibility to other responses (McCay 1978:410; see also McCay 2002:375-378).

However, many Texel fishermen switched from fishing altogether when they deemed the returns on their capital and labour insufficient. There is usually a point long before complete resource depletion when fishermen turn their backs on the sea. One unintended consequence of such behaviour may be a recovery of the marine environment's carrying capacity.

In conclusion, the sum-total of conscious adaptations sometimes brings about an adaptive process that operates beyond human awareness. Though it would be a mistake to consider these cybernetic processes evidence of control over nature or automatic homeostasis, they are nonetheless important in understanding how natural resources are exploited. Accusing fishermen a priori of being inclined to reckless exploitation is as short-sighted as apologetically viewing them as 'noble commoners'. Fishermen are neither noble nor nefarious. The use of communal natural resources in complex and dynamic socio-ecological systems cannot easily be explained by such simplistic and deterministic models as the tragedy of the commons, but should, as I have shown, be interpreted in a broader contextual framework. Overexploitation is often brought about by a complex interplay of processes in the wider society and their articulation with actors at the local level. Ecological transformations (whether the result of human adaptations or not), demographic changes, developments in transportation and communication, the opening up of new markets, and the integration into a global economic system, for example, will affect the efforts fishermen deploy to exploit marine living resources. This was clearly the case in the Texel oyster industry. Given the resilience of Texel fishermen's adaptations, there was no inevitable route towards a tragedy of the marine commons. We should be careful, however, of interpreting their adjustments as unequivocal signs of conscious strategies that aimed at achieving ecological sustainability. Clearly, they were not. Yet under certain conditions the chances of successful exploitation of the commons may be considerable. This applies if: the number of resource users is limited and not subject to sudden increases; territoriality and rights-to-use among local fishermen are firmly established; rule-breaking or cheating within the user-group and incursion by outsiders present no problems; alternative niches and resources are available and provide for flexibility, and diversification is economically viable (see also Wade 1987). In the next chapter, we shall see that the construction of a dam had devastating consequences for the inshore fishing industry and undermined the inshore fishermen's resilience and adaptability.

Figure 1. The early days of Oudeschild, 1609 (courtesy of Zuiderzeemuseum, Enkhuizen).

Figure 2. Harvesting oysters near Texel, 1670s (engraving by Wenzel Hollar, courtesy of Rijksarchief Noord-Holland, Haarlem).

Figure 3. Oudeschild shipyard, 1835 (courtesy of Rijksarchief Noord-Holland, Haarlem).

Figure 4. Oudeschild harbour, 1890s (courtesy of Zuiderzeemuseum, Enkhuizen).

Figure 5. Village scene from Oosterend, 1909 (courtesy of University of Amsterdam).

Figure 6. Village scene from Oudeschild, 1909 (courtesy of University of Amsterdam).

Figure 7. Texel fishermen hauling a mussel dredge, 1917 (photograph C.J. Hofker, courtesy of Spaarnestad Photo).

Figure 8. Texel anchovy fishermen in Enkhuizen, 1920s (courtesy of Maritime and Beachcombers Museum, Oudeschild).

Figure 9. Unloading eelgrass in Oudeschild, 1920s (courtesy of Zuiderzeemuseum, Enkhuizen).

Figure 10. Wieringen eelgrass mowers, 1925 (courtesy of Zuiderzeemuseum, Enkhuizen).

Figure 11. New steel-hulled boat (TX 37), 1920s (courtesy of Jan van der Vis).

Figure 12. Texel's first motor-powered cutter (TX 12), 1927 (courtesy of Sam van der Slikke).

Chapter 3

Between the Devil and the Deep Blue Sea

As we have seen in the previous chapter, Texel fishermen diversified their operations such that a complex annual round of fisheries came about. Some opted for specialization or intensified exploitation, while others withdrew from fishing temporarily or seasonally alternated onshore jobs and fishing pursuits. Exploiting various niches in the marine ecosystem, the fishermen adapted to the erratic forces of nature, the market and state intervention. They proved to be genuine opportunists, who were ambivalent concerning self-imposed or external restrictions and cooperation among themselves and with other parties in the fishing industry. Depending on such factors as, inter alia, resource availability and accessibility, competition for extraction of particular species, customary rules and restrictive external regulations, marketing opportunities, seafood prices and alternative sources of employment, they selected specific modes of marine resource utilization. Their individual decisions were usually intentional, but the sum total of the actors' choices brought about an adaptive process, the outcomes of which no one had envisioned or planned. In order to cope with the gamut of uncertainties in their métier, fishermen often turned to specific modes of labour organization, remuneration and mutual cooperation. Doing so enhanced their adaptability and resilience. A precondition for their adaptive performance was that resources were held in common and that access to and use of the marine domain was relatively easy. Ecological perturbations and economic cycles to a great extent explained why actors either entered or exited the fishing industry, as state intervention in the fishing industry was rather limited, usually pertaining to legal seasons and gear. The exception to the rule was the enclosure of parts of the marine commons for oyster cultivation – which was short-lived – and for eelgrass harvesting. Other than that, fisheries policy was largely based on a laissez-faire philosophy.

All this changed when the state initiated a project of coastal engineering with little consideration for either the natural environment or the fishermen's interests. In addition, the state intervened in the market through production and price regulations. In the 1930s and the first half of the 1940s, the island's fishing industry was hampered by a triple crisis. Firstly, there was a severe *ecological* crisis. The construction of a Closure Dam in the Zuider Sea would have devastating consequences for various stocks and – by extension – for inshore fishermen, including those from Texel. The Dam's completion in 1932 turned the Zuider Sea into a freshwater lake, but also transformed the ecosystem and morphology of what was henceforth dubbed the Wadden Sea, leading to habitat destruction for various species. Secondly, the ecological crisis was compounded by the global *economic* crisis that assailed the capitalist world. Market gluts and tumbling catches led to a situation in which artisan fishermen could barely survive economically. Texel fishermen became involved in a struggle to gain recognition as an interested party that had suffered losses due to the hydraulic works, which would entitle them to at least some financial compensation. After a fierce and prolonged battle they were finally successful in 1938. Thirdly, the economy was slightly improving again when the local fishing industry was hit by a *political* crisis: the Second World War. The occupying German *Wehrmacht* severely restricted fishing operations and impounded the best vessels. Perforce, the North Sea cutter fishermen resorted to fishing with sailing and rowing boats, while the inshore fishermen continued to work with old wooden-hulled sailing vessels. While the war's effects continued to influence the fishing industry for a few years after the liberation of the Netherlands, the Closure Dam's ecological consequences were lasting and caused Texel's inshore fishing segment to seriously decline and almost evanesce in the span of a few decades.

This chapter explores the events of the 1930s, 1940s and 1950s and the ways in which the Texel fishermen responded to the problems that confronted them. It will become apparent that not all categories of fishermen were equally affected by the triple crisis. For example, there was a marked distinction in this regard between the inshore and the offshore segment, which to a considerable extent overlapped the occupational communities of fishermen in Oudeschild and Oost on the one hand, and Oosterend on the other hand. The distributional implications of the Zuider Sea's closure were such that Texel's artisan fishermen unexpectedly suffered serious blows, which were exacerbated by the 1930s' economic depression. Their time-honoured adaptive strategies of diversification, intensification and pluri-activity were insufficient to cope with the problems. They were trapped in a downward spiral, from which there seemed to be no escape. Cooperation and political bargaining intensified, but initially did not yield results, and when they finally did, could not provide a way out of the entanglement in which the petty owner-operators found themselves. The offshore fishermen did not face the kind of problems that assailed their inshore compatriots. On the contrary, a specific set of incentives and motives and a lack of restraints encouraged rapid modernization and growth. The differential success seemed to be self-perpetuating. Are the reasons for this dichotomy of an economic, social or cultural nature? How and to what extent did the socio-cultural divergence of the two communities hamper collective action and supra-local cooperation?

The Closure Dam and Its Consequences

For decades, there had been discussions regarding the building of a dam in the Zuider Sea and the reclaiming of land from this inlet of the North Sea. It took a 1916 storm surge that led to dyke breaches and flooding to convince politicians that an improvement of the area's safety was necessary. A law was passed in 1918 that approved the construction of a thirty-kilometre-long Closure Dam. It would connect the provinces of North Holland and Friesland and transform the Zuider Sea into a freshwater lake – Lake IJssel (*IJsselmeer*) – of about 3,500 square kilometres. Opponents of the Closure Dam warned of the devastating consequences for the fishing industry and the loss of unique local cultures should it be built. However, their protests were in vain. Environmental and cultural considerations were not then integral to contemporary governmental planning regimes. Prior to the work on the Closure Dam, the government in 1925 decided that people whose livelihood would be negatively affected by the Closure Dam should receive partial compensation. This marked an important change in state involvement in social affairs. Hitherto, the state had clung to a laissez-faire policy, leaving issues of social security to churches, charities and local government. However, it was a half-hearted policy, as government and parliament alike denied the fishermen any rights of ownership. The state had merely consented to their use of a common pool resource that was in fact considered to be state property (Bossaers 1987). The Zuider Sea Relief Act (*Zuiderzeesteunwet*) of 1924 stipulated that only those who, prior to the 1918 decision, owned a fishing vessel or other fishing-related businesses – such as shipyards, sail-makers and fish processors – would be eligible for indemnifications or other support measures. Support included compensation for depreciation, schooling, credit facilities, job mediation or a licence to fish in Lake IJssel. First and foremost, the measures intended to re-educate people who had been earning a living in the fishing industry so that they could find alternative employment. A state bureau, the Zuider Sea Council (*Zuiderzeeraad*), was established to implement the measures, while local committees would review applications. The former was rather unforthcoming. Worse still for Texel fishermen was that the Council stated that it defined as Zuider Sea the area south of the imaginary line where the Closure Dam was planned. The waters north of this line, which had previously been known as northern Zuider Sea, would henceforth be designated as Wadden Sea. Importantly, only people working in the fishing industry south of the Dam were to be entitled to at least some indemnifications and state support. The government did not regard Texel fishermen – or any other fishermen working from fishing ports in the northern Zuider Sea area – eligible to any assistance from the Zuider Sea Relief Fund.

In 1920, Texel fishermen's associations informed the local authorities that they feared their interests would be harmed following the imminent closure of the Zuider Sea. They believed that mussel, oyster and particularly anchovy fisheries would decline or become impossible. Moreover, they expected that hundreds of fisher families from around the Zuider Sea coast would settle on the island and compete in shrimp and North Sea fishing. In the early 1920s, when the fisheries crisis made itself felt again, Zuider Sea fishermen were already able to apply for temporary support measures. This was not possible for Texelians, who were not considered to be Zuider Sea fishermen even though they had been fishing its waters for centuries, albeit seasonally. The local administration deemed this unfair since the Texel fishermen's financial situation was generally similar to that of their Zuider Sea compatriots. As share fishermen, they were exempt from unemployment insurance. The Minister responsible thereupon stated that they should then call on poor relief. What ensued was a prolonged battle concerning geographic definitions. The government refused to give way and continued to stand by its definition. This would have severe consequences for the Texel fishing industry. Initially, the islanders grudgingly accepted the situation. However, once the coastal engineering project was completed in May 1932, it soon became apparent that the Texel inshore fisheries were suffering and declining rapidly. As expected, the closure had drastic ecological effects on the Zuider Sea's fauna and flora, as it was completely shut off as a spawning and nursery area for a plethora of species. It turned into a brackish and later freshwater lake. Many people lost their employment in the fishing industry and scores of towns and villages saw their traditional livelihoods and cultures changing rapidly.

In what was henceforth dubbed the Wadden Sea, the ecological impact was also devastating: 'The closure drastically distorted the autonomous behaviour of the inlet system and large-scale effects in the hydrodynamics and consequently the morphodynamics were observed' (Elias et al. 2003:630). For example, the tidal prism increased by about twenty per cent, water velocity increased, currents changed and sedimentation occurred. Traditional knowledge on where to set stationary gear and fishing positions more generally became useless. Species that had been thriving there before began dwindling swiftly or disappeared altogether. Prior to the Closure Dam's construction, the area had been important for spawning herring and anchovy stocks. After the dam's completion, these stocks vanished, ending the seasonal anchovy fisheries. For a few years, herring continued to migrate to the Wadden Sea and Texel fishermen had exceptional catches. However, finding the spawning grounds blocked, the herring never reappeared after 1935. Eelgrass also vanished, halting its exploitation. Scientists concluded that the cause was a wasting disease, not the damming off of the Zuider Sea. The approximately 150 Texelians who had directly or indirectly depended on eelgrass harvesting stood empty handed. With the eelgrass beds' micro-ecology ravaged and the wider ecosystem thoroughly disturbed, knock-on effects soon became apparent. Turbot, ray, flounder, eel, oysters, garfish, periwinkles and brill, once important target species for some Texel fishermen, all suffered a serious decline.

Unexpectedly, shrimp catches also diminished sharply. Whereas shrimp had previously been caught in the winter and spring, following the closure the season in which they could be netted shifted to the summer. In that season, the crustaceans spoiled quickly and demand was at a low. Subsequently, most shrimp catches went to the local fishmeal plant where a fifty-kilo basket sometimes fetched a guilder. Only small amounts of boiled and heavily salted crustaceans could be exported to the French market for human consumption. However, it was often saturated as early as the beginning of April. As a consequence, the fishermen and dealers agreed on a quota system. The auction only took fifty or a hundred kilos of boiled shrimps for human consumption per boat per day. From 1928 to 1932, shrimp landings at the Oudeschild fish auction averaged 425 metric tons per year, while during the following five years they tumbled to less than fifteen per cent of that quantity. In mid-summer, the auction even closed down for a few weeks. From a mean of over 74,000 guilders between 1927 and 1931, the annual turnover of Texel's fish auction dropped to less than half that between 1932 and 1937. The nearby Den Helder fish auction experienced a similar decline. In 1937, the quantities of seafood landed in Oudeschild and Den Helder were less than a quarter of what they had been in 1931 (Hildebrandt 1952: X-85). Some fishermen worked with intermediaries and sent shrimp to the French market directly, but this was to little avail. Whereas Texel fishermen specializing in shrimp fishing had earned from 1,000 to 1,500 guilders annually in the years prior to the Zuider Sea's closure, their income plummeted to a mere 250 to 300 guilders in 1933 (Schreur 1953:21-22). The figures are indicative of the crisis that had assailed inshore fishing. Although 1933 was an extremely bad year, in the remainder of the decade the incomes of small fishermen were about a half to a third of what they earned before the Closure Dam's construction. The Zuider Sea Relief Fund deemed a weekly income of twelve guilders for a married couple sufficient, to be supplemented with an extra guilder for each child under the age of fourteen (Plomp 1989[1940]:79). Texel inshore fishermen's incomes were often considerably lower than that.

With nearly all inshore fisheries being struck by a severe crisis, the fishermen had few opportunities to adopt their usual strategy of switching target species. Under the new ecological conditions, only whelks thrived. Many fishermen – among them several from Wieringen – began pursuing this species during the legal season from October until April, which quickly led to overcrowding. Their number increased from a total of forty in 1934 to sixty-one four years later. High landings were offset by extremely low prices. The members of Oudeschild fishermen's association voluntarily agreed to limit fishing and sell boiled and salted whelks to a Belgian merchant directly. Three petty fish traders from Oudeschild, including Abraham Boon, were not amused and protested. Four members of the association continued delivering whelks to them but after being threatened with expulsion from the association they conformed to the deal. This once more illustrates the fishermen's ambiguity with regard to mutual cooperation and relationships with traders. The ethic of fairness still applied, however. Whelk fishermen would occasionally take turns to supply the auction. The concentration on whelk fishing was from sheer necessity. Having insufficient capital to invest in larger vessels, the only feasible alternative – North Sea fishing – was unattainable for most fishermen.

The ecological crisis was compounded by the devastating economic depression that assailed the capitalist world in the 1930s. Most fish landed by Dutch fishermen were exported and this also applied to the landings of Texel fishermen. Prices plummeted, however, and national protective measures were taken. For example, Germany introduced trade barriers, import duties in Great Britain were high, France began levying an import tax of thirty per cent and a devaluation of the British pound and the French and the Belgian francs made things even worse. The Dutch state intervened in 1934 because it believed supply to be in excess of demand. It established the Dutch Fishery Board (*Nederlandsche Visscherij Centrale*) to cope with the problem and to regulate the market: that is, restrict the supply of specific species, including shrimp and whelk.[1] To obtain a fishing licence, membership was compulsory. A blackboard on the auction building specified the quantities each vessel could supply. Inshore fishermen who had engine-powered boats were allowed to supply twice as much as those operating sailing vessels. Several had engines installed in their vessels for that reason, although some fishermen refrained from actually using them. Whelk fishing had to be done without engine power and a maximum of four dredges. When the whelk season began in October, the fishermen went to the shipyard's slipway to take off the propeller and the propeller-shaft. This restrictive measure did not, however, ameliorate the fishermen's destitute situation and in 1938 it was repealed. In combination, the ecological and economic crises dealt a serious blow to the Oudeschild and Oost inshore fishermen. They became more and more impoverished. Agriculture, accounting for about half of the island's employment opportunities, also suffered from the Great Depression and many farmhands were laid off. The usual hiring of fishermen for seasonal farm work in the summer pretty much turned into something of the past, virtually blocking this escape route. Unemployment rose to unprecedented levels and hundreds of islanders became dependent on municipal employment relief – including well over a hundred fishermen – or ended up on the dole. As the local authorities pointed out, the fishermen 'perceive unemployment benefits as a pittance and used as they are to being self-employed, unemployment relief goes against the grain for them'.[2] The owners of fishing boats exhausted their own capital before applying for any kind of support. Savings that were badly needed for maintenance and reinvestment evaporated.

The consequences for those Texel fishermen and their families who had been particularly dependent on inshore fishing and eelgrass harvesting were severe. The shipyard's owner, the sail-makers, the blacksmith, the fish auctioneer, the fish packers, the fish traders and others who directly or indirectly depended on the fishing industry suffered along with them. Those shrimp fishermen who tenaciously held on to fishing could barely weather the bad economic times:

> In many fisher families, incomes are below those of men working in unemployment relief projects. Needs are pressing indeed. Shopkeepers are complaining tremendously. Ship's maintenance is suffering. ... If the expectation that the circumstances will deteriorate rather than improve comes true, the future of the fishermen and Oudeschild and of the petty Oost and Oosterend fishermen looks bleak (*Texelse Courant*, 5 January 1938).

Fewer and fewer skipper-owners were able to maintain their boats properly, and they used the vessel share for an income and avoided the shipyard to save on expenditures. Once in a while they would put their vessels aground during ebb-tide, scrape off barnacles and tar planks on what was dubbed the 'beggars' slipway', a shallow foreshore near Oudeschild. Several fishermen refrained from sailing at all, as the costs of fishing sometimes surpassed the proceeds. With their solvability on the decline, fishermen encountered difficulties finding creditors, who were increasingly reluctant to deliver goods on tick. Poverty was looming large in many fisher families. Fisherwomen, who generally managed the household budget, encountered problems in making ends meet.

Initially, the fishermen had feared an invasion of Zuider Sea fishermen, whom they thought would be likely to migrate to Texel to continue fishing. Although this expectation proved to be false, the Texelians who depended on inshore fishing for their livelihood did face the direct consequences of the Zuider Sea's closure. Resenting the fact that they were not entitled to any kind of indemnification from the Zuider Sea Relief Fund, they united and sought recognition as victims of the Zuider Sea's closure and as rightful claimants to the fund's benefits. Through their organizations, the fishermen launched a prolonged and bitter fight with the state. The Oudeschild fishermen's organization was *Door Eendracht tot Vooruitgang* (Progress through Unity), originally founded in 1918, while the Oosterend North Sea skipper-owners were organized in a Fishery Cooperative – the Co-op for short – which was established in 1931. It also functioned as a voluntary interest association. The organizations and their members had regular joint meetings to discuss the situation. Attendance was usually high, as much was at stake. The boards sent letters and telegrams to various authorities, including the local and provincial administrations, the Minister of Water Affairs, the Minister of Social Affairs, the House of Representatives, individual parliamentarians, the boards of political parties and – as of 1934 – the Dutch Fishermen's Union (*Nederlandse Vissersbond*). They explained that inshore fishing had suffered from the coastal engineering project and requested compensation. The mayor, aldermen and councillors of Texel immediately supported the local fishermen, deeming the Closure Dam the cause of the deterioration of inshore fishing and the local extirpation of eelgrass. It was partly from interested motives that they did so. They feared that unemployed fishermen would become dependent on municipal poor relief. The municipal council – consisting of thirteen persons – included three fishermen from Oudeschild (Social Democrats Jan Henkes, Teunis Dogger and Jacob Bruin) and one fisherman from Oosterend (Jan van der Vis, of the Christian Historian Union). Together, they tirelessly supported the fisherfolk's cause and at local level they had considerable political clout.

In January 1933, the fishermen's associations submitted a joint request to the Ministry of Water Affairs concerning the local fisheries' bad state after the Closure Dam's completion. It was signed by 161 Texel fishermen and eelgrass workers, who demanded that they be eligible for Zuider Sea relief. The municipal council and the mayor and aldermen supported their address. The then responsible Minister, Paul Reymer, rejected the request, arguing that Texel fisherfolk fished in the Wadden Sea not the Zuider Sea. Apparently, the government was loath to see more people than had been planned apply for money from the Zuider Sea Relief Fund, as it would exhaust the budget early. With the severe economic crisis coming down harshly on the country, the government's policy was to cut expenditure. Understandably, the Texelians experienced the Minister's decision as a grave injustice. They had been fishing the Zuider Sea's waters seasonally and they were assailed by the closure's devastating consequences. Many requests would follow in the early 1930s. In July 1933, fifty unemployed fishermen demonstrated in front of the town hall in Den Burg. A delegation of four was allowed in to talk to the mayor, W.B. Oort, who sent a telegram to the Minister that very day. His plea fell on deaf ears, however, and the same happened a month later with another telegram. The mayor and a delegation of fishermen visited the ministry, but their voyage was futile as the Minister did not give an inch. The whole affair started all over again when a new Minister, Jacob Kalff, was appointed in 1934. In June that year, he decided that the Texel fishermen could not apply for Zuider Sea relief because they did not fish in the Zuider Sea and eelgrass had disappeared as a consequence of the wasting disease not the Closure Dam. The fishermen were back to square one. The mayor and aldermen of Texel protested and pointed out that the Minister was not acting according to the spirit of the law. They achieved the inauguration of a special committee (the committee Defoer) to investigate the reasons for the disappearance of eelgrass and to subsequently advise the Minister. The results were disappointing. The committee concluded in June 1936 that a wasting disease not the closure had brought about the problem. Although two aldermen of Texel and Den Helder, who were on the committee, opined that the closure had had a negative effect, the government did not heed their minority advice to compensate the eelgrass harvesters. It refused to accept Texel fishermen as being eligible for benefits. Nonetheless, the fishermen never tired of petitioning the Minister of Water Affairs, claiming that the dam had caused the fisheries' deterioration and stating that the Texel fishermen should therefore be entitled to Zuider Sea relief. They were prepared to drop their claim if they could get permission to fish mussels or gain entry to the Lake IJssel fisheries. The state rejected this option as it had just adopted a policy of limited access and restricted production to contain the fisheries crisis.

The Oudeschild inshore fishermen, who were hit especially hard by the crisis, began using their Labour Party contacts to lobby in the House of Representatives. In 1936, especially, Labour parliamentarian Willem Drop put pressure on the Minister to come up with the committee's results. He repeatedly submitted written questions to the then responsible Minister, Otto van Lidth de Jeude. Drop also visited the island on a number of occasions to consult with Texel – and particularly Oudeschild – fishermen. They made plans for justice to be done. On the basis of detailed information concerning the experiences and proceeds of 112 fishermen, they sent a long and thoroughly underpinned request to the House of Representatives in June 1936, calling for new scientific research into the causes of the decline. Another research committee was formed: the Reigersma committee and in addition, a Parliamentary Committee – which included Drop – would investigate the consequences of the closure for the Texel fisheries and report on the island's fisher families' social situation. The former again concluded that a causal link between the closure and the decline of fish stocks was unproven and that the extinction of eelgrass was solely due to the wasting disease. The Parliamentary Committee did deem the Zuider Sea Relief Act applicable to Texel fishermen, because they had been fishing in the Zuider Sea, and the dam had significantly affected the Texel fisheries and eelgrass exploitation, even though the primary cause of the eelgrass's decline was the wasting disease. In February 1937, the committee advised the entitlement of Texel fishermen who had been working in the fishing industry prior to 25 July 1918 to Zuider Sea relief. Following this advice, the mayor and aldermen of Texel insisted that the House of Representatives accept the committee's conclusions, supported by yet another petition of Texel fishermen. The House of Representatives indeed heeded the advice on 12 March 1937. The efforts of Texel fisherfolk appeared to finally be yielding some results. In November, however, the Minister rejected the proposal to allow Texel fishermen under the terms of the Zuider Sea Relief Act, still arguing that the Wadden Sea was not the Zuider Sea and that the wasting disease had led to the extinction of eelgrass. The Oudeschild and Oost fishermen were outraged, desperate and deeply disappointed in the right-wing government, a coalition of confessional parties. The destitute circumstances of many Texel fishermen and the Minister's obstinacy led to several joint meetings of Progress through Unity and the Co-op. Many fishermen attended such gatherings, which often resulted in new petitions being sent to the Minister and the House of Representatives. The associations' leaders and parliamentarian Willem Drop pointed out that it was important to stick together and that any kind of fission –

which had previously often sapped sustained collective action – would be dangerous now. A local merchant, Abraham Boon, had attempted to strike a separate deal for the eelgrass workers, but in response to the plea stated that he understood what was at stake and that a separate action would be 'insane'.

A new Parliamentary Committee (again with Drop as a member) was inaugurated. Its report was due in February 1938. Like its predecessor, it advised the Minister of Water Affairs to support Texel fishermen through the Zuider Sea Relief Fund. At long last, on 1 April 1938 the new Minister, Johannes van Buuren, decided that the boundary of the Closure Dam would no longer be decisive in applying the Zuider Sea Relief Act. Van Buuren was the fourth Minister with whom the fishermen had had to deal since 1932. All along, changes of government had delayed decision making. Without voting, Parliament accepted the Minister's decision a few days later. This meant that Texel fishermen could finally call on the Zuider Sea Relief Fund. By the end of the year, ninety of them received weekly allowances amounting to between 1.50 and 12 guilders (with a mean of 6.50 guilders). However, admission was slow due to bureaucratic procedures. The measure was not retroactive, so that Texel fishermen were not compensated for their losses from 1932 to May 1938.

There were more reasons to be dissatisfied. It still took another year before the rules were relaxed somewhat to include eelgrass harvesters and certain categories of fishermen. In addition, the Zuider Sea Relief Act stipulated that only those working in the fishing industry prior to 25 July 1918 were entitled to compensation. Being unaware that the closure would have such negative consequences, Texel inshore fishermen had nevertheless continued to invest in their businesses whenever possible. Their sons had taken up the occupation, never expecting that they would be confronted with the rapid and unanticipated decline of inshore fishing. They only discovered the consequences after 1932. Therefore, the Texel fishermen's associations continued to petition the government and pointed out the unfairness of sticking to the '1918' rule. The fishermen also deemed it unfair that in certain cases money was deducted from their allowances, for example when they were working in unemployment relief projects, received interest from savings or had relatives living with them who earned an income. Without support, it was nearly impossible to earn a living in the inshore fisheries – unless skippers had young sons who did not have to be paid a full share. Many inshore fishermen therefore refrained from fishing altogether. They could not even afford the costs of wear and tear. The small-scale fleet was on the decline quantitatively as well as qualitatively. Along with that, it became harder and harder to find creditors, entrapping the inshore fishermen in a downward spiral.

Much bitterness was felt and the Oudeschild and Oost inshore fishermen's experiences would colour their attitude towards the authorities for years to come. The government's protracted rigid stance and its narrowmindedness brought about a surge of bitterness on the island, the ripples of which were still noticeable when I did research decades later. Several retired eelgrass harvesters expressed no doubt whatsoever: the Closure Dam had caused the eelgrass to vanish. Later research would largely confirm this view. The wasting disease was important, but so were higher velocities of tidal currents and sedimentation following the Closure Dam's construction (Giesen, van Katwijk and den Hartog 1990). Even after the wasting disease had subsided, the eelgrass did not return. In the 1930s, the fishermen's opinion did not count for much. State representatives and scientists patronized them as ignorant people. The dam caused an ecological disaster, but back then ecology did not count for much either. The dam brought misery upon Texel's inshore fisherfolk. By the end of the decade, agriculture had luckily recovered from the crisis and in the summer many fishermen – including owners – could work as farmhands again. This temporarily alleviated the inshore fisherfolk's situation, but in the extremely cold winter of 1939-1940 inshore vessels could not sail for months due to ice and ice-drift. The inshore fishermen fought to keep body and soul together and many led a hand-to-mouth existence. According to an Oudeschild skipper, inshore fishermen were earning 'just enough to keep the hunger away, but definitely not more than that'. However, the segment of the local fishing industry that did not depend on exploitation of the inshore zone was doing rather well.

In general, Oosterend fishermen were still more successful than their Oudeschild counterparts. In spite of the global economic crisis, the former motorized their vessels and as of the late 1920s introduced steel-hulled vessels of a type dubbed cutter (*kotter*). Cutters were comparatively cost-effective and efficient and could be handled by a relatively small crew. Their owners obtained excellent financial results. The first such vessel, equipped with an 80 h.p. engine, was added to the Texel fleet in 1927. Due to financial problems the vessel was sold to Den Helder after a year, but soon other Texel fishermen followed suit. Although motorization made fishermen less dependent on the weather, landings and gross proceeds still fluctuated heavily. A 1930 account book of an Oosterend skipper makes this abundantly clear. That year, the two owner-operators and three deckhands grossed an aggregate of 17,616 guilders with their steel-hulled boat, which was equipped with a 45 h.p. engine. Less than a third of the gross proceeds went to operating costs and share wages for non-related deckhands. The crew did not fish for six weeks due to inclement weather. Their highest gross revenues after a week's fishing trip amounted to 1,616 guilders, the lowest to a mere 85 guilders. Still, the figures proved that investment in cutters – initially requiring a sum of 20,000 to 30,000 guilders but rising to 45,000 guilders by the end of the decade – reaped rewards. This in turn changed the balance of forces of production from labour being more important to capital becoming ever more important, although the local fishing industry's social organization remained based on family firms. A time of rapid modernization ensued. In 1931, Texel's fishing fleet boasted four steel-hulled cutters with 100 to 150 h.p. engines owned by Oosterend firms: eight years later there were twelve such vessels, eleven of them owned by Oosterenders. Five families were instrumental to this modernization: the Drijver, Vlaming, van der Vis, and Ellen families from Oosterend and the Krijnen family from Oudeschild. These families constituted the local 'fishing elite'. A councillor noted that the Oosterend North Sea fleet was the country's best and that despite the difficult circumstances its fishermen exhibited 'entrepreneurial spirit, pluck and initiative' (*Texelse Courant*, 4 December 1935).

It was not mere local partisanship. At the time, the Texel North Sea fishermen were indeed the vanguard of the Dutch skipper-owned cutter fleet's development, owning a third of the total number of this type of vessel in the country. Their pioneering role is generally acknowledged in the Dutch fisheries literature (see, for instance, Hildebrandt 1952; Schreur 1953, 1966; Kranenburg 1977). The example of Texel fishermen who successfully used cutters was swiftly followed by skipper-owners hailing from other Dutch fishing communities. They had observed or heard of the catches the Texelians landed. However, in addition to looking at landings the Texelians also took account of costs, something that others occasionally seemed to forget. A sociologist who conducted research on the island of Urk – a community that is currently completely dominated by the fishing industry, with the largest fishing fleet and the largest fish auction in the Netherlands – wrote in 1940 that the Oosterend North Sea fishermen compared favourably with their Urk compatriots. They grossed more, they were on top of new developments, replaced engines long before they were completely worn out and were not heavily indebted. The Oosterend fishermen, he wrote, 'are very active and energetic and form a close-knit organization with, amongst others, a cooperative credit-bank' (Plomp 1989 [1940]:67). In 1937 and 1938, the mean annual gross proceeds of Texel North Sea fishermen were over 18,500 and 17,000 guilders, respectively, which was considerably more than what Urk and Den Helder fishermen working with similar vessels grossed.

Texel skipper-owners and crewmembers working in the offshore segment obtained good annual incomes with a mean of 2,400 guilders in 1937 (Plomp 1940:381). Cutter fishermen often earned considerably more than skilled labourers during the years of economic crisis. Since investment in steel-hulled cutters required considerable sums of money, the share division between vessel and crew was fixed at fifty-fifty. With a crew of five, this implied that the skipper and full-blown deckhands would obtain a ten per cent share each.[3] With a fuel price of three cents per litre and a fuel consumption of about 400 litres per week, operating costs were low. Even though herring fishing had suffered, sole and plaice prices were generally rather good. Their motorized vessels enabled the North Sea fishermen to market their fish at the auctions that yielded the best prices, for example IJmuiden. Weeks in which the gross revenues amounted to 700 guilders were no exception in the late 1930s. If operating costs amounted to a hundred guilders, the vessel share – which was used for depreciation, insurances and so on – would be 300 guilders and the same amount would go to the crew share. Each of the crewmembers, including the skipper, would then have a remuneration of sixty guilders, which was extremely good money. Of course, there were weeks in which the boats could not sail due to inclement weather, but in general, the North Sea fishermen were rather well off. Thus, the economic depression did not hit the North Sea fisheries to the same extent as it did the inshore sector and on the whole, the Oosterend fishermen have been considerably more successful than Oudeschild fishermen. The question is: why?

Worlds Apart: 'Jerusalem' and 'The Red Village'

The differential success of Oudeschild and Oosterend fishermen dates back to the late 19th century, when their fishing economies began diverging after a brief spell of convergence (see Chapter 2). The Oosterend fishermen's tendency was to orient themselves on North Sea week-trip fishing, whereas Oudeschild – and Oost – fishermen were mostly inclined to day tripping on the Zuider Sea and the Wadden Sea. The Closure Dam's construction therefore had devastating consequences for Oudeschild and Oost fisherfolk in particular. Not only did they lose access to a vast area of fishing grounds, the closure also had a strong negative impact on commercially important species of fish, shellfish and crustaceans in the Wadden Sea. Eelgrass exploitation, which was still important in the early 20th-century seasonal cycle of Oudeschild and Oost fishermen but to a considerably lesser extent in that of Oosterend fishermen, ceased to exist in the early 1930s. Those who mainly plied the North Sea did not face similar problems, although those who seasonally switched to shrimp, herring and anchovy fishing in the Zuider Sea of course also felt the effects. With their type of vessels, it still proved to be much easier for Oosterenders to find alternatives, enabling them to avoid the crisis. The events of the 1930s deepened the differences between both occupational communities of fishermen. Even though they shared a pervasive labour ethos, Oudeschilders were predominantly content to eke out a living, whereas Oosterenders were much more inclined toward accumulating profits and staying ahead of their competitors. The former perceived themselves as labourers, the latter as entrepreneurs. In the long run, only those who modernized their operations succeeded in maintaining a profitable business. Early on, the economic successes of Oosterend fishermen were connected with ideological factors, most notably religion (see Chapter 2). So, let us take a closer look at the communities and cultural conventions of Oosterenders and Oudeschilders and see how their respective ideologies might have influenced the fishermen's behaviour, goals, attitudes and labour ethos.

The vast majority of the inhabitants of Oosterend were Protestants of various denominations: Dutch Reformed, Calvinist Reformed and Mennonite. There was a tiny Catholic minority. With four local churches, Oosterend was visibly the most religious community on the island, hence its designation as 'Jerusalem'. With hardly any exception, the fisher families were Protestant, belonging to either an orthodox congregation or a latitudinarian one. As we have seen, the heterodox and the orthodox Protestants were nicknamed 'coarse' and 'fine', respectively. The terms refer to the degree of strictness with which people observed the Protestant creed. On Sundays, fishermen refrained from sailing and attended church services. The Calvinists went twice, the heterodox once. On this day, religious allegiance became particularly manifest:

> In general, the feud between fine and coarse fishermen is not so grave. They meet each other at the quay, at sea, at the fish auction, in the fishermen's association, on Saturday evenings in the house of head teacher Daalder. They brave the same dangers, have the same fears and the same expectations and they confide in and pray to the same God. On Sundays alone this all changes completely. Then they go to different churches, have their faces in a particular fold that excludes any communion, then the coarse are allowed to do more than the fine and the religious differences are strongly emphasized (Daalder 1978:38).[4]

At face value, the differences between the various Protestant denominations were over obscure exegetic details that easily gave rise to dissension. This became evident once more in the mid-1920s when a conflict in the Calvinist Reformed Church led to a schism. As of 1926, there would be a fifth church in Oosterend: the Restored Union (*Hersteld Verband*) Church, which had hived off from the Calvinist Reformed (*Gereformeerde*) Church. The whole affair was more or less coincidental. It so happened that in 1924, following a six-year vacancy, a young and charismatic minister, Jan Buskes, was appointed to Oosterend's *Gereformeerde* Church. Buskes was quite popular, but became involved in a conflict when a Calvinist theologian claimed that the Bible story of Genesis 2 and 3, referring to the speaking serpent in paradise, should be interpreted metaphorically. This led to a clash when the national *Gereformeerde* Synod stated that the Bible ought to be taken literally and that ministers must sign a statement that they subscribe to the Synod's viewpoint. Buskes read the statement in a Sunday service, but confessed that he had great difficulty with it. A conflict was born. Buskes was suspended and he left the island, but his followers established the new Restored Union Church and about a fifth of the Oosterend Calvinists enrolled as members. A woman, born in 1914, recalled:

> My father was the Mennonite Church's deacon. There would be a service once every three weeks. His cousin had joined Restored Union, and he came to ask whether it would be possible to use our church if there was no service. My father agreed. After that, the Calvinists refused to greet us, although they had always done so previously.

Within the religious community of Calvinists, relationships between those who opted for Restored Union and those who continued to be members of the *Gereformeerde* Church grew quite tense and even disrupted entire families. The members of Restored Union built a new church in 1928 and a year later, Buskes returned to the island to become its minister. This would only sour relationships further. The conflict had a strong moral dimension in that it hinged upon a discourse as to who adhered to the 'right' interpretation of the Bible. It realigned the local community and stirred emotions precisely because the opponents had much more in common than what divided them dogmatically.

The main dogmatic difference concerned the view on predestination. The orthodox were convinced that they were chosen to behold paradise, whereas the heterodox expressed more doubts. Yet at the level of social relationships, there was strong moral construction and constriction to adhere to one's own church. 'In the old days, you *had* to attend church services and confirmation classes, otherwise you would be in trouble,' related a man born in 1921. Social control was tight:

> They simply made you. You didn't have any opportunities for an outing, so you stayed in the village. You were a member of a Christian youngsters association or a Christian glee-club. That was it. Later on, though, [i.e., in the 1960s] it got much more liberal.

The villagers were born into a particular religious community and would not easily switch to another. They were disciplined into 'the right belief' not only in church, but also in a variety of other institutions such as voluntary associations and the school. Religion divided the village's communal life. To a large extent, it determined whom one married, whom one befriended, where one bought one's groceries and so on. Mixed marriages between orthodox and heterodox Protestants were taboo and if they did occur, conflict would usually run in the families concerned. However, marriage partners of different latitudinarian denominations (for instance, Mennonite and Dutch Reformed) were seldom a problem. Protestant children who went to the secular school and their orthodox counterparts who attended the Protestant school would often get into fights with each other. The latitudinarians deemed the Calvinists sanctimonious and Pharisaic. As a Dutch Reformed man, born in 1923, related: 'When they [the Calvinists] were in the village, they behaved in a saintly way. Once they were off the island, however, they would go and see prostitutes and indulge in drinking. That happened during the herring season. We never did that.' From the other standpoint, the orthodox considered the latitudinarians to be insufficiently sound in their faith. Still, religious denomination was not important in regard to crew recruitment or with whom one pair trawled. Heterodox Protestant deckhands could be found aboard an orthodox Calvinist skipper's boat and vice versa. As time went by, the relationships between the denominations relaxed considerably as was the case in Dutch society as a whole.

The religiousness of orthodox Oosterend fishermen showed among many other things in vessel names (for example, '*Deo Volente*', '*Soli Deo Gloria*', '*Ora et Labora*') and in the maxims that could sometimes be found in the wheel-houses ('How vast is your sea, O Lord, and how tiny my boat'). For the orthodox Protestant majority of fisher families, bicycling on Sundays was taboo. Several fishermen were on church councils or acted as church elders – this applied to both the Dutch Reformed and Calvinist Reformed congregations. Reference to biblical language was quite common and the orthodox were versed in Scripture. For all outer signs of religiousness, several ministers still deemed the Oosterenders' faith rather superficial. They appreciated the latter's civility but at the same time felt that their belief was not submissive but rather moralist and rationalist (Buskes 1946:36-37; Saal 1948:9-14; Janse 1955:266). According to these clergymen, a sense of sin was lacking. The emphasis was on abiding by biblical or commonsensical yardsticks. One of the ministers asked an old fisherman whether he had been experiencing God's greatness and majesty at sea. He referred to Psalm 29:3 – 'The voice of the Lord is upon the waters: the God of glory thundereth: the Lord is upon many waters.' The fisherman replied: 'I've been sailing for sixty years, sir, but I've never heard it' (Buskes 1946:19-20). For the villagers, church services first and foremost signified *social* occasions, not *religious* ones, the ministers claimed. A minister was appreciated for his oratory gifts and social talents, not his evangelical message. Indeed, a Dutch Reformed skipper-owner, born in 1923, pointed out that for him personally, the Sunday service offered a moment of reflection, adding: 'Following the service, you would come together with the family for a cuppa. You were together then and that was congenial.' However, the same interlocutor also related the following:

> My father was a deeply religious man. There have been times when I thought: why don't you curse everything to hell now. He never did. I remember that once we were sailing together in really foul weather. My father said: 'If you help a little bit, and He does too, we'll manage.' He never used angry words.

Personal experiences and anxieties could indeed lead to a thorough sense of sin and salvation. A Calvinist Reformed man, born in 1904, was on board a cutter that capsized in a 1938 storm. Miraculously, no one was killed in the accident. The next Sunday, the minister preached Jonah 2, verse 3: 'For thou hadst cast me into the deep, in the midst of the seas; and the floods compassed me about: all thy billows and thy waves passed over me.' My interlocutor, who unlike most of his contemporaries was still not confirmed at the time of the mishap, took the accident to be a sign of God. He was confirmed and he turned into a pious believer: 'God did not want me yet, but I have always believed that He had an intention with this event.' As a 94-year-old, he related:

> That night was the great night of my life; a milestone, a night in which the clear light of God's mercy went up on my tormented soul, and the points were shifted to the right track. ... I feel that I had to go through this, that as a notorious sinner I had to experience this to enter the right path to truly get to know Him like once Jacob did on the banks of the Jabbok River (*Visserijnieuws*, 28 May 1999).

Of course, this concerns an individual experience. Accidents did not determine the measure of religiousness. Other interlocutors who had also had narrow escapes rationalized why they survived by pointing out the particular conditions under which mishaps occurred, including the boat's seaworthiness and so on. Their creed could offer comfort, though. In addition to being religious linchpins, the Protestant churches generally were social foci in the community.

Oudeschilders maintained quite another ideology. Thirty to forty per cent of the population had no religious affiliation, while the Dutch Reformed were mostly nominal church members. The congregation often did not have a minister for years, and if a visiting minister led a service, attendance was negligible. It was only the Roman Catholic minority – very few of whom were involved in the fishing industry – that was tightly integrated and adhered to the church. Like Oosterenders, Oudeschild fishermen refrained from fishing on Sundays. To a large extent, the latter were embedded in capitalist relations of production and often felt exploited. In Social Democracy, they saw a means of achieving social justice. It harked back to the late 19th century, when poverty had driven many villagers into wage labour on the mainland. In the port of Amsterdam, in particular, many found work at the shipyards and as dockworkers. Soon they discovered that they were used as scabs. Dockworkers were on strike and told the islanders in no uncertain terms what their work amounted to. This experience made the men susceptible to Socialist leaders who went to the island in order to recruit a following there. With six others, fisherman Jan Henkes had established a local department of the Social Democratic Labour Party in 1908. He was an independent skipper-owner of a small boat. Henkes would be a local councillor for thirteen years and was strongly committed to elevating the working class. For a quarter of a century, he was also the local fishermen's association's chairman. Several self-employed boat dischargers in the harbour also supported Labour. Oudeschild turned into a stronghold of Social Democrats and was dubbed 'the Red Village'. The Oudeschilders used a socialist jargon, referring to 'exploitation' by the 'capitalist class' and demanded government intervention to tackle the economic depression. They celebrated Labour Day on the 1st of May. At local level, however, class relationships were mitigated by a moral economy. The former fish auctioneer, born in 1916, characterized Oudeschild as a rather poor and more or less classless community:

> Some were slightly better off but there were no great class distinctions. They [the villagers] mainly opposed government, not so much capitalists, for there weren't any here save for some petty entrepreneurs. Nor did religious differences matter. The majority was indifferent, some were Dutch Reformed and you had the Catholics who did attend Sunday mass. The Dutch Reformed only went on Christian holidays or not at all.

All local children went to the local secular school. The Oudeschilders opted for self-help: 'No-one had money, so you just had to help each other.' It was an egalitarian and close-knit community. A shrimp fisherman, born in 1915, said: 'Skipper or deckhand, it didn't matter. You were equals among each other. We didn't have any differences.' Indicative of their communitarian ideology is the following. Shrimp fishermen landing their catches first usually received the lowest prices. Therefore, they agreed on having a fixed price, so that it was not disadvantageous to come first. The Oudeschilders' solidarity and neighbourliness were acknowledged and admired by Oosterenders, who usually mentioned it in interviews.

The ideological divergence of Oosterend and Oudeschild was rooted in structural and contingent factors. At national level, the emergence of socialism and orthodox Protestantism impacted what happened at local level, but – to a considerable extent at least – it was a coincidence that people who subscribed to these respective ideologies gained so much influence locally. For example, the fact that the Dutch Reformed of Oosterend grew more orthodox by the turn of the century was linked to the fact that an orthodox minister was called to the pulpit. The Christian and Social-Democratic ideologies had an impact on the respective villagers' worldviews. They suited the different experiences of Oudeschilders and Oosterenders and provided models on how to act in the situations they confronted. The former restricted their operations to the inshore domain, were impacted by capitalist relations of production and perceived themselves as labourers, whereas the latter plied the dangerous North Sea and regarded themselves to be entrepreneurs whose this-worldly success would be an indication for other-worldly salvation. Oosterenders were inclined to take greater risks than Oudeschilders and in Calvinism they found a way of mentally coping with potential perils. It reminds one of Malinowski's anxiety–ritual proposition, stating that the greater the risks, the more rituals and taboos (Malinowski 1955). Elaborating on this theme, one might say that the Oosterend fishermen's religious orthodoxy was perhaps linked with their métier, being as it was full of danger and uncertainty. Oudeschilders took fewer risks and were much more inclined to self-help and solidarity. Because the Reformed church had stood aloof in adverse times and had increasingly estranged the local populace, the Social Democratic movement with its promise of future emancipation of the labour class rapidly gained influence. As national Labour politicians showed an interest in the destitute situation of Oudeschild fishermen following the Zuider Sea's closure, and in parliamentary debates supported their fight to be recognized as victims, the local support for the Social Democratic movement only increased.

It would seem plausible to link differential economic success with the diverging politico-religious ideational systems of the two villages. This has indeed been done by the sociologist Schreur, who in the early 1950s conducted research into the Texel fishing industry. He points to Weber's thesis on the 'Protestant ethic' and the 'capitalist spirit' (Schreur 1953:53; Weber 1969[1920]), arguing that the Oosterenders' diligence, thrift, restraint and rational economic orientation explain why they were successful entrepreneurs. Their leading was 'due mainly to their economic rationalism that is linked to their Calvinist ethos' (Schreur 1964:222). However, this is a rather vulgarized version of Weber's thesis. The German sociologist made it abundantly clear that there was no unilateral causal relationship but a mutual influence, that the capitalist spirit sprouted from an ascetic way of life but that economic rationalism was hardly restricted to Protestants and had gradually penetrated all 'modern culture'. Oudeschilders had certainly been influenced by Protestantism, even though a process of secularization had affected their outlook on life. Instead of a religious ideology, they opted for a secular political ideology as their compass in life. This does not mean, however, that they were any less diligent and thrifty or had less self-control than their northern neighbours. Quite the contrary: the Social Democratic movement stimulated such 'civil virtues' as a road towards emancipation. Oosterend skipper-owners were often keen on hiring Oudeschild deckhands, who were reputed to be highly skilled net-menders and to work hard to earn their share. Whereas in Oosterend individual profit maximization and Protestantism were in perfect harmony, in socialism profit-making and an ethic of equality seemed to be at odds, since profits would be made at the expense of colleagues. The Oudeschilders distrusted capitalist entrepreneurship as they perceived themselves as a kind of seagoing proletariat – despite the fact that they were petty entrepreneurs. First and foremost, their orientation was towards fulfilling household needs.

Another factor, directly linked with the different ideologies, was also important. Oosterenders had larger families than Oudeschilders. Although there were more mouths to feed, this enabled them to recruit family crews, enhancing their versatility. As a former Oudeschild fisherman related: 'In Oudeschild, there were more coffins than cradles. Sons were lacking, making it difficult to weather hard times. ... Here, we only had one skipper with a couple of sons and his firm has been quite successful.' Demographic statistics to some extent confirm this view. The demographic composition of the two communities originated in politico-religious differences. Churches stimulated large families, the Social Democratic movement smaller ones. So in addition to affecting labour ethos, the two ideologies impacted the ethos of human reproduction. Oudeschilders indeed had smaller families, whereas Oosterend fishermen generally had more sons who contributed importantly to reducing labour costs and accumulating capital. Interestingly, in the 1910s and 1920s many Oudeschild fishermen had acquired smaller boats so that they did not need a crew of three but could handle the vessel alone or with one deckhand. It was precisely this boat type that was unfit to sail the North Sea and as we have seen, those who depended on inshore fishing were hit hardest by the 1930s ecological and economic crises. This in turn diminished the Oudeschild fleet's viability, whereupon recruiting crew became increasingly difficult. Even occupational succession was no longer a matter of course as it had previously been. All in all, their greater economic and social assets gave Oosterenders a considerable edge over their northern neighbours. However, those Oosterend fishermen who also depended on inshore fishing were hit equally as hard as their Oudeschild compatriots, while the only Oudeschild skipper-owner who owned a cutter was quite successful. We should, therefore, avoid generalizing too starkly.

The inhabitants of the respective villages themselves usually argued in terms of 'mentalities' to explain the differences between the occupational communities of fishermen. An Oosterend fisherman said: 'It is a matter of mentality: more entrepreneurial spirit, more viciousness, we had more of that here than in Oudeschild. They were more relaxed, more congenial, gossiping about the neighbours. Here we are a stubborn lot, and religion also plays a role.' Many of my interlocutors – regardless of whether they hailed from the one or the other village – alluded to 'the fact' that the Oudeschilders 'never lost sight of the local windmill' or that they 'always wanted to be able to see their chimney smoke'. The Oudeschild fishermen allegedly wanted to return to 'Cape Bum' (meaning their wife) each night. The former fish auctioneer said: 'In Oudeschild, people were more relaxed. They would say: "If it doesn't happen today, it'll happen tomorrow".' By contrast, the Oosterend fishermen were said to be 'more active', 'entrepreneurial' and 'keen'. To some extent, the images were appropriated as self-images. A retired Oosterend owner-operator, born in 1904, contended: 'In Oudeschild they had a big shrimp-fishing fleet. Fine shrimpers, but they wanted to be home each night! It was another mentality. The Oosterenders' outlook was more progressive: keep up with the times, acquire a bigger vessel with a more powerful engine.' Of course, the proximity to the fishing grounds enabled the Oudeschilders to return home each night and as shrimp spoil quickly, it was paramount that they be marketed as quickly as possible. An Oudeschild skipper, born in 1918, felt that the Oosterend North Sea fishermen were disdainful of Oudeschild shrimp fishermen:

> They thought they were fine fishermen if they had a big boat. That's not true. Unlike us, they were not very good with nets. Nonetheless, some Oudeschilders looked up to those Oosterenders because they earned much more. Anyway, we had little to do with each other, partly because our labour rhythms varied.

Although he clearly resented what he assumed to be the Oosterenders' stance, he continued, saying:

> We did like to be home, though, and traditionally Oudeschild fishermen were not North Sea fishermen. The Oosterenders were. For their part, they disliked shrimp fishing because it was a laborious kind of fishing. It involved a lot of work. You sailed short distances and made brief tows.

Such stereotypical views not only impacted the perceptions of each other, but also their mutual relationships. A retired Oudeschild skipper-owner, born in 1915, related the following:

> To be honest, we can't stand each other. No, we can't. They are a different kind of people. Here they were... Well, religion didn't matter in Oudeschild. I am a Roman Catholic and for a long time I've been the only Catholic skipper in the fleet. Still, I've been the chairman of the [Oudeschild] fishermen's association for twenty-five years. I mean it didn't matter. For the Oosterenders it did matter. Then again, we did have a Catholic and a Protestant bonfire. However, my father was a carpenter, and he used to say: 'If I only have to live off Roman Catholics, I might just as well put a rope around my neck.' He could only make a living from the entire village. He also made us go to the secular school here, not to the Catholic one in Den Burg. I am of the same opinion. It would harm the village community if I had sent my kids to Den Burg.

The perceived differences between the two fishing villages continued to be considerable. A symbolic boundary rather pertinently divided and is still dividing both communities. Their different outlooks on life in general, and their métier in particular, were reinforced by their diverging ideologies. These in turn provided anchorage for their specific modes of operation, the risks they were willing to take and their labour motivations.

Their separate worldviews and modes of production also affected occupational organization from its very onset (see Chapter 2). Fishermen from both villages continued to maintain their own associations. The pioneering skipper-owners of five family firms who constituted the island's fishing elite established the Cooperative Purchasing Association for Fishery Necessaries – as the Co-op's official name read – in May 1931. These rather successful firms owned and operated the innovative steel-hulled cutters with powerful engines. The Co-op's first board reflects the importance of these families: Biem Vlaming was elected chairman, Aris Ellen became its secretary, Willem Drijver was its treasurer and the other board members were Fup Krijnen and Jan van der Vis. The Co-op was born out of necessity in times of grim economic recession. It started off with thirteen members, representing nine cutters, who had to chip in a hundred guilders each. The Co-op's aim was to bypass intermediary traders and directly buy fishing materials and fuel from wholesale dealers at reduced prices. This went against the grain for several local tradesmen, among whom were the sailmaker, the shipyard and the firm that supplied fuel and lubricants to the fleet. They rightly feared that they would lose customers. The Co-op bought its own warehouse to store fishing materials and a tan-vat to preserve nets and sails. The Co-op's scope went beyond that of buying fishing materials at reduced prices. It also was a voluntary association that acted on behalf of its membership in general fisheries matters. On the basis of the minutes of Co-op board meetings, one gains the impression that mutual relationships were rather genial. Crosscutting ties of kinship, friendship and community membership contributed to this. Most members even lived in the same street in Oosterend. Although the Drijver and Vlaming families were Calvinist Reformed and the Ellen, van der Vis and Krijnen families were Dutch Reformed, the emphasis was on togetherness. For example, the treasurer – who could invariably report good financial results during the initial stage of the Co-op's existence – encouraged the members in 1936 to continue supporting it and wishing it well by maintaining it together. Business went so well that a year later, the Co-op was able to buy its own small oil tanker to fetch fuel from Amsterdam. The vessel's capacity was sufficient to supply the entire cutter fleet with red diesel for three weeks.

In contradistinction to the Co-op, both skipper-owners and non-propertied deckhands could become members of Oudeschild's Progress through Unity association for the modest fee of five cents per week. Generally, the Oudeschild fishermen resented the Co-op's efforts to bypass the intermediary dealers. They called the Co-op 'the hand-that-strangles institution' (de Wolf 2004:28). At the root of this perception was the conviction that fishermen, suppliers and tradesmen were interdependent and that their fates were inextricably interlinked. The fishermen supported the interests of the small suppliers and traders by remaining loyal and expected them in reciprocal fashion to support the fishermen, for example through acting as guarantors, supplying informal credit or upholding prices to a certain level. In other words, the Oudeschild fishermen and the local petty traders and suppliers constituted a strong community of interest based on a moral economy (see also Chapter 2). Economic transactions were to some degree guided by norms, values and mores and face-to-face relationships helped mitigate power differences between fishermen and merchants. The fact that the association was led by local leaders of the Social Democratic movement – including a fish dealer – is important in understanding this viewpoint. Moreover, both fishermen and middlemen had joined forces in their fight to be acknowledged as rightful claimants to the Zuider Sea Relief Fund.

Although organized separately, the island fishermen worked together if and when they deemed doing so beneficial. Mutual problems, goals and desires brought them together in an attempt to achieve common objectives. For example, the Co-op wholeheartedly supported the Oudeschild fishermen's association's struggle to become entitled to Zuider Sea Relief.

Both the Co-op and Progress through Unity joined the National Union of Fisheries Associations that was established in 1934 and later renamed Dutch Fishermen's Union. It was the first such organization that horizontally integrated the interests of disparate petty producers that did not disappear from the fisheries arena within a couple of years. It was no coincidence that fishermen finally had arrived at sustained supra-local collective action by this time, as the Zuider Sea's closure and the economic depression had a severe impact upon the Dutch fishing industry. However, by the time the Union was gaining some political clout, it had to suspend its activities due to dramatic geopolitical events in Europe, beginning with Germany's invasion of Poland in 1939. Dutch fishermen soon felt the effects. Prices of fuel and fishing materials soared, while it became increasingly difficult to ship fish because foreign markets became inaccessible. For some time, fishing in the North Sea was prohibited due to the danger of sea-mines.

The War and its Aftermath

German troops invaded the Netherlands in early May 1940. Texel cutter owners moored their vessels near Oost. They deemed the Oudeschild harbour unsafe, as Den Helder was a naval port that was heavily shelled. Fishing was impossible until the end of October. For some time, the fishermen were allowed to participate in day-tripping fisheries with Scheveningen and later IJmuiden as their port of call. German navy vessels accompanied them to the fishing grounds. The Co-op was disbanded in December 1940. The German army had requisitioned the Co-op's oil tanker a few months earlier, and due to the war it was increasingly difficult to acquire fishing materials. After that, the Co-op stocks were redistributed to the members. The Oudeschild fishermen's association's work also came to a standstill. As of February 1941, the *Wehrmacht* began impounding the best vessels and converted them into tug boats and patrol vessels. Soon, all Texel cutters were confiscated. 'It was a terrible time,' one of the skipper-owners recalled. 'You had been working so hard for that boat.' The fishermen were allowed to strip the fishing equipment off their vessels and take it home to Texel. The occupiers were not interested in wooden-hulled rowing boats or sailing vessels equipped with small engines. As of May 1941, day tripping from the harbour of Oudeschild with such vessels was allowed with a special permit. The fishermen could leave between 6 and 8 am after reporting to the Military Post and showing their *Ausweis* in the port office. They were not allowed to call at other ports, had to be in possession of a German fishing licence and should return between 14 and 16 pm and report back to the authorities.

Nearly all skippers and deckhands who had become unemployed following the cutters' impoundment acquired new flats or an old sailing vessel. The local fleet eventually comprised 174 small sailing vessels and rowing boats. Only thirty-four had auxiliary engines, with predominantly less than 20 h.p. Until September 1944, the owners of the requisitioned cutters received a small monthly allowance in compensation. Some of them – in particular those who had bought new vessels shortly before the outbreak of the war – were deeply indebted and had to redeem their debts. Most former North Sea fishermen began fishing dab with hoop nets and, to a lesser extent, flounder – species that had not been particularly sought after previously. Still, under the prevailing conditions the fishermen could make rather good money landing them. The local auction even received a fillip during the occupation. Its annual turnover increased sevenfold relative to the five years before the war. A lot had to do with the fact that Texelians were not allowed to land their catch in the naval port of Den Helder or elsewhere as many had previously been doing. In addition, fish was a valuable food item, and the German army readily bought most of the supplies. There were many extra mouths to feed on the island. Fish prices were fixed but generally, the fishermen earned a good living. As a trap netter, born in 1920, recalled: 'Inshore fishing was good because there was no fishing in the North Sea. Species such as mackerel and scads had been gone for years, but they returned. We caught huge amounts and everything you landed was worth gold.' Several contemporaries confirmed this view. 'With a hoop net, I sometimes caught six to eight hundred pounds of dab in a tide. You would leave the harbour when high tide began, and with the turning of the tide go back with the flow. The auction bought the dab for a fixed price, I think it was a quarter a kilo,' said a fisherman, born in 1914. Having a small boat was a means of obtaining an income but with increasing scarcity, it was also used to acquire food for one's own household and an exchange item in a developing informal economy: 'We used to sort the big dab and cycle to the farmers and swap them for butter.' In the summer, however, shrimp and dab migrated to other waters and most fishermen had to find jobs as farmhands. Whelk fishing came to a standstill during the war, since the Belgian, French and English markets could not be supplied.

Although they resented the occupation and the occupiers, the fishermen could have fared much worse. They were exempt from the *Arbeitseinsatz* if they reported to the Industrial Board of Fisheries Products, an institution established by the German authorities. However, as the war progressed, equipment and fuel were rationed and distributed. The island's fishmeal plant closed down in 1942. For a long time, the devastating practice of landing undersized by-catches had yielded a substantial part of the gross revenues and contributed importantly to the incomes of many fishermen. By 1943, fuel was no longer available. When the occupying forces also impounded the fishermen's bicycles, the Oosterenders had to walk the six-kilometre distance to the Oudeschild harbour. The situation further deteriorated when the Germans in November 1944 commanded hundreds of Texel male civilians aged 18 to 35 to conduct forced labour, among whom were several fishermen. They only returned to the island in March the next year. A fortnight later, Georgians who were in the *Wehrmacht* rose up against the Germans. The insurrection – locally known as the 'Russians' War' – had devastating consequences as the belligerents' hostilities took weeks and the civilian population was caught in the crossfire. The event exacted a heavy toll in human lives (see Chapter 1). Several vessels were badly damaged when Oudeschild and the harbour came under fire. A fisherman, born in 1914, remembered the first day of the Georgians' uprising:

> I went to the harbour that morning, and the gate was still locked. I said to another skipper: 'I don't know what is going on, but they are shooting like madmen.' It later turned out that the Russians' War had begun. I was lucky not to have entered the harbour quay. There was a dab fisherman who had and he was taking cover near a building that was shot to fire. He got badly burned. I wanted to go home, but the bullets were whistling all around me so I had to duck for cover.

The Russians' War would last for weeks and kept the island's populace in its grip. Texel was one of the last areas in the Netherlands to be liberated. In the final stage of the war, heavy damage was inflicted upon the island that only a few months earlier had seemed to be escaping relatively unscathed.

As the Netherlands emerged from the Second World War, the fishing industry was in a state of severe disorder. The modern fishing vessels had been impounded; others were demolished or badly damaged. The country was in dire need of animal protein and the post-war government's reconstruction policy prioritized food production and economic recovery. Immediately after the conclusion of hostilities on the island in late May 1945, Texel fishermen whose vessels had been confiscated by the German occupying troops longed to resume their means of livelihood. However, they could only guess as to the whereabouts of their cutters. Search parties were formed and several vessels – some of which had served as far away as the Baltic Sea – were recovered. Three cutters were never retrieved. They were eventually substituted with new-built craft with the same registration numbers, but meanwhile their owners had to crew for others for want of a boat. The vessels that were repatriated to Texel needed substantial repairs. The state indemnified the owners and as soon as the boats were ready to return to sea, the Texel fishermen began fishing again. With catching efforts in the North Sea having been at a low for five consecutive years, fish stocks thrived. Once back in business, Texel fishermen discovered that large sole, plaice, cod and other species could be caught in abundance. A retired skipper-owner recalled:

> Right after the war, we fished with a flat-bottomed sailing vessel with an 18 h.p. engine. We otter trawled, using a thirty-pound footrope only, and we caught plenty of big flatfish and cod. Within thirty-six hours, we grossed

> 2,500 guilders in Den Helder. As they say: big fish eat small fish. Initially, we hardly caught smaller specimens.

Many of my interlocutors confirmed this view. The fishermen made good money, even though the Dutch government introduced a system of maximum prices. Markets at home and abroad were craving for fish and prices rocketed. Bumper catches and fairly good prices led to unprecedented gross proceeds and net incomes. For example, after three days and nights of fishing, the crew of a Texel cutter landed such quantities of fish at the Oudeschild auction in October 1945 that it took hours and hours to unload, sort and pack the catch, which grossed a record of about 9,000 guilders. Bumper catches and windfalls encouraged risk-taking. That fishing immediately after the war could be hazardous became evident when in 1946 and 1947 three Texel vessels hit sea-mines. Two vessels were heavily damaged, while one boat was lost and a crewmember – the skipper-owner's son – badly injured. In solidarity, the Texel fishermen collected money to provide the uninsured owner with a substitute vessel. It could have been much worse: when fishing just north of Texel, a cutter from neighbouring Den Helder sank in July 1945 after it netted a sea-mine. The device exploded, killing a crew of six.[5] But Texel's North Sea fishermen did not regard such perils and mishaps sufficient reason to quit fishing. They were keen on maintaining their leading position and dared to invest in improving their vessel's performance and safety.

Texel's North Sea fishermen confronted the question of whether or not to revitalize the Co-op. It was only in February 1946 that the first post-war meeting was held. Mutual relationships had suffered from the fact that a few skipper-owners had been able to keep their boats for some time and continue fishing, whereas most fishermen had lost their vessels to the occupying forces for all those years. Meanwhile, some operators were making good money and this went against the grain for those whose cutters were never found again and whose owners had to crew for others until they could acquire new boats. It took some heated discussions to convince everyone to uphold the Co-op, including its tannery and heavily damaged warehouse, which badly needed repairing. A new board was elected, but restoring mutual trust was not easy. During the next meeting, the Co-op's secretary hammered on the need to let bygones be bygones:

> We should support our cause together and in unison because it is in our mutual interest. This requires that all of us mean well and refrain from cold-shouldering each other or indulging in slander. Just frankly say what fault you find with each other. For how can we build up something if mutual relationships are corrupt?

This emotional appeal struck a chord. Becoming a Co-op member seemed easy enough: it still required a symbolic contribution of a hundred guilders. The skipper-owners began cooperating as of old, although it took many years for ill feelings to finally subside. The Oudeschild association Progress through Unity, however, did not face problems of internal strife and was re-established as soon as the war ended. A new board, succeeding the old generation, was elected and would stay in office for many years. Generally, board members of Texel's fishermen's organizations were re-elected by acclaim.

Despite the fact that the local fishermen organized, their self-image was of an individualist and independent lot whose differences of opinion were legendary. 'You won't ever find two fishermen who agree,' many remarked, usually referring to past and present experiences and examples. In the immediate post-war era, this folk model of a deep-seated penchant for individualism was backed by outside views on Dutch fishermen. In a speech for the 1946 Fisheries Days, a social geographer propounded the following view:

> The fishermen's inclination to associate is poorly developed; the women do not play a role in this whatsoever. Organizations and cooperatives, which are thriving in other parts of society, are hardly known. Only hesitatingly do our fishermen become members of local occupational associations, even though it is currently evident that this is desirable. ... They persist in their individualism. This is an unfavourable disposition. Leaving aside the important fact of externally promoting interests with the state, they should organize to obtain better social security (Visser 1946-47:315).

Other contemporaries also referred to the alleged individualism of Dutch fishermen and their disinclination to organize (Hildebrandt 1947-48).[6] Fishermen 'take great pride in their occupation and if at all possible they solve their own and their family's problems. If they are unsuccessful, they turn to the unorganized spontaneous help of their own group' (Hildebrandt and Visser 1947:64). Contrary evidence as in the Texel case notwithstanding, auto-images and external images seamlessly overlapped. To be sure, it was difficult indeed to encompass the entire heterogeneous occupational community of fishermen at local and national levels, but this did not mean that fishermen were unorganized or under-organized. The Dutch Fishermen's Union – on whose board a man from the Vlaming lineage represented Texel fishermen – even reinforced its bargaining position in post-war times.

Before the war, the island's North Sea fishermen had avoided the Oudeschild auction because they could obtain much better prices in IJmuiden. However, with prices fixed at maximum levels, it did not matter where fishermen landed their catch as they would obtain similar prices everywhere in the Netherlands. Forced by the German occupying authorities to focus on inshore fishing, Texel cutter fishermen had grown used to landing their fish in Oudeschild. Inshore fishermen were initially also experiencing good times. Due to a shortage of coal, which was needed to boil shrimps aboard just after capture because they spoiled quickly, they initially did not go into shrimp fishing. Mackerel, scads, whelks, sole and plaice were now their prime target species. They were earning well and so were petty fish traders. The dozen or so fishermen utilizing stationary trap nets, who had been landing (part of) their catch in Den Helder, also opted for the Oudeschild auction. However, it was mainly due to the fact that the local North Sea fishermen landed their catch in Oudeschild that the auction – a municipal venture – thrived for five consecutive years after the liberation. Its annual turnover amounted to a mean of around a million guilders (with only fifteen to twenty per cent coming from the inshore segment). Demanding a three per cent fee on sales, the municipality could employ four to five men, including the auctioneer. In addition, about twenty persons were involved in unloading, sorting and packing the landings. A handful of traders bought the fish, shrimps and shellfish and shipped the merchandise to wholesale dealers at home and abroad, mostly in Belgium and France. Since prices were fixed, there was no real sale by auction. The dealers had agreed among each other what percentage of the landings they would buy. The retired auctioneer said, 'There were a few fish dealers, but everything was so petty, you can hardly imagine. They would buy small quantities of dab, flounder and shrimp; it was really, really small-scale. It is hard to explain. They all worked on commission for larger traders elsewhere in the country.'

The good times for the petty fishermen were short-lived. With the exception of trap netters, the inshore fishermen turned to shrimp fishing again as soon as sufficient coal was available. Since first-comers usually obtained low prices, the drawing of lots decided who had to land shrimps first at the auction. By October 1946, shrimp prices fell mainly because the exchange rates of Belgian francs were unfavourable. The Dutch Fishery Marketing Board (*Nederlandsche Visscherij Centrale*) still regulated prices. In 1947, a category of inshore fishermen working with eleven medium-sized vessels, mostly from Oudeschild, obtained gross revenues ranging from almost 8,250 to a little over 16,000 guilders. Another fifteen small boats, whose owners hailed from the hamlets Oost and Horntje, grossed less than 6,500 guilders. The majority of these fishermen only fished seasonally or part-time and used rowing boats or sailing vessels with small auxiliary engines. Some hardly sailed at all because the gross proceeds would not even cover the operating costs. Even for the full-timers, many fishing days were lost due to ice, inclement weather and sundry other reasons. In the second half of the 1940s, for example, they only fished 140 to 150 days on average, as opposed to 210 to 220 days in the offshore segment.

In 1947, there were fifty-seven owners or co-owners of boats in a total of 114 active fishermen. Thus, a considerable part of the occupational community was propertied, although perhaps about twenty (co)-owners had retired and were not counted in the 'active fishermen' statistics. In the vast majority of cases co-owners of a vessel had agnatic kinship ties (see Chapter 4). Part of the problem in the inshore sector was, however, that fathers who owned small boats did not encourage their sons to become fishermen any more. In this section of the fishing fleet, only nine sons of skipper-owners crewed vessels, whereas in the offshore segment they numbered thirteen. The petty inshore fishermen were rather pessimistic about the fisheries' future. The Zuider Sea relief prevented some older fishermen from poverty, but fewer and fewer young fishermen entered a fishing career. The financial results of the inshore sector proved to be so bad that many owners of inshore boats complained that it was hard to recruit young deckhands. The reason was that few young adults saw future prospects in the Wadden Sea fishery, as it could hardly provide an income. By the late 1940s, the devastating pre-war depression of the local fishing industry's inshore segment had returned in full force. Occasionally, inshore fishing was so bad that fishermen preferred working for farmers: agricultural jobs provided well over forty per cent of employment on the island, fishing less than four per cent. Other petty skipper-owners worked as farmhands in the summer months. The local shipyard got fewer and fewer customers. Lacking the means, fishermen again did not maintain their vessels properly and postponed highly necessary repairs.

On the contrary, the offshore fishing fleet continued to perform extremely well. In 1947, eight North Sea cutters with 100 to 150 h.p. engines grossed between 52,000 and 82,000 guilders at the Oudeschild auction alone. On account of not having fished during a part of the year, two did less well – grossing a mere 28,000 to 29,000 guilders. However, this constituted only a part of their proceeds as Texel offshore fishermen discovered that they could earn more in herring fishing than in flatfish fishing. From mid-October until early March, they followed the herring into the southern North Sea, steaming as far south as the French coast between Calais and Dieppe. They landed 'the silver darlings' in the Zeeland port of Breskens. Their catches were so plentiful that nets were frequently torn by their weight. In his cashbook, a Texel skipper-owner jotted down on 24 January 1948 that his landings of herring fetched a phenomenal 7,761 guilders – a grossing that was almost equalled a week later. For deckhands, this implied that they made extremely good money. On average, they earned three to four times as much as the deckhands of the most powerful inshore vessels. Bigger boats with powerful engines and a crew of five or six men of course implied higher costs that had to be subtracted from the gross revenues than smaller ones with only two or three crewmembers. Yet even taking this into account, the financial results of the North Sea cutters were much better than those of shrimp and other inshore fishermen. The North Sea fishermen's rule of thumb was that depreciation and profits should be kept aside for investment in a new cutter. In addition, owner-operators did not take their full crew share but instead made do with a weekly allowance and used the remainder to redeem debts and save money (see also Chapter 4). Although by that time it was common practice to take out a bank loan, they were of the opinion that their own capital should provide the greater part of the initial expenses for a new boat.

Whereas the offshore fishermen were able to steam to remote fishing grounds, the inshore skippers lacked such opportunities, and at times when shrimp prices improved, North Sea fishermen would also switch to shrimp fishing in the Wadden Sea. As of 1945, the Progress through Unity board had been attempting to get access to Lake IJssel (the former Zuider Sea) for Texel fishermen who wished to participate in eel fishing there. However, the state wanted to check the number of fishermen operating there and bluntly rejected the Texelians' request, as it had previously dismissed a Texel fishermen's petition to be admitted to mussel fishing. The same happened when Zeeland mussel farmers gained entry to traditional shrimp-fishing grounds in the Wadden Sea. In 1950, a parasitic copepod killed a large proportion of mussels in Zeeland waters, where ninety-five per cent of the country's total mussel production was farmed. Some enterprising Zeeland mussel planters obtained permission to cultivate plots in the Wadden Sea, which was not infested by the parasite (see van Ginkel 1990, 1991). Soon all Zeeland musselmen relocated parts of their production areas to the western Wadden Sea. This was to the detriment of Texel and other shrimp fishermen who much to their dismay saw 'their' territory radically reduced while they themselves were still coping with the consequences of the Closure Dam. Realizing that mussel cultivation was more profitable than shrimp fishing, several Texel fishermen considered switching to farming mussels. They applied for licences, but the Ministry of Agriculture and Fisheries refused to provide licences. It had reserved these grounds for Zeeland musselmen who faced a problem at home and it did not want to expand the number of mussel farmers. The ministry put the interests of the successful Zeeland mussel industry over those of the petty inshore fishermen whose troublesome ventures it neglected. Repeatedly, the Texel inshore fishermen – with the support of the local and provincial administrations and a Labour parliamentarian – applied for plots. The state, however, refused to budge.

Understandably, the Texel – and particularly the Oudeschild – fishermen deemed it a grave injustice and were outraged. 'We fished mussels here until the 1920s,' Jan Boom, the local fishermen's association's chairman from 1945 until 1971, contended. He continued saying, 'But they didn't grant us access even though the mussels were farmed right in front of our door. It was a grave injustice, a scandal!' The issue obviously rekindled the inshore fishermen's resentment of state institutions. The Oudeschild and Oost fishermen certainly had reason to be suspicious and spiteful. Boom related that he had had several meetings with state representatives to discuss the matter. On one of these occasions, a civil servant riposted his desperate remark that the petty fishermen had a *right* to participate in mussel farming with the following words: 'Well, mister Boom, there is much more that you are entitled to, but that you nonetheless won't get.' Apparently, the arrogance of 'might is right' applied. The inshore fishermen continued to be entangled in a ruthlessly negative spiral: they had neither the means nor the opportunities to relocate their fishing activities; financial institutions would not provide loans; for want of credit they could not acquire a modern cutter that would require an investment of 180,000 guilders. They had to continue working with ageing boats, and due to their meagre gross revenues they failed to attract crew.

Winners Take All, Losers Stand Small

In 1950, fishing accounted for approximately four per cent of employment on the island, with a turnover of 1.4 million guilders. Agriculture and tourism yielded twelve and two million guilders, respectively. The fishing fleet consisted of a mixture of cutting-edge steel-hulled cutters with up to 200 h.p. engines, old wooden-hulled flat-bottomed sailing vessels that were equipped with auxiliary engines, and a small number of flats used mainly in seasonal trap-net fishing. The socio-economic differences between expansive entrepreneurs and petty self-employed fishermen had been intensifying for decades and this would continue to be the case for some time to come. The divergence was also still evident at the organizational level. The Co-op expanded its activities from merely acquiring commodities at reduced prices to other ventures. It became the co-owner of an ice-making factory, had a new warehouse built and bought a new oil tanker to supply the fishing fleet with fuel. It also established a fishmeal plant that reduced fish offal and discards into animal feed. However, due to fierce mainland competition, the plant proved a failure and closed down within months. With the Co-op's expansion, profits increased. Half of the profits were shared out to the members – the cutter owners – and put on an account at a five per cent interest rate. Members could take out money from this account for investments or borrow money from the Co-op at a modest interest rate. This state of affairs aroused the curiosity of a news reporter, who believed the local fishermen's 'progressive business outlook' and 'good mutual relationships and cooperation' to be unique in the Dutch fishing industry. One of the Co-op leaders responded by saying: 'Look, we may live on a remote island, but this certainly does not mean that we are backward in development' (*Het Vrije Volk*, 6 March 1951). Partly through the Co-op, the local fishermen kept up to date with new developments. For instance, it organized courses in the use of radio communication equipment and echo sounders, it discussed and arranged collective insurances for vessels and accident and health insurances for crew, and it generally acted as an interest association for its members. It employed several wage labourers. All in all, the Co-op constituted an important linchpin for the North Sea fishermen, who still hailed mostly from Oosterend.

The vast majority of Progress through Unity members continued to stand aloof from the Co-op. With their small boats, the Oudeschild fishermen had little to gain from Co-op membership. For them, the cuts on the costs of fuel and commodities were relatively small and they received only a fraction of the profits because every member had to contribute his share to the Co-op's operating costs. Being strongly communitarian, some Oudeschilders doubted its 'cooperative' character. As one of them explained:

> The Co-op had a peculiar arrangement. Fuel supply made for its profits, but the warehouse always had deficits because wage labour was expensive. All members were liable for deficits. So if there were twenty members, shortages would be divided by twenty. However, the fuel profits would be shared out according to the litres you had consumed. We owned a cutter with a small engine, so we received virtually nothing. It was not a cooperative idea. I therefore terminated my membership. My sons became members again, but by then it had all changed.

Lacking the means to ameliorate their situation, most inshore fishermen exited the fishing industry or simply muddled through.

The situation was much the same in other Dutch fishing communities. A substantial part of the inshore fishing fleet was ageing and was believed to be in need of modernization. The state established a committee to look into the matter. It reported that family firms encountered difficulties obtaining external capital for investments; that the fishing fleet needed to expand and that state support for newly built vessels was necessary (Commissie Sanering Zeevisserij 1952:5). Based on the committee's advice, the state introduced special grants for decommissioning and development of the fishing fleet. Texel fisherman Jan van der Vis responded in a March 1953 issue of the weekly *Visserijwereld*. He deemed expansion and state support unnecessary and even undesirable as well as irresponsible. He regarded the extant fishing fleet as sufficiently large and growing exploitation and increasing competition were not in the fishing industry's interest. Early on, the Dutch Fishermen's Union also expressed concern about the cutter fleet's expansion. It defended vested interests. However, the committee argued that since foreign states protected and supported their national fishing industries, the Dutch government should follow suit. The Social Democrats, who were in the 1950s coalition governments, came up with a report of their own and backed the idea that the fishing industry's modernization and expansion were in the national interest (Anonymous 1956). Hence, subsidies became available for the modernization of the fisheries, not because fishermen called for them but because politicians indulged in economic nationalism. As we will see in the following chapters, this would have far-reaching consequences.

Many older boats were decommissioned or scrapped, while new ones continued to be added to the fleet. The process was also evident in the Texel fishing industry. By the end of 1952, Texel's inshore fishing fleet consisted of twenty vessels, three-quarters of which were old wooden-hulled boats with less than 25 h.p. engines. Their owners hailed from Oudeschild (seven vessels), Oost (seven), Horntje (four – all of them trap netters), Oosterend (one) and Den Burg (one), while the twenty-two deckhands mostly lived in the same locales as the skipper-owners. Apart from the trap netters, shrimp fishing was the predominant occupation of these fishermen, while some also fished whelks, cockles, flounder, eel and sardines. Only two skipper-owners of somewhat larger vessels fished flatfish in the summer. Nearly all the inshore fishermen were day-trippers. Five fishing firms in this category grossed between 10,000 and 20,000 guilders; the others grossed less than that. Several of these fishermen worked as farmhands during a part of the year or were on the dole. On average, the inshore fishermen were older than North Sea fishermen. Sons of crewmembers often opted for working aboard North Sea cutters (this applied to ten out of fifty crewmembers in the North Sea fleet; seven of them were from the hamlet of Oost). Only a few sons of skipper-owners in the inshore segment followed in their father's footsteps (Schreur 1953:33-37). The old wooden vessels required a lot of maintenance and it was increasingly difficult to recruit crew. By the mid-1950s, most owners had sold their outdated boats. Likewise, the trap netters could no longer exist from fishing alone and some gave up, while two men combined fishing with fish hawking and odd jobs. One of them, born in 1920, remembers: 'You had to work hard, but you were free even though it was a meagre livelihood.' His colleague, a year older than him, recalls: 'You did not want to give up, you were always hoping for better times.' However, by the end of the decade both men had switched to onshore jobs. By then, there were only six shrimp-fishing boats left. Two of their skippers found new ways to use their vessel in connection with the growing importance of tourism: during the summer season, they began taking holidaymakers aboard on shrimp-fishing trips for a fee of two guilders per person. In the months of July and August, it was too warm to preserve the shrimps, so this proved to be a good alternative. 'You were a kind of entertainer,' said a retired shrimp-cum-tourist-boat operator from Oudeschild, 'In those months we were pretty busy and we earned well. We did three two-hour trips a day. During the shrimp-fishing season we earned a rather meagre living.' Four wooden-hulled sailing vessels hardly ever came off the mooring ropes and were left to rot in the harbour.

The miserable state of affairs in the inshore fishing industry had knock-on effects in the local economy. As soon as the price-fixing system was abolished in 1951, the bigger boats began marketing their catches in IJmuiden again. Dealers there offered higher prices and the additional transportation costs of ferry and train made it difficult for the Texel fish traders to compete. The extent to which the Oudeschild auction had depended on the landings of Texel's North Sea cutter fleet soon became evident. Its 1951 turnover was half that of the previous years. Within two years, the auction personnel were reduced to one, the team of sorters and packers was downsized and only a few fish traders remained active. Trap netters went predominantly to Den Helder again and only obtained approximately twenty per cent of their gross revenues on Texel. The other inshore fishermen continued to land their catch in Oudeschild and received eighty-five per cent of their grossing there. Although prices in the nearby ports of Den Helder, Wieringen and Harlingen were usually higher, the latter remained loyal to the local auction because it saved time and they preferred to be home each night (Schreur 1953:47-48). Across the board, though, shrimp prices were low, hardly ever surpassing fifty cents per kilo. The national Fish Trade Organization therefore introduced minimum prices for shrimps as per April 1952, but for the Texel shrimp fishermen this measure barely improved their situation. With their number dwindling, the auction's importance was on the decline. Hardly any fishermen from outside visited the island to dispose of their catch in Oudeschild. Its peripheral location and the additional costs of ferry transportation proved to be an obstacle. The consequences of these developments were that the auction's turnover plummeted to a mean of 134,000 guilders per year in the remainder of the decade. By 1956, the local authorities began to consider discontinuing the auction but as yet they merely resorted to reorganization. The main reason for keeping the auction open despite it not being cost-effective was that the municipality did not want to sack its manager, but preferred to wait until his retirement.[7] Suppliers of fisheries goods also suffered, as offshore fishermen acquired most of their necessaries via the Co-op. In 1958, the shipyard shut down. According to its owner, it had become impossible to make a living because too few boats showed up for maintenance and repairs. In Oudeschild, the island's ferry company TESO increasingly provided employment for former fishermen, particularly when its schedule intensified following the upsurge of tourism. By the end of the decade, nearly two score villagers were working for TESO. Oudeschild's labour force of active fishermen had dwindled to less than twenty-five; most of them were crewing on the North Sea cutter fleet. Contrary to the inshore segment, it was still thriving and expanding.

In 1948, there were eleven cutters of over 50 gross register tonnes (GRT) and equipped with up to 150 h.p. engines. A decade later, this number had increased to sixteen, thirteen of which were built after the war, while engine power had increased to between 150 and 250 h.p. The owners belonged to the island's fishing elite lineages, although two in-laws had become co-owners. With the exception of the Krijnen family, they lived in Oosterend. Depending on the stage of the family cycle, family members crewed the boats. About half the offshore fleet's vessels had a core of three or more agnates, often partly supplemented by in-laws. In addition, quite a few Oudeschild crewmembers and even men from the mainland were recruited. Based on the share system, remunerations were excellent as in the 1950s sole and herring catches were plentiful and prices were generally high. Oosterend fishermen were relatively wealthy: 'Exceptionally good money is earned in the [offshore] fisheries and it is no exaggeration to say that by Texel standards the Oosterend cutter fishermen are affluent. However, they have experienced different times' (Janse 1955:260-261). A few figures can illustrate the proportions the offshore segment had assumed. Fishing contributed 700,000 guilders to the local economy in 1950, rising to over 2.5 million guilders in 1959. Per boat, the gross revenues rose from a mean of just over 100,000 guilders in 1950 to around 150,000 guilders by the end of the decade. With such gross proceeds, expansion and modernization were feasible. Moreover, parts of the investments were tax-deductible as of 1954. Wheelhouses made way for bridges, while ship-to-ship and ship-to-shore radios and echo sounders – instruments needed in the herring fishery – became standard equipment. Of course, decisions to invest were based on any number of personal considerations, for example depending on the stage of the family cycle; age and health of the owner-operator; number and age of (potential) successors; their interest and ability in pursuing a fishing career; availability of credit and so on. Along with the offshore fisheries' modernization went a demand to professionalize. As of 1947, certificates were required for certain categories of crew – skippers, steersmen and engineers – of vessels of 50 GRT and over. Previously, the acquisition of knowledge and skills was exclusively based on 'learning by doing' – an epistemology of practice. The state began monitoring the Sea Fishing Certificates Act more tightly as of the mid-1950s and, subsequently, vocational training in a Fishery School turned into a matter of course, particularly for sons of owner-operators.

Although the fishing fleet's modernization was a major step forward, the work aboard was arduous. Shipboard facilities were rather basic, and it was hard to preserve milk and bread. Due to a lack of vitamins, poor hygiene and the chafing of oilskins, many fishermen developed sores. An eyewitness told:

> You should have seen the hands of those men! Of course, hygiene was not as good as today and there was no good soap either. The fishermen's disease was 'sleeve eaters', that's how we called them. They were tiny sores they had on their wrists. The oilskins would chafe them open continually. They would also get them in their neck. God, I've seen old fishermen who had them all over. Tears would spring to your eyes. Of course their hands were wet and cold and in the winter when they had to pull the nets... They didn't have hands but claws. Obviously there was a lack of hygiene and they couldn't do anything about it. They would rub some ointment on their hands and put a cloth on them, but, well, they just had to go to sea. They just had to earn a living.

A deckhand, born in 1934, said about the 1950s fisheries:

> We fished with otter trawls. It was mostly manual labour. You hauled the nets by hand. Only the cod end line and the otter doors were winched in. [...] There was a small stove in the accommodation. In the wheelhouse it was cold and draughty. I always wore long underwear in the winter and you really needed that. You had mittens, but they didn't help much when you were hauling the nets. You were always exposed to all sorts of weather. We would not get out of our clothes the entire week; you didn't even change your underwear. We had to wash at a small tap, using a pump. When it was cold, it was sometimes frozen. You hardly ever used it anyway and you

> didn't shave. We didn't have a toilet either. We defecated on a barrel or, if the weather permitted, right across the rail.

His wife adds, 'When he got home, he stank to high heaven. That fish smell and the engine room stench!' Step by step, however, shipboard conditions improved, as skippers continued to invest in their boat. Continual reinvestment prevented the owner-operators from having to hand over a considerable part of their profits to the tax inspector. They preferred to keep the money in the firm. It was precisely this kind of attitude that the state sought to stimulate through its fisheries development policy.

When pursuing flatfish and other demersal species from May to November, the Texel offshore fishermen used otter trawls, as they had been doing for decades. The technique is based on the use of two heavy wooden or steel boards – the otter boards or 'doors'. The boards work like a kite and extend the net's mouth horizontally using the vessel's motion. In the first few years after the war, otter trawling was also used in the herring fishery from November to April. However, after a few good seasons, bumper herring catches with otter trawls were a thing of the past. Then the fishermen began hunting herring using pelagic nets made of nylon. Two vessels dragged the net in between them. A 'head rope' connected them and kept the distance between the boats at approximately sixty metres. The technique called for considerable expertise and close cooperation, particularly because herring fishing was mostly done at night, which increased risks. To get the catch aboard, the boats had to come alongside each other and this was possible only in relatively calm weather.

Knowledge of the fishing grounds was rather basic. Following the compass, the skippers determined a course and after a specific steaming time they knew where they were – by approximation at least. Using lead lines, they fathomed the depth and by putting some grease on the lead, they collected samples of the type of soil. In conjunction, the information gave fishermen an indication of their position. In flatfish fishing, this usually sufficed to steam to locations where they expected concentrations of flatfish to be. Herring being a migratory species, the herring hunt required another approach, to some extent based on hunches. To discover where the herring shoaled, skippers would look for gulls, gannets and seals. 'The herring gathered in shoals. If you were just by yourself, you had to be lucky to hit upon a school. So usually we searched with the entire fleet. When you saw that a pair (*span*) shot their nets, you would go there and also try your luck,' said a retired skipper-owner, born in 1904. Competition was fierce, however. A colleague, twenty-eight years his junior, claimed that skippers would 'almost try to sink each other's boats': 'They begrudged each other everything. My father was a persistently successful herring skipper, and his colleagues could drink his blood for that.' An important improvement in catch efficiency was the introduction of echo sounders, although there were teething troubles. Texel skippers proved to be agile and quite successful herring hunters. The demand for herring was high, particularly in Germany. The herring was shipped to the country by truckloads and the Germans paid handsomely. The good outcomes attracted many islanders into the herring hunt. In the second half of the 1950s, nearly the entire local offshore fleet participated in pair trawling. To some extent, it led to further integration of the island's occupational community of North Sea fishermen as Texel owner-operators pair-trawled with each other, and together they usually cooperated in the initial search for herring. They shared experiences not only in fishing, but also in being away from the island for part of the year. Still, tensions always lurked in the shadows. A retired skipper-owner, born in 1928, intimates, 'If you pair trawled, the skippers had to get along with each other. There were weeks when you earned hardly anything at all ... When you didn't gross anything, there was a tendency to blame each other.' Furthermore, differential success within the local fleet gave rise to jealousy (see also Chapter 5).

Since the distance to the fishing grounds was considerable, many Texel cutter fishermen decided to take up temporary residence in the Zeeland port of Breskens during the herring season. They rented summer cottages and those who were married with children brought their families over, transporting the furniture with their boats. When the herring season ended sometime in March, the families returned to Texel. Hence, a particular kind of transhumance came about. From mid-March until the end of October many Texel fishermen conducted the flatfish fishery from their island; in the remainder of the year they moved over 300 kilometres south to Breskens for the herring pair-trawl fishery. Some crewmen preferred travelling to and fro between Breskens and the island by bus or train for the weekend, a trip taking many hours. Many interviewees had fond memories of the Breskens episode in their lives. They experienced the relations with their hosts as congenial. Several Texel children were born there and five young Texel fishermen found a bride in Breskens. 'She was my best catch,' joked a deckhand married to a woman who was born and raised in the Zeeland town. All the newlyweds settled on Texel. Since herring catches were often so plentiful as to necessitate landing the catch two or three times a week, the women saw their husbands, albeit briefly, more often than they were used to on Texel. Moreover, pair trawling could only be done in fair weather, so the boats remained in port for more days than during the flatfish fishery. 'Sometimes we went home for a cuppa,' a skipper-owner born in 1921 said. 'Come Saturday, we told the deckhands to leave early as they had a long trip ahead. Those staying behind would clean the boat and mend the nets for the upcoming week.' The Breskens population welcomed the Texel fisher families who temporarily lived in the town, spent their money there and provided work for many labourers. It was an era of affluence for the Texel colony and Breskens retailers and ship chandlers alike. The riches of the seasonal herring fishery contributed greatly to the wealth of Texel owner-operators and deckhands. However, it is a well-known fact that herring are a notoriously fickle fish, sometimes showing up in abundance in specific areas for a number of consecutive years or decades, only to disappear again for prolonged lengths of time (see Corten 2001). In 1959, several Texel cutters returned from Breskens in early February, as the herring were too far south. Subsequently, the importance of herring fishing declined and the Breskens episode in the Texel fisherfolk's lives would gradually come to an end (see the next chapter).

'Frogs in a Wheelbarrow'

The petty inshore fishermen's experiences with state intervention fuelled distrust in government. Firstly, they regarded it a grave injustice that for six years, the state had not acknowledged that the Closure Dam had had a devastating impact on their economy, and had only belatedly recognized them as an interested party that was entitled to indemnification, which subsequently arrived slowly and in a limited supply. Secondly, they resented the fact that they were not granted access to Lake IJssel while their Wieringen compatriots could fish in both the inland lake and target shrimps in the Wadden Sea and so compete for scarce resources and market opportunities. Thirdly, they loathed the fact that Zeeland mussel farmers could rent plots in what they perceived to be 'their' territorial waters, while Texel fishermen were not granted access to these fishing locations for mussel cultivation on the account that they had not been fishing mussels for years. These events severely restricted the small-scale fishermen's possibilities to obtain a livelihood from fishing, whereupon this segment of the local fishing industry suffered a serious decline. The petty owner-operators found themselves trapped in a whirlpool that sucked them down and did not let go. Creditors looked away, the government seemed to be glad to get rid of an outdated fleet in times of modernization, non-propertied and unrelated deckhands voted with their feet, and – having already eaten into their capital for a considerable length of time – the skipper-owners of inshore boats lacked the financial and social resources to pull themselves up by the bootstraps. Most of them attempted to muddle through for as long as possible, either because they desired to remain independent or because they had no alternatives.

The offshore segment, on the other hand, was in quite another maelstrom. The Closure Dam did not impact the fishing elite to the extent that it impinged upon the inshore sector. With their vessels' range of action, the cutter fishermen did not depend on exploiting the Zuider Sea or Wadden Sea. Nor did the economic depression hit them significantly. Good outcomes accomplished with cost-effective boats enabled continual reinvestment, stimulated by state subsidies and tax policies. In hindsight, the war was an intermezzo that temporarily interrupted the North Sea fishing fleet's development. Upon the war's conclusion, the offshore fishermen did not have to start entirely from scratch but in most cases could soon begin operating their cutters again. With the excellent post-war results the skipper-owners continued to reinvest in their boats and, if feasible, expand. In addition to extrinsic incentives to modernize, there were several intrinsic motivations to keep up with the times. It would be tempting to also involve the divergent ideational systems and mental dispositions in explaining why differential success so clearly marked the communities of Oudeschild on the one hand, and Oosterend on the other. However, this might boil down to social and cultural reductionism. No doubt, different outlooks on life and varying practices importantly impacted the ways in which the distinct categories of fishermen went about their business, but it would be wrong to assume that cultural factors alone could provide a satisfactory explanation. Above all, it was the particular articulation of various forms of capital assets – ecological, economic, social, cultural and psychological – at individual and firm level that made for success or failure. As we shall see in the next chapter in considerable detail, the family firm's social and cultural importance and the way in which it tied together these capital resources explain a lot in this respect.

The divergence of the two segments – inshore and offshore, to a considerable extent localized in the distinct communities of Oudeschild and Oosterend – had important consequences for collective action. It may perhaps seem self-evident that shared interests should bring entrepreneurs together in voluntary associations, cooperatives and other organizations, but this is not so 'logical' as it might appear to be. Historians, sociologists and ethnographers have presented myriad examples of failed fishermen's organization from various corners of the world. This lack of sustained collective action is often attributed to an interrelated cluster of economic, social, cultural and psychological factors deemed typical of the fishing industry. Some argue that their frequent absence from shore and their inexperience in dealing with officialdom prevent fishermen from organizing (Smith 1977:5; Andersen 1979:3). Others surmise that non-cooperative behaviour is inherent in the capitalist mode of production where perpetual competition for scarce common pool resources impedes concerted action (Thomas, Johnson and Riordan 1995:143; Jentoft 1986:199; Libecap 1989:74). Skipper-owners are each other's rivals and compete for prime fishing positions, bumper catches, markets and prices. In their folk models, fishermen often value the conviction that they are independent or autonomous personas (Thomas, Johnson and Riordan 1995:150-151). Indeed, in the Texel fishermen's self-image, the view that they were and are an independent and individualist lot was and is strongly embedded. 'It is hard to make all noses point in one direction. Fishermen are like frogs in a wheelbarrow: they usually jump in all directions,' an Oosterend fisherman said. He used this metaphor to refer to the problem of sustained cooperation among his colleagues at the local and supra-local level.

Fishermen greatly value the notional idea of 'autonomy' and the corresponding identity as 'independent individualists'. In many fishing communities, an ethos, rhetoric and idiom of freedom, self-reliance and autonomy prevails (Vestergaard 1996; McGoodwin 1990, 2001). In social science analysis, the predominant – yet most unconvincing – argument for failed organization is that competition and economic and socio-cultural autonomy lead to the psychological character trait of individualism and a 'need for independence' that allegedly inhibits the creation of associations and cooperatives.[8] These explanations are summarized in the following proposition:

> the mode of production associated with commercial exploitation of a resource held in common engenders an atomistic organization of labor, which results in a culture and psychology of independence. This independence, in turn, constrains the ability of fishermen to act collectively (Thomas, Johnson and Riordan 1995:144).

In my view, such circular arguments hold little water in the case of Texel's fishing industry and indeed probably more generally also. They are tautologies claiming that fishermen do not cooperate because they are individualists and that they are individualists because they do not cooperate. It was Emile Durkheim who stated: 'every time a social phenomenon is directly explained by a psychological phenomenon, we may rest assured that the explanation is false' (1982:129; also see Durrenberger 1992b:153). Although fishermen may often cooperate reluctantly, institutionalized collective action did and does occur. It would be impossible to explain their attempts at organization if they were really 'independent individualists'. Moreover, independence and individualism do not necessarily confound cooperation; they do not exclude each other, as Lawrence Taylor shows with regard to fishermen of Dutch descent working on the Great South Bay in the USA. In this connection, he uses the concept of 'contractual community' (Taylor 1983:9ff.), a configuration of individuals who freely choose to establish and maintain social ties.

It is important, then, to understand under what conditions fishermen seek to organize themselves; what kind of outcomes they attempt to achieve through organization; which type and degree of organization they deem necessary; and what causes success or failure. As early as the 1870s, Texel's occupational community of fishermen acted as a formal interest group while previously informal collective action for specific ends had also come about. For various reasons, the island's fishing communities clung for a long time to their own organizations and their activities waxed and waned. This particularistic stance needs explanation. Why did Oosterend and Oudeschild fishermen usually refrain from establishing a joint voluntary association? The answer is rather complex. Given the fact that at face value Texel fishermen had common interests and confronted similar problems, one would expect them to cooperate supra-locally. Often, the local organizations discussed similar issues and pursued similar goals and in general, the boards of these voluntary associations met frequently and developed joint plans of action. As a rule of thumb, they also addressed authorities together. The fishermen acted collectively if they deemed it necessary to do so. The same applies to cooperation with fishermen from outside. Every now and then their associations joined a federative fishermen's organization. Usually this happened when a spokesman from such a regional or national corporation visited the island and expounded the view that 'union is strength'. More than once this kindled the Texelians' enthusiasm. However, it usually vanished equally quickly because they were not really interested in the wider and long-term objectives of a federation, which of course could not simultaneously deal with the specific problems of all categories and communities of fishermen. The Texel fishermen were preoccupied with their own immediate interests, for which a local association was best suited. They did not want to give up their own organizations and autonomy. This also applied to the situation on the island and was a consequence of socio-economic differentiation within the occupational community. So the reasons for collective action problems are social, not psychological. The sociologists Bechhofer and Elliott maintain that within the class of small independent entrepreneurs, institutionalized collective action is hampered by 'the disparateness of the various elements and the intense individualism that pervades it' (1981:190; also see Verrips 1989 on Dutch bargemen). More generally, Mancur Olson contends that 'unless there is coercion or some other special device to make individuals act in their common interest, *rational, self-interested individuals will not act to achieve their common or group interests*' (1971[1965]:2). However, he also states that the prospects for voluntary collective action are high for small interest groups. This is so because frequent interaction and face-to-face contact facilitate social pressure and social incentives to maintain cooperation, for example by ostracizing non-cooperative individuals and rewarding cooperative actors (ibid.:61-62). It is much harder to mobilize and organize large-scale social configurations, particularly if they are economically and socially heterogeneous.

The fishermen indeed constituted a heterogeneous economic grouping and differed widely with regard to their social positions, fishing strategies, investment policies, adoption of innovations, work attitudes and worldviews, and these differences were in turn linked with demographic and ideological differences, among other things. The majority of Oudeschild fishermen specialized as shrimp and shellfish fishermen who mainly operated in the Wadden Sea. Most Oosterenders were North Sea fishermen, catching flatfish and herring and in the harsh winter season some fished shrimp in the Wadden Sea. Besides common interests, both categories also had their own specific ones. Pursuing different species led to divisiveness, not because fishermen were independent and individualist competitors per se, but because their group interests did not converge. Although Texelians adhered to the subjective belief and rhetoric of independence and individualism, their actual behaviour showed that this 'folk model' had little to do with practice, since they organized time and again and acted collectively if the need arose. The fact that they cooperated – and still cooperate – in associations belies their image of being independent, individualistic, non-cooperative and politically inert human beings. This is not to say that sustained solidarity came about automatically and that it was entirely unproblematic. Collective action problems did occur. Above all, it was the economic heterogeneity with its concomitant divergence of sectional interests – not some innate penchant for individualism and independence or a collective behavioural disposition leading to social atomism – that impeded cooperation and sustained solidarity. In addition, to some extent the organizational localism also had roots in socio-cultural contrasts between the island's two fishing villages. Conflicting ideologies and the creation of symbolic boundaries divided the communities and fanned a slumbering antagonism. However, as we shall see in later chapters, faced with problems of scale and increasing intervention from outside, the Oudeschild and Oosterend fishermen would eventually overcome their differences.

Chapter 4

Booming Business: The Rise of Beam Trawling

The offshore fleet's gradual growth and the inshore segment's rather rapid decline characterized post-war developments in the Texel fishing industry. It was evident that owner-operators needed to modernize to stay in business. Those who were unable do so began to lag behind and in most cases were ousted from the fishing industry or had to muddle through and accept substandard incomes. In hindsight, modernization in the offshore cutter sector was rather modest. In many respects, the 1960 boats resembled the ones that dated from the 1930s. Generally, engine power had only slightly increased, while several skipper-owners continued to operate pre-war boats into the late 1950s. Otter trawling for flatfish and pair trawling for herring did not require enormous pulling power. With the advent of new technology, this would change radically as of the early 1960s. It was the beam trawl's reinvention, in particular, that brought about a headlong expansion and innovation of the fishing fleet, leading to an unmitigated 'horsepower race' in the Texel and Dutch fishing industries. Catches increased extraordinarily and so did the fishermen's incomes. It was not a strictly local or national phenomenon. Globally, the development of fisheries since the Second World War has been dramatic. Reported landings more than trebled between 1945 and 1972, an expansion caused by technological developments and the human demand for fish (Garcia and Newton 1994). In most countries, the emphasis of fisheries policy was on fleet development and increased production to set off a Blue Revolution, with relatively little concern for the dangers of over-fishing. International negotiations and agreements on the exploitation of marine living resources, however, did lead to the foundation of the North East Atlantic Fisheries Convention in 1959. Having extremely limited powers, the convention remained a paper tiger for quite some time, even after the establishment of the North East Atlantic Fisheries Commission in 1964 to further articulate and implement it. That year, sixteen Western European states introduced a 12 nm (nautical mile) zone at the European Fisheries Convention, giving these states jurisdiction in specific coastal areas. Foreign fishermen could only fish within this zone if they had historical rights to do so. In this case, too, protecting national interests rather than fish stock management was the predominant goal.

As of the early 1960s, the Dutch family-owned fishing fleet expanded rapidly and the fisheries thrived. The dual fisheries policy of decommissioning and development seemed to be paying off. In addition, the Treaty of Rome increased the trade potential for fish landed by Dutch fishermen. The European Economic Community – established in 1957 by six nation states, including the Netherlands – created a common market without trade barriers and tariffs that had previously hampered free trade to a large degree. The Dutch fishing industry benefited as the Netherlands was the most important fish-exporting country of the original member states (Kranenburg 1978:293). The Dutch seafood-processing industry and wholesale dealers also profited from these European policies, while in addition the national fishing fleet assured a steady supply of fish because auctioning in the Netherlands was mandatory from 1959 onward. The objective of mandatory auctioning was still to regulate the market to some extent. The obligation for Dutch fishermen to sell fish through Dutch auctions was abolished in January 1976, because it appeared to be at odds with European Community rules pertaining to the common market. Meanwhile, however, the Netherlands had reinforced its position as one of Europe's leading fish-exporting countries. This position was partly based on transit trade.

Like their compatriots in other fishing communities, the owner-operators and crewmembers of the Texel offshore fishing fleet benefited from the new opportunities that arose in the 1960s. From the mid-1970s onwards, however, the fisheries management regime would change profoundly. In this chapter, I will cover the period from 1960 to 1975 when the sky seemed to be the limit. It was a time when fishermen gained a reputation of being successful petty commodity producers: 'The Dutch fisherman is an improviser, perseverant with entrepreneurial gusto and zest for work. He is an individualist, certainly, but he also has an eye for the power of cooperation' (de Boer and van der Meulen 1976:8). With few restraints and many incentives to invest, the Texel – and the Dutch – fishing fleet expanded at a fast pace, both in number of boats and in engine power. The present chapter seeks to address the underlying causes for this unprecedented growth. As we shall see, they were not of an economic nature alone, as there were several other reasons to expand. In part, these were of a social and cultural character, with the logic and dynamic of the family firm being at the forefront. Although owner-operators were well aware that collectively their decisions to expand had negative consequences, as individual actors they had their own rationalities – or, rather, 'reasonable strategies' – to participate in the game of keeping up with the competitors. The particular system of remuneration and the labour ethos it stimulated were also important. So, what was the tangle of factors that improved the Texel fishermen's situation?

The Reinvention of the 'Wondyrchoun' and the Horsepower Race

Beam trawling presently dominates offshore flatfish fishing, with important fleets of beamers in several fishing ports in the Netherlands. Beam-trawling effort is concentrated in the southern and south-eastern North Sea. Until the late 1980s, use of this gear was mostly restricted to the Netherlands and Belgium. Later, it also spread to Germany and the United Kingdom (Rijnsdorp and van Leeuwen 1996:1201). The beam trawl became highly popular in the 1960s, but the basic principles of this fishing technique are much older. A predecessor of today's beam trawl was already in use in late 14th-century England. The net was spread open by a heavy ten-foot-long bar and had a leaded rope weighted with large stones attached to the net's lower part. A group of hook-and-line fishermen from the Thames estuary worried that 'this new and subtly contrived instrument' – dubbed the 'wondyrchoun' – could seriously damage fish stocks (Kennelly and Broadhurst 2002). Two centuries later, the Dutch prince William the Silent (William I, Prince of Orange) launched similar complaints and cautioned that catches could decline rapidly if fishermen continued applying the device (de Groot 1984, 1988:7). Throughout the history of its utilization, the beam trawl has been highly controversial. Due to its disturbance of the benthos (organisms living on or near the seabed), in recent times ecologists and environmentalist organizations have emphasized the alleged harmful effects of beam trawling (see Chapter 6). Early criticism arose especially from fishermen deploying other types of gear, particularly stationary nets. For centuries, fishermen working with mobile gear and those using fixed nets have resented and clashed with each other. There have been occasional spells when beam trawling has been prohibited, but it has always been reintroduced. For example, in the late 19th century, Texel fishermen used this type of trawl net to catch North Sea flatfish. However, in the days of sail, no more than three relatively light tickler chains could be used and the beam's width was usually less than eight metres (Kranenburg 1977:83). It was only with the diffusion in the 1910s of the otter trawl, a net that was easier to handle and more efficient, that the beam trawl temporarily lost its popularity.

In the late 1950s, the beam trawl was reinvented. New technology enabled the use of two such beam trawls simultaneously, one on each side of the vessel. They could be operated by means of derricks or outrigger booms and winches. The double-rig beam trawls were towed by means of warps passing through blocks at the end of the outriggers. Following these and other modifications, it became much easier to shoot and haul beam trawls than otter trawls. Moreover, due to their twin character the trawl path was wider than that of otter trawls while in addition beam trawls could be used at low speeds. These advantages led to a rapid distribution of beam trawls in the early 1960s. The relatively specialized beam-trawl flatfish fishery would soon turn into a particularly Dutch concern. By August 1963, save for two cutters, the entire Texel offshore fishing fleet had switched from using otter trawls to beam trawls. The technology's basic principles are still much the same as in the original version. A heavy steel tube spreads the net open horizontally. The apex of triangular trawl heads ('skids' or 'shoes') are mounted at the beam's ends and keep the cone-shaped net open approximately half a metre vertically, thus creating an oblong 'mouth'. The shoes' lower part moves across the seafloor. Sets of tickler chains (*wekkers*) are suspended along the bottom of the mouth and run in front of the ground rope, disturbing the seabed's top layer and activating flatfish species such as sole, plaice, dab, turbot and brills lying on or partly buried in the seabed. The fish start to swim, but due to the towing speed and the top net, they are funnelled into the cod end (see figure 1). The headline of each trawl is less than a metre from the bottom to avoid by-catches of cod and whiting. The net's bottom sheet is made of heavy twine to resist abrasion and, in addition, nylon strings protect the net's underside from wear and tear. A fishing rope is attached to the cod end; when the gear is hauled in, only the cod end is lifted aboard by means of the rope and emptied on deck.

Beam trawling was hugely successful in terms of catching efficiency. The rather shallow depth, the extensive areas of relatively smooth seafloor and the productive ecosystem of the North Sea make it a particularly good area for intensive beam trawling. Before long, the beam-trawl fishermen discovered that the more and the heavier the tickler chains, the larger the catch, particularly of sole. The tendency was to use more and heavier chains. A retired skipper-owner (born in 1922) explained: 'We sometimes wanted to fish with lighter gear, so that we could make longer tows and get more sleep. However, if another fisherman, who used heavier gear, caught considerably more, you swiftly deployed heavier chains yourself.' More and heavier chains required greater pulling power and hence engine capacity or, rather, thrust of the propeller. To increase pulling power and to improve propulsion efficiency, in the late 1960s most cutters were fitted with nozzle propellers. More horsepower also allowed fishing at greater speeds of up to seven knots to cover much larger areas. Soon the development towards more powerful boats was evident in the entire Dutch offshore fishing fleet. The fishermen were quick to make the necessary investments, leading to a steep growth of engine capacity. In less than twenty years, the aggregate engine power of the Texel fleet increased tenfold to 17,500 h.p. by the end of 1969. Vessels became longer, wider and had more draught. Older technology made way for new. The use of synthetic fibres became widespread in the late 1950s and early 1960s. Nylon nets replaced cotton and hemp ones. They improved the handling characteristics and efficiency of fishing gears and were less susceptible to wear and tear than cotton nets.

Due to these technological developments, sole catches increased. Prices paid for sole were several times those paid for plaice, and the fishermen's gross proceeds were generally considerable (see below). But the success of

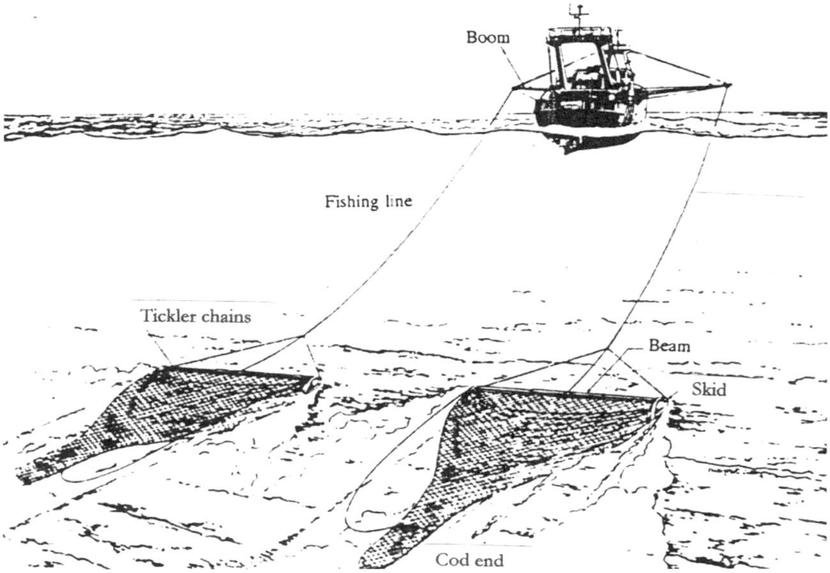

Figure 1. *Beam trawling.*

beam trawling came at a price. It proved a rather dangerous fishing technique that exacted a heavy toll in human lives. From 1960 up to and including 1970, twenty-eight Dutch beam trawl cutters toppled over in fair weather. During storms another six beamers capsized. In these accidents, fifty-seven beam-trawl fishermen were killed (Vissersbond 1994:52), including five from Texel. For various reasons, the beamers' stability was poor. Underwater streamlining of hulls to increase speed and heavy superstructures had negative consequences for stability. For vessels under 50 gross register tonnes (GRT), no seafaring certificates were needed. Since many fishermen had not had any formal vocational training, several skippers did not have such diplomas. When owners ordered new boats, they would sometimes limit the tonnage to 50 GRT, yet have powerful engines installed. Again, this was detrimental to vessel stability, but a requirement to undergo stability tests was lacking for new-built boats up to 200 GRT. The technical problem was that if gear snagged on a wreck or another obstacle, the momentum and the engine power would cause the boat to be pulled over sideways by the boom. If the cable did not snap in time, the cutter would capsize. A new Vessel Order was introduced in 1965. It demanded stability tests for all sea-going vessels to obtain a 'certificate of soundness'. The measure was immediately effective for new-built vessels. Between September 1967 and April 1969, all Dutch cutters – including older ones – had to undergo such tests. No less than 175 boats failed to pass. Many had to be replaced or lengthened and would consequently surpass the 50-GRT limit. For vessels built before March 1967, the Shipping Inspectorate was lenient. It dropped the diploma requirements if the vessel had passed the stability tests. In 1968, the Inspectorate, in conjunction with the Fishermen's Union, came up with additional safety measures. It demanded the application of safety-release systems in the boom stays or winch breaks to minimize the vessel's lever action if the gear snagged. It also came up with procedures on how to handle adverse and hazardous situations. These regulations have benefited fishing-industry safety (Vissersbond 1994:65-68).

The measures to improve stability brought about another wave of rapid modernization in the late 1960s, on Texel as well as elsewhere in the Netherlands. In 1967 and 1968, there was a veritable investment boom in the island's fishing industry. Despite increasing prices, in these two years alone ten cutters were launched or lengthened and equipped with new engines. By then, to be modern was to own a vessel with a length of 26 metres and a 600 h.p. engine that required a crew of five or six. Bigger boats with increased draught and powerful engines enhanced seaworthiness, allowing for more days at sea, greater pulling power and a wider radius of action. Between early 1965 and early 1969, the number of Texel cutters expanded by forty per cent (from 35 to 49), while aggregate engine power more than doubled from approximately 7,500 h.p. to 16,000 h.p. The expanding fishing fleet necessitated the construction of a new harbour in 1969. Texel fishermen boasted that they had 'the most modern fishing fleet of the Netherlands', but in other fishing communities similar developments occurred. The inshore segment of the Texel fishing fleet, which had been declining for decades, also slightly recovered. In the 1960s, five to eight owner-operators had shrimp-fishing licenses, including two that still took tourists on shrimp-fishing trips in the summer. However, by the end of the decade, shrimp fishermen encountered difficult times again due to low prices connected with the import of cheap shrimps from Germany. Structural state support as of late 1970 led to a considerable improvement. To a modest extent, the shrimp fishermen also modernized, mainly through the subsidized acquisition of rinsing and sorting machines. The inshore segment also included a small number of boats whose owners pursued other species, such as cod, eel or cockles. However, the inshore sector was very small in comparison with the offshore fleet. In 1970, out of 186 fishermen a mere twenty worked in the inshore fisheries. Clearly, expansion was most evident in offshore fishing, where there seemed to be no limit to the growth in engine power.

Several beam-trawl fishermen showed concern about the 'horsepower race'. They believed that it would ultimately result in unmitigated competition and possible bankruptcy, while the effect on flatfish stocks, in particular, would be devastating. In June 1971, Texel's Progress through Unity and several other local fishermen's associations in the Netherlands demanded a 800-h.p. limit for sole fishing. They stated that increasing fishing capacity endangered sole stocks. Texel fishermen were already experiencing problems, because catches lagged behind what could be expected with more engine power. The Progress through Unity board maintained that such a measure would protect fishermen against themselves. The horsepower limit would apply to sole fishing only. However, by then there were already three beam-trawl cutters with engines over 800 h.p. The association's leadership suggested that these should be either exempt or financially compensated. Eighty per cent of the members were in favour of a legal limit at 800 h.p., twenty per cent favoured capping it at 900 h.p. The Texelians realized that to be effective, such a measure would have to be taken at international level, not just in the Netherlands. However, they believed that if they set an example, they would have 'reason to speak vis-à-vis our foreign colleagues', as someone said at a meeting of the fishermen's association. 'But we should act now, for otherwise we will miss the boat, just as before with the 500 h.p. limit.' Another skipper-owner opined, 'All of us are fishermen and if we want to continue being that we should be in favour of capping engine power.'

The Dutch Fishermen's Union also discussed plans to limit engine capacity to 800 h.p., but by October 1971, it was clear that the plans to protect the sole stocks would not become effective, mainly because by this time many vessels were already equipped with engines that were more powerful. Subsequently, the Union suggested introducing a 1,000 h.p. limit. All of its local departments subscribed to this proposal. In January 1972, the Union again urged fishermen to restrain themselves, since more powerful engines would endanger the sole stocks and at the end of the day would be detrimental to the fishermen themselves: 'You will not only oust others from the market, but also yourself.' The problem was that the fisheries sector could not arrange this voluntarily. The Union emphasized that the fishermen were caught up in a devastating competition and that the government should intervene. Once again, however, some fishermen had already ordered more powerful vessels. Expecting that the state would heed the Union's desperate plea, their strategy was to stay at least a step ahead of potential restrictions. Tragically, the state had no plans whatsoever to put a lid on the booming fishing industry. On the contrary, it still extolled modernization and stimulated its achievement through various support measures (see below). Policymakers and advisors argued that bigger boats would mean higher catches, lower costs and larger profits, which they considered to be in the national interest.

A fishing-industry meeting discussed the issue of capping engine power in February 1972. A fisherman made a dramatic call on the state: 'We are busy destroying each other. If my neighbour has a 1,000 h.p. engine, I want one of 1,200 h.p. Government, please control us.' However, the administration refrained from intervention, as it did not want to curb the growing industry. A senior official said that he regarded an engine power limit as a measure of 'distributive justice that cannot work in the long run'. The government favoured a stop on fishing in certain areas during certain times. This would maintain fair competition and could be arranged through the recently established Producer Organizations. However, the European Economic Community would have to agree upon such a deal. The state did not want to regulate engine power because it did 'not have any means to intervene', as the head of the ministry's Fisheries Directorate, Th.J. Tienstra, explained. He pointed out that the 1963 Fishery Order did not provide for possibilities to limit engine power, and that the government – fearing foreign competition – would continue its liberal fisheries policy. This message would have drastic consequences. One of my interlocutors told me: 'Immediately after Tienstra had said this, many fishermen went home, grabbed the phone and lodged an order for a new vessel.' Indeed, during the following two years no less than sixteen new thirty-two to thirty-six metre-long beamers with flake-ice machines and engines ranging from 1,200 to 1,500 h.p. joined the Texel fishing fleet. In part, they replaced older vessels, while several other boats were completely overhauled, lengthened and equipped with more powerful engines. In addition, Texel fishermen bought three old but powerful side and stern trawlers from Katwijk and Scheveningen and had them converted into beam trawlers.

Similar developments occurred in other Dutch fishing ports. These were hectic times for shipyards. They launched 282 new cutters between 1967 and 1974. The fences seemed to be down. The fishery was still mainly controlled by input measures such as minimum mesh-size and fish-size regulations, but at the international level, discussions were meanwhile underway to limit the almost unbridled fishing effort through output measures. The buzzword of the day was 'quotas'. Based on recommendations of the International Council for the Exploration of the Seas, the North East Atlantic Fisheries Commission by 1970 began discussing the concept of quota-based catch restrictions. In the Netherlands, the expectation was that international quota measures were about to be introduced soon. A leading Dutch fisheries biologist, Pieter Korringa, was one of the keynote speakers at the 1972 Bergen meeting and proved to be an advocate of closed seasons and quotas: 'Fishermen themselves can decide whether they want to catch a quota with either the trawl or a dip net.' He added, 'Long before the fish is exterminated, the point is reached where fishing is no longer rewarding and owing to the sea's enormous reproductive potential, recovery will come soon.' With state officials and fisheries biologists seemingly rather unconcerned about growing engine power, one could hardly expect self-restraint from fishermen. However, they did not abandon all hope. Reviewing the state of Texel's fisheries in 1972, the chairman of the fishermen's association, Ben Daalder, wrote in the local newspaper, 'The fleet's expansion implies a further taxation on the already "overfished" North Sea. I do hope that next year we, in fruitful cooperation with the organizations and the fishing trade and industry, can arrive at a sensible management of the fishing grounds' (*Texelse Courant*, 29 December 1972). His words were ignored, at least concerning engine capacity. The Texel fleet's aggregate engine power more than doubled to over 36,000 h.p.

between early 1970 and early 1975 (see Appendix B), while in both years the number of vessels was about the same (forty-six and forty-eight, respectively). Aggregate landings were similar to ten years earlier, but effort had increased tremendously. That the fishermen were well aware of the dire consequences of the sum total of their individual behaviour shows in the remarks of a Texel owner of a brand-new vessel. After stating that he was not happy with the powerful engines because there was staggering over-fishing already and nursery areas needed protection, he said: 'But as an individual, you have to stay abreast when other fishermen continue to increase their engine power' (*Texelse Courant*, 17 July 1973). The horsepower race also went on unabatedly elsewhere in the Netherlands. The Dutch cutter fleet grew spectacularly and in the span of two decades, aggregate engine capacity increased tenfold to around 250,000 h.p. in 1971, steeply growing to approximately 380,000 h.p. by 1975. Reversing a prolonged trend of decline, employment on the Dutch cutter fishing fleet increased from more than 1,300 crewmembers in 1958 to over 2,400 crewmembers in 1964, subsequently growing slightly to approximately 2,500 fishermen in 1974.

Why, then, did Texel and other Dutch offshore fishermen continue to invest so heavily in the early part of the 1970s – even though they appeared to be aware of the fact that collectively, their decisions might lead to ruin? There were several incentives to do so. As we have seen, one reason was the race to stay ahead of a possible introduction of an engine power limit. Most fishermen seemed to argue as follows: 'It is fine by me to restrict engine capacity, as long as my vessel will be the most powerful one, for otherwise I will be in the category of losers.' They were entangled in what is a typical example of a prisoners' dilemma. The best solution was to collectively limit catching effort and thus engine power. Flatfish stocks would not be depleted and – all other conditions being equal – the beam-trawl fishermen would earn fairly good incomes. The worst solution was to collectively increase engine power in an unmitigated way. This might lead to a collapse of flatfish stocks, ultimately bringing ruin to the entire occupational community of fishermen. From an individual viewpoint, the highest gain could be had if all others restricted themselves save for oneself. Having the most powerful vessel would mean taking the highest proportion of the aggregate catch without the flatfish stocks being endangered. The worst situation from the individual viewpoint would be if a fisherman chose to restrict himself, but not all others followed suit. His share of declining catches would be small, and the fishery would be heading for disaster. With the first solution depending on state intervention that failed to come and dismissing the fourth possibility as foolish masochism, the fishermen opted for the third, which, because it was opted for by them all, in fact led to the undesirable second 'solution': theoretically the worst of all. Of course, the real situation was further complicated by the fact that the Dutch fishing fleet did not operate in isolation. Both fishermen and the Dutch state were firmly embedded within the dynamics of fishing industries and fisheries policies in a much wider context. Even so, when the government refrained from intervening, keeping up with the Joneses was the norm. This penchant also entailed an 'irrational' aspect that was linked with occupational culture. Fishermen took profound pride in their vessels, owners as well as non-propertied crewmembers (see also Chapter 5). Having the latest equipment at their disposal had an intrinsic value. In their perception, 'new' usually equalled 'good' – until proven otherwise – and being able to purchase state-of-the-art gadgets to become a top-ranking skipper was part and parcel of the fishermen's competitive game. In nearly all interviews, my interlocutors pointed out that they had been 'the first' in the Texel or Dutch fishing fleet to introduce this or that technique, instrument or piece of equipment. Having done so provided a source of self-esteem – although older fishermen would often state that dependency on electronic devices had led to a loss of traditional skills.

In order to avoid lagging behind, fishermen needed the financial resources to modernize their vessels. Between 1970 and 1975, the mean price of a new beam trawler belonging to the most powerful segment doubled from one million to two million guilders. It was no sinecure for family firms to come up with this amount of money. Thus, there must have been other reasons why the offshore fishing industry expanded at such a swift pace. Because of certain tax measures, it was wise to reinvest in vessels and equipment. This stimulated the expansion of the fleet and aggregate engine capacity. Like most people, fishermen were not fond of paying taxes. If they were making a profit, they had to pay tax. If they reinvested, however, this 'fiscal profit' would often evaporate and stay in the firm. Most fishermen therefore opted for a forward retreat. They had to continue doing so to escape from the tax collector. In addition, the state strongly encouraged modernization through its fiscal policies: in the mid-1950s, it had introduced tax-deductible investment reductions of twenty per cent during a five-year period (see Chapter 3). As of 1964, fishermen who modernized their vessels could obtain a twenty-five per cent state subsidy. Similar 'structural measures' leading to subsidized overcapacity applied to the shrimp-fishing fleet, where decommissioning and replacement schemes enabled fishermen to buy bigger and better boats – including beamers. Still, the Fishermen's Union pointed out that the heavily subsidized fishing industries of Germany and the UK had led to larger supplies of fish and lower prices, particularly following the development of the common market and the abolishment of protective measures. With fishing equipment becoming increasingly expensive, the organization called for more state support.[1] Indeed, in 1972 tax arrangements turned even more lenient: fishermen could subtract a quarter of investment costs over a five-year term and could write off the cost of their vessels in five years' time. Moreover, bank loans were easily available. Financial institutions had long been hesitant to provide fishermen with capital, since a vessel did not seem to be secure collateral. For this reason, it had been hard for inshore skipper-owners to work their way up into the offshore fisheries. Members of the fishing elite lineages usually made sure that they were possessed of a large part of the capital needed for a new vessel: they were often well beyond sixty per cent solvent. However, against the background of previous successes in the share-based offshore fisheries, the banks cast off old prejudices. As a retired skipper-owner from Oudeschild, born in 1915, told:

> For my first boat, I had to mortgage my house. The *Boerenleenbank* [a cooperative farmers' bank] would not give loans on vessels. Later, they gave me 300,000 guilders on my bare bum. Everything had changed by then, and they did provide mortgages on vessels. Previously, however, that bank shunned fishermen.

Fishermen proved to be very dependable when it came to paying interest and redeeming debts. Shipyards were of course more than willing to build new vessels with bank guarantees. Moreover, to obtain such orders they paid handsome prices for second-hand boats, which they sold abroad or at home. With the entry of scores of newcomers – usually ex-crewmembers – into the Dutch fishing industry, the demand for used vessels was high. Often, they were sold at prices that equalled the initial expense – or even more than that. Again, taxes, subsidies, bank loans and high trade-in prices are only part of the story. The risks fishermen ran in becoming heavily indebted had to be offset by optimism about the opportunities to redeem loans and mortgages, account for depreciation, pay all operating costs and crew shares and still make a profit.

Indeed, based on the fishermen's experiences in the recent past and despite the growing concern that they might out-compete each other and ruin flatfish stocks, in their view the future looked rather bright. Beam-trawl fishermen had had some extremely good years. Those hailing from Texel grossed an aggregate of three million guilders in 1960, steadily climbing to twenty-one million guilders in 1972.[2] The most powerful vessels obtained the highest proceeds; one vessel grossed over a million guilders in 1972. By then, mean gross incomes of Texel share fishermen amounted to 47,500 guilders. Corrected for inflation, this implied a doubling of income in fifteen years, making fishing an attractive occupation. Beam trawling brought prosperity. Many local young men desired to be on a crew, so there was no shortage of hands. There was even a pool of substitute crewmembers. Only a decade earlier, two score of fishermen from the mainland had had to work aboard Texel cutters for want of islanders willing to crew the boats of the expanding local North Sea fleet. In 1963, for example, approximately forty to fifty of the 175 men and one woman who crewed the Texel fishing fleet hailed from the mainland. By then, there were thirty-one fishing firms, operating twenty-six big cutters and ten small cutters. Due to the fleet's growth, more and more hands were required. Initially, it was hard to come by crewmembers, especially qualified crew holding certificates. In the 1960s, new legislation made education compulsory for youngsters up to sixteen and no longer allowed fourteen-year-olds to hold full-time jobs. Henceforth, boys entered fishing careers when they were fifteen or sixteen years of age. As of the mid-1960s, young Texelians became increasingly interested in getting a berth aboard a beamer. With weekly gross revenues of over 10,000 guilders unexceptional, many were lured into fishing. By the end of 1967, only three mainlanders were still crewing on the Texel fishing fleet. Things had obviously turned for the better. In 1969, only seven per cent of Texel's occupational population worked as fishermen, but they generated no less than seventeen per cent of the gross income of the island. In 1973, there were 227 active fishermen crewing forty-seven boats. The easy recruitment of locals was inextricably connected with the attractive share remuneration (see below). Bigger beamers required no more labour effort than smaller vessels and had considerably better gross proceeds and living conditions. Therefore, crewmembers attempted to join the crew of the bigger vessels. To avoid losing experienced deckhands to other skippers, this created another incentive for owners to expand. As a Texel fisherman argued, 'I dare say that some skipper-owners had cutters built to keep their crew. They said, "If I don't go along, I won't get good personnel." And it worked out fine.' The beam trawl fleet's impressive results were due to larger catches – although not for all species – and better fish prices.

In the winter season, a number of boats still switched to pair trawling for herring, operating from the Zeeland port of Breskens, as many had been doing since the end of the war (see Chapter 3). In the early 1960s, the Texel herring hunters began the season as early as August, starting to fish off the coast of Flamborough Head and gradually following the herring south along the English coast into the Channel until early spring. In 1963, there were record catches, but prices were so low that most herring went to fishmeal factories. By 1965, the seasonal herring fishery conducted from the port of Breskens ended. Higher quality demands inhibited transportation on motor trucks to Germany. The herring had to be packed in wooden boxes before being supplied to the traders and Breskens had neither the space nor the facilities to switch to the new mode of operation, and so the Texel fishermen's transhumance came to an end. The dwindling number that continued to seasonally pursue herring would from now on operate from Texel and land their catches in such ports as IJmuiden and Scheveningen. More powerful and faster boats also enhanced their radius of action, diminishing the need to operate from other locales during the herring-hunting season. With the good results in flatfish fishing, there seemed to be no reason to switch to herring fishing, but some skipper-owners nonetheless continued to do so. Texelians had focused mainly on spent herring (herring that has ejected its spawn), which was processed into pickled herring. In 1971, six Texel pair trawlers participated in fishing fat herring or '*maatjes* herring', which was processed in brine. The skippers had to steam for thirty hours to the Scottish coast. As the Texelians were inexperienced concerning the necessary fat content of herring and the salting process, each pair had a Katwijk herring fisherman aboard. Catches were disappointing, however, and later that year, a temporary herring ban was promulgated, which was repeated the following year. Despite this, in the early 1970s, several Texel fishermen had their newly built vessels equipped with net drums aft so that they could continue combining beam trawling with pair trawling or otter trawling. They did not want to put all their eggs in one basket, even though it proved increasingly attractive to concentrate on flatfish fishing.

In terms of the contribution to gross revenues, sole was the offshore fleet's most important target species. Three-quarters of Texel's fishing fleet focused on sole fishing for most of the year. On average, sole usually made up less than fifteen per cent of the total catch in weight, but it often accounted for more than half of the gross revenues. Aggregate landings in the Netherlands went up from around 4,200 metric tons of sole in the late 1950s to between 7,000 and 11,300 metric tonnes in the early 1960s. Owing to the beam trawl's efficiency and widespread distribution, 1966 saw a record high of 23,300 metric tonnes. From then on, sole landings steadily declined to 15,000 metric tonnes in 1972, levelling off at around 14,000 metric tonnes in the next three years – despite growing engine capacity and catching effort. In other words, it took ever more energy and more days at sea to catch a dwindling amount of sole. This tendency was believed to be a fair indication of overcapacity and overexploitation of sole stocks. The secretary of Texel's fishermen's association emphasized that this could not go on. The fishermen's worries about the state of the stocks in connection with increasing engine capacity did not come out of the blue, but were firmly based on their experiences.

Steadily increasing market prices easily made up for diminishing sole landings. Between 1957 and 1963, mean sole prices had been fairly stable at around 2.87 guilders per kilo, deviating by no more than eighteen cents. In 1964, however, prices rocketed because sole catches were exceptionally low and demand remained high: the extremely harsh winter of 1962-1963 had wiped out a large percentage of the sole stock's younger year classes. On average, a kilo of sole fetched no less than 5.32 guilders. Although prices fell by twenty-five to thirty per cent in the following four years, they eventually climbed again to 9.13 guilders in 1973, remaining fairly high from that time on. Thus, in the time span of a decade sole prices rose quite considerably – even with correction for inflation. Prices of other important species such as plaice and herring also went up, although less spectacularly. The total landings of plaice averaged 28,639 metric tonnes per year between 1966 and 1969, increasing to 44,576 tons over the next four years. In 1972, the Dutch cutter fleet's aggregate gross proceeds were almost sevenfold that of 1958 (and if corrected for inflation: 3.4 times as much). It rose from 32 million to 217 million guilders (Rijneveld, Smit and de Wilde 1973:7). Plaice and particularly sole were highly valued, prime fish species for which there was a buoyant export market, especially after the establishment of the European Community with its ideology of free internal trade. Overall, the fishermen's experience was that both catches – with the exception of sole catches – and fish prices increased and their expectation was that this would continue to be the case. In addition, vessel operation costs depended largely on fuel prices. Although engine capacity and thus fuel consumption had increased slowly in the 1950s, when the fleet developed at a much quicker pace, red diesel prices went down for most of the 1960s, then slowly increased again until 1972. Fuel accounted for less than ten per cent of operating costs. Calculating from their account books, the fishermen certainly had reason to be optimistic.

Then the global oil crisis of 1973 struck, hitting the fishing industry extremely hard. Arab oil-producing countries boycotted the Netherlands, along with the United States, for its support of Israel in the Israeli-Palestinian conflict. The embargo severely affected the Dutch economy, not least the fisheries. Heavily indebted due to their recent investments, most beam-trawl fishermen faced serious difficulties when red diesel prices began soaring. Within a year, they rocketed from nine cents to thirty cents per litre. The most powerful vessels by then consumed up to 25,000 litres of fuel per week, weighing heavily on exploitation bills. Adding to the fishermen's misery, sole and plaice landings decreased compared with previous years, despite the growth of the fleet and engine capacity. Plaice prices rose slightly but sole prices fell and so could not make up for higher fuel costs. On aggregate, the mid-1970s yielded negative net results of many millions of guilders for the North Sea fishing fleet. Instantaneously, the fishermen's optimism made way for pessimism and in some cases even panic. Investments halted. The leader of the local fishermen's organization, Ben Daalder, told his compatriots: 'We are in the midst of the overfishing problem.' The association's secretary noted in his annual review:

> All of us should cooperate with the recovery of fish stocks. ... If everyone is honest, we have to admit that all of us have been involved in bringing about the current situation. Of course, the state is also to blame because it did not intervene [that is, cap engine power] despite the organizations' repeated calls. Our individualist attitude makes doing so difficult but not impossible.

A majority of the fishermen were in favour of catch reductions, provided that the state compensated the fishermen. Again, they turned to the government for assistance. This time, it did not leave them in the lurch, as it had done in regard to the engine capacity limit. Rationing and distributing red diesel and temporarily providing fuel subsidies, the state sought to prevent the fishing industry from facing mass bankruptcy. It also introduced a decommissioning scheme in an attempt to cope with overcapacity. With the risks of investments now obvious, the owners began to negotiate with crew concerning new divisions of the share remuneration (see below). For the time being, fishermen refrained from lodging orders for new vessels or yet more powerful engines. Some even began to economize on maintenance. Most fishing firms proved to be resilient, not in the least because generally they were family-owned and operated. In the post-war era, the specific dynamics of family firms had also been significant in bringing about the Dutch fishing fleet's growth and modernization. The owners did not solely reckon in terms of net losses or profits and future returns on investments or other economic incentives. Nor were status considerations alone sufficient reason to desire growth. Social reasons – continuity of the family firm and equal opportunities to all sons – were also important in their motivations. Fishermen thus had their own rationalities to expand; rationalities that might deviate from pure business economics.

Blood Runs Thicker Than Water: The Family Firm

Kinship plays a prominent role in the fishing industry's social organization, not just on Texel or in the Netherlands, but also elsewhere (see, for instance, Byron 1975, 1986; McGoodwin 1990; Menzies 2003). A number of economic, social and psychological factors are significant, as are several adaptive characteristics. Anthropologist Reginald Byron emphasizes the socio-psychological function of family crews: 'the model of a family partnership makes possible the conversion of the kinship values of trust, solidarity, the equivalence of siblings and mutual aid on the private level into the organizational norms of voluntary cooperation, equality and mutual reliance on the team level' (1975:155). Close social ties enhance teamwork aboard a fishing boat. Rules of inheritance are also important (Löfgren 1972; Norr and Norr 1978). A vessel constitutes an undividable asset. Consequently, relatives – and especially agnatic kin (father-son(s), paternal uncle-nephew(s) or brothers) – are strongly interdependent; they have to stick together to continue fishing. More importantly, a family firm is a common pool of resources. Relatives share forces of production, labour, capital, knowledge and expertise. In so doing, they spread risks and reduce vulnerabilities. The earnings remain within a narrowly defined core of kinsfolk, enabling the quick accumulation of capital in good seasons. All participating family members ultimately benefit. Therefore, the family firm is a system of self-motivated labour (Thompson et al. 1983:156). Conversely, in bad times a particular economic outlook is highly significant. If revenues are low due to meagre catches or disappointing fish prices, fishermen who are (part-)owners tend to work longer hours and/or cut back on their expenditure and on their income from the ship's and crew share. They will postpone reinvestments in equipment and possibly even eat into their own savings to weather a depression. Generating profits is not their only objective. Thus, the family firm is also a form of self-exploitation (Jorion 1983:10). Their pervasive labour ethos and the specific definition of profitability enable fishermen to keep their firms afloat even in the face of serious economic crises: 'Family involvement gives small capital a flexibility that ... alters basic assumptions about rationality in economic behaviour' (Apostle et al. 1992:321).

The economic logic and social dynamic of the family firm contributed importantly to the Texel and Dutch fishing fleet's expansion. Highly significant to this is an ethic of deferred gratification. Relatives are prepared to cut their own incomes for some predefined or implicit future goal. With the greater capital outlay needed in the North Sea cutter fisheries, it was important to acquire the necessary funds to buy seaworthy boats. Whereas labour used to be more important than capital in the era of wooden-hulled sailing vessels, this changed as of the 1930s and more so after the Second World War. Although skipper-owners were usually able to redeem debts, account for depreciation and still save money to buy new boats, after the war vessel prices were such that they usually needed to contract higher loans. Using a boat – and often a house too – as collateral, banks would be prepared to provide part of the money, particularly if an owner had sons. Sometimes, bank loans would not entirely suffice. The fishermen therefore also had to raise funds through private loans that had to be redeemed. Usually, they called on relatives whom they knew or believed to be affluent. In order to quickly redeem such loans, their most commonly used strategy was to take out 'week money' (*weekgeld*) for a fishing trip. On paper, there would be the vessel share and the crew share (see below), the former utilized for depreciation and interest and the latter providing the money that crewmembers – including crewing co-owners – earned. Whereas non-propertied deckhands would obtain their full share, owners and their crewing offspring usually only took a modest predetermined sum of cash per week. This would amount to less than a deckhand's share. Young unmarried crewing sons of owners would generally receive 'spending-money' only. Thus, family labour was often underpaid. An Oudeschild man who began his fishing career aboard his father's shrimp cutter in the 1950s said: 'My brother and I were teenagers then. Father gave us just a few guilders a week and that was it. We did not care at all. We just wanted to fish and thought of nothing else.' 'With your own people [meaning kinsfolk], you had fewer costs,' summarized a skipper-owner. The money thus saved would go straight back into the firm so that debts could be redeemed quickly, enabling investment in a new boat at some point in future. This mode of self-restraint and self-exploitation stimulated rapid modernization and growth. 'It cuts both ways,' stated a retired skipper-owner, born in 1923: 'Because you had sons, you could progress and later on, you in turn would help those lads.'

Relatives working as crewmembers provided an important social asset. Depending on the stage in the household's life cycle, a father and son(s) or brothers constituted the crew's core. If skipper-owners had male offspring, their sons would often be impatient to assume their father's role, work aboard and eventually become skipper-owners themselves. When a fisherman's young son showed an interest in becoming a fisherman, he would occasionally join his father on a trip to sea, simply being aboard and lending a hand if and when possible. This way, the father got to see whether his son had 'a turn for it'. There was a pervasive sense of the significance of occupational inheritance and handing down of fishing traditions, knowledge and zeal. It was every owner's dream to have at least one successor to continue the family firm, putting a strong moral obligation on the male offspring of owner-operators to follow their father's example. Fishing was an important source of identity and an occupation that was highly valued by those in the industry, something skipper-owners in particular wanted to pass on to the next generation. Better still was to have more sons and to provide each of them showing an interest in fishing with a vessel of his own. Thus, one of the driving forces underlying the growth of the Texel (and Dutch) fishing fleet was the fishermen's desire to become independent and, ideally, to set up each owner's son with his own vessel to skipper. When the father retired and became a 'shore skipper' – often maintaining a position as paterfamilias for a considerable time – brothers usually continued the firm and cooperated until their sons joined them aboard ship. There would sometimes be insufficient positions to accommodate them all. In addition, patrilateral cross cousins often did not get along regarding fishing matters and other decisions and wanted to work with their own father on their own vessel. Consequently, schisms of family firms usually occurred during the stage when two or more brothers each had sons aboard. Conflict and subsequent fission could also occur if one of the co-operating brothers had the feeling that he contributed disproportionately to the firm. An Oudeschild skipper-owner (born in 1940), who turned independent in the early 1970s, related the following:

> We ran a single firm, my father and my brothers, with four cutters. I was the only one who sailed the North Sea. They were home every evening, while I was away for a week, from Monday until Saturday. I was earning the greater part of the firm's proceeds while they all shared equally. The thing that bothered me most was that my brother's wife told my wife that she could suffice with less week money because I was on the cutter the entire week and had my board there. She shouldn't have said that, it went against the grain. It spoke volumes about her lack of understanding. I was the one who was earning most of the money for the firm! Therefore, I bought the cutter and began working alone.

This was certainly not the only example of brothers splitting up owing to disagreement on skewed contributions to the firm or – often equally important – incompatible characters. Occasionally, brothers did not get along. This could be a reason for a firm to hive off into two or more firms. An older informant, born in 1905, gave a concrete example and said the boys' mother had had many sleepless nights. Finally, the parents decided to have a second cutter built so that each brother could skipper his own vessel. 'They just did not get along, until the death of the eldest brother. They simply couldn't tolerate each other!' A father had another solution for such a case, as a retired deckhand, born in 1928, intimated: 'The father of two boys chalked a line in the wheelhouse and told them that they shouldn't cross it.' Conflicts could also arise concerning who would be the skipper, as an Oosterend owner-operator (born in 1923) related:

> Working with cousins aboard is difficult, although even brothers sometimes have problems. There is bound to be someone who is obstinate and it easily kindles feelings that the one does more than the other does. You should contribute equally, but a skipper does less work with his hands, and if his brother is a deckhand and does manual labour, he might say 'you don't do anything'. The fat would be in the fire then...

Both centripetal and centrifugal forces (Faris 1973:93) consequently characterize the family firm. Schisms generally occur during a particular stage in the family cycle. If brothers have sons who are old enough to join a crew, they often break away from the existing firm and establish their own. Due to this dynamic, the Texel fishing fleet expanded, particularly in times of boom, making family firms even more common. Occupational inheritance and – where several agnates (or cognates) were co-owners – fission or expansion of family firms meant that the number of vessels increased. As in other Dutch fishing communities, the fishing family firm dynamic was an important reason for the local fleet's growth. Continuity of the family firm was deemed extremely important – for most owners even a *raison d'être*. Obviously, it depended on the social contingency of having at least one son who, in addition, had to be physically and mentally up to the job. 'If you don't go out to sea with heart and soul, it won't work, even if you come from a fishing lineage,' an informant said. Daughters were – and are – regarded as unsuitable successors. If they married in community of property, a deceased skipper's widow owned the cutter, but her son(s) usually ran the firm. If she had no boys or if they were too young to crew, she might employ a hired skipper and deckhands. Unless he had taken out life insurance, a deckhand's wife would stand empty-handed if her husband died. This was not the case with the spouse of an owner, however. For example, the widow of an owner-operator who suddenly died from a heart attack age 47 continued the firm and had the relatively new cutter skippered by an unrelated deckhand. As in this case, most couples were married on equal terms, so that the wife in theory owned half her husband's property.

The fact that in the Texel fishing industry women generally did not crew is not exceptional. In many places across the globe, there is a clearly gendered division of labour: men work at sea while most women stay at home and manage the household and the household budget and take care of child rearing.[3] In fact, this required continual role-switching: 'While their men are at sea, women must become "reluctant matriarchs"; when their men are at home, women must turn into dutiful wives' (Davis and Nadel-Klein 1992:139). Underlying the division of tasks are cultural and symbolic constructions and constrictions. This does not mean that women's work is less important than men's work: 'not infrequently it is their economic contribution that underwrites or provides the risk fund necessary to sustain fishing activities' (Davis and Nadel-Klein 1988:19). The family firm usually rests on close conjugal cooperation and the fisherman's wife's organizational, economic and emotional contribution to the fishing household is often crucial for its flexibility, versatility and resilience.

The role of women in Texel family firms was usually considerable. Some women played an important part in running the financial side of the firm. For instance, an Oudeschild North Sea cutter owner's wife, born in 1914, did all the bookkeeping, handled the bank accounts and paid the deckhands from the moment her husband took over the firm from his father in the early 1950s. She said her spouse lacked the time to do so during the weekend – 'and the banks would be closed anyway'. When her husband passed away, she inherited part of his share. While a widowed septuagenarian, she was still actively involved in her sons' firm. In the North Sea segment of Texel's fishing fleet, there were more examples of women who were a firm's financial brains. In contradistinction, wives of shrimp fishermen were not involved in bookkeeping or similar tasks. Their husbands were home often enough to take care of this chore. Generally, spouses would discuss financial decisions concerning the firm. A woman, born in 1921, recalled that when her husband was still a deckhand, she was only remotely involved. This changed when he became a co-owner:

> When the firm was ours, I managed all sorts of business. You contacted an accountant who had to take care of things, you made phone calls, wrote letters, remitted money and so on. Sometimes I knew more about what had happened than he did. You always had to arrange all sorts of things. I did it myself, but my sister-in-law never took care of that. It was my job; I had the authority to take it on. My husband used to call home via the radio and say something like 'we will be in port by that and that time and you should make sure that so-and-so or such-and-such is there'. Bringing up the kids was my task. You just did it alone, punishing them too. My husband never had to punish them. He didn't have to be the bogeyman. I never told them 'you just wait until your father gets home.' My sister was different. She always told her husband [a fisherman] what the kids had done wrong and he would give them a smacking. However, when my husband was home it was always a party for the kids. You had to do everything by yourself, though. Sometimes it was difficult, for instance when the children were ill. Their care always rested on my shoulders. When they got a bit older, you had to make sure that they did their homework for otherwise they would do it haphazardly.

Her husband interjected: 'Yes, a fisherman's wife has a great responsibility. She is a mother and a father at the same time.' She continued:

> He would come home late on Saturday afternoon. The weekend was brief. A fisherwoman was already emancipated back then. You took care of everything yourself. It demanded a particular attitude. I could never share my worries with him, so I had to carry the burden alone. Oh well, you were aware of that if you married a fisherman.

Although they were mostly indirectly involved in the fishing industry, the women's – and particularly the skipper-owners' wives' – supportive role enabled the men to go out to sea and earn a living. Hence, there was a clear-cut division of labour, with men's tasks being at sea and women's tasks being on shore. In the 1960s, only one girl worked aboard her father's boat for several years. They took tourists on shrimp-fishing trips. The skipper had sons, but they did not like sailing with holidaymakers and they went on to skipper their own vessels. This was an exceptional case. In their role of rearing the children, mothers were important for ingraining the values boys needed at sea. They were just as keen as their fathers were to prepare them for their future role in the firm, as occupational inheritance in skipper-owner families often was a matter of course. Ideally, skipper-owner's sons would be skipper-owners themselves in future.

Living economically and being thrifty often enabled expansive growth. Women had a significant part to play in accomplishing this goal. Their spending patterns reinforced the firm's financial strength and facilitated redeeming debts and paying interest on loans in time. If they did not internalize this kind of behaviour early on, there would sometimes be a mother-in-law to tell her son's wife she should not be spending too much. This was not always appreciated: 'At some stage, I had a job and if I bought a coat or something like that, my in-laws couldn't say anything about it for it was *my* money,' said a woman, born in 1931, still slightly indignant. By making do with a rather modest weekly allowance for running the household, she enabled her husband to reinvest or to keep up with his brother(s) in acquiring his share in the firm. An Oudeschild skipper-owner, born in 1940, put his mother on a pedestal:

> She was always toiling to make ends meet. My father was terribly thrifty, almost stingy. Nearly all the money he earned went straight into the firm. That is how we could expand, but my mother made clothes on an old sewing machine and did the laundry for twelve children by hand. She had to deny herself a lot, for my dad would always put the firm first.

Some women were keenly aware of the fact that their austerity contributed to a firm's solidity: 'I helped build up the firm,' a skipper-owner's wife (born in 1921) contended. The men acknowledged – and appreciated – this fact. A retired owner-operator, born in 1923, related the following:

> In Oosterend, family ties have always been tight. The fishermen's wives played an important role, not just with regard to raising the kids and run-

> ning the household. If something needed to be taken care of for the boat and such, it always had been done at the time we returned home. They had to talk to agents and manage all sorts of business affairs. The Oosterend women were squarely behind their husbands. They never complained and always said, 'You just go ahead, I'll make sure that everything is in order here.'

In honour of his wife or their mother, owners in several cases gave their vessel her name. 'When we named our boat after my wife, she got even more involved and made sure that everything was thoroughly cleaned on Saturday,' a skipper-owner intimated.

The fishermen cherished their weekend, which they tried to spend with their wife and children. Usually, however, they had several additional commitments. First of all, after returning to the harbour of Oudeschild on Saturday at around noon, the boats had to be made shipshape for the upcoming fishing week, ice had to be stored and drinking water bunkered, nets mended, fishing necessaries replenished and so on. The weekend thus usually began only late on Saturday afternoon. Birthdays were always celebrated at the weekend, and for many skipper-owners there would be work on committees and boards of voluntary associations, either fisheries-related or connected with school, church and so on. Oosterend fisher families were commonly sabbatarian and attended Sunday worship. A woman, born in 1916 and married to an Oosterend skipper-owner, related:

> We had seven kids. My husband went away on Monday morning, and returned from sea by Saturday noon. Then he usually had meetings of the fishermen's association and on Sunday morning there would be the Church Council and the religious service. It was only by Sunday afternoon that we could finally say: 'At long last, we'll have father to ourselves for a brief while.' Sometimes he would sail again by Sunday midnight. Well, you just had to accept that.

Fishermen's wives perceived themselves primarily to be homemakers. They made sure that when their husbands returned from sea, they had done all the shopping and the house was spick and span. On Saturday, many skipper-owners' wives cleaned the boat's pantry, wheelhouse and accommodations and did the firm's laundry. In addition, they would often prepare meat for the crew's upcoming fishing trip.

When their husbands were out at sea, fishermen's wives occasionally turned to each other for moral support. Some would also meet in a women's club, but for those with small children it was not easy to leave home. News from relatives who were away on a fishing trip was scant. Until the 1980s, there was only ship-to-shore communication. At specific times, women had to listen in to a special radio frequency that enabled them to hear their husband and kinsfolk at sea conveying short messages that could also be picked up by others. Communication was therefore one-way and public. For this reason, the fishermen kept their 'calls' brief and rather general, often saying little more than that everything was alright and that catches were satisfying or disappointing. 'I always listened to the radio', a woman born in 1908, said, 'and I still do.' Particularly during foul weather, there used to be considerable anxiety and uncertainty. With a wind force of seven or eight, the fishermen would attempt to reach a safe haven or refrain from sailing. 'You didn't sleep well when they were out at sea in a storm, your thoughts were with them and you prayed,' an owner-operator's wife, born in 1921, confessed. A deckhand's wife said:

> You sometimes worried. There was a terrible storm once and I was anxiously listening to the radio. Apparently, a huge wave hit the cutter and I heard the skipper say 'goddamn'. The man never ever cursed. Then the radio went silent. I was convinced they had capsized. It was only after three-quarters of an hour that I heard them again. Save for this once, I have never been afraid, although you did worry from time to time.

In general, women born into fishing lineages were more used to coping with the men being away most of the time than women with a 'landlubber' background.

All in all, the family firm's economic logic and social dynamic – including the role of women – enabled the expansion of production units, ultimately leading to processes of fission. Newcomers from non-fishing lineages also contributed considerably to the local offshore fishing fleet's growth. In the 1960s, especially, several deckhands aspired to become independent and bought their own vessel. 'It is better to be a small boss than a big servant,' as they expressed it. Having served as deckhands for a number of years, they had learned the ropes of fishing. If they had earned a certificate as a helmsman, there were few obstacles to beginning for themselves, apart from financial ones. Not all deckhands desired to become independent owner-operators. Several of the retired crewmembers I interviewed told me that they refrained from taking on the responsibility of starting a firm of their own because they thought they would be unable to handle the responsibilities. 'I did not dare, I was afraid of being in debt,' one of them, born in 1912, disclosed. 'I preferred being a deckhand.' Another (born in 1938) said, 'I was given the opportunity, my skipper offered to help my brother and me to set up a firm. I had a good life as a deckhand though, and perhaps I didn't have the guts or I wasn't enterprising enough.' Several crewmembers did not have such hesitations. As we have seen, they often earned considerable sums of money under the share system of remuneration (see also below), part of which they could set aside for investment in a vessel of their own. In addition, banks provided loans and credit more easily than in previous years. One of the newcomers, who in 1967 as a twenty-six-year-old began a fishing firm with his brother, remembered:

> The manager of a certain bank would not even discuss the option of providing a loan, but the manager of another bank was less reluctant and said, 'You just buy that cutter.' The man apparently had faith in our abilities. I never gave a thought to the idea that it might be a financial risk. We were self-assured and thought we could run a cutter with a bank loan just as well as others could run one with family capital. The skipper-owners of the traditional fisher families were sceptical until we proved them wrong. Initially, they told deckhands who began sailing with us: 'Are you working for them? You must be mad. They are as poor as church rats. You will have to take your own toilet paper with you.' Later, they had to admit that they were wrong and they never mentioned it again. They simply forgot that we had had years of experience before we began skippering. They firmly believed that only sons of skipper-owners could become skipper-owners. My father was a farmer. Later they would say, 'The farmers are catching the fish now.' They were also sceptical because originally, our family was Roman Catholic – even though only nominally so – and nearly all fishermen here are Protestant.

Another newcomer, who as a twenty-two-year-old started a firm with his nineteen-year-old brother, at the time considering this 'nothing special', said similar things about the initial scepticism of descendents from traditional fishing lineages: 'They regarded us as intruders; at least this is what we sensed. It didn't take long though before we were accepted and belonged to the club, but you certainly had to prove yourself.' Some newcomers were assisted financially by their former 'boss' or by other wealthy fishermen. Shipyards and suppliers were also important in providing loans, on the condition that aspiring owners bought their vessels and equipment (for example engines) from them. Several rather young Texel fishermen started a firm this way, some being heavily indebted initially. Due to the excellent financial results in these years, they were able to 'fish their firm out of the water', as they expressed it, meaning that they were able to redeem their bank loans and debts. The upward social mobility of some pioneers stimulated several others to follow suit. Among the newcomers were several fishermen from Oudeschild. In 1966, for example, Oudeschilders operated six big beamers, while in addition the village boasted six or seven shrimp-fishing boats. Inshore fishing also provided opportunities for newcomers, particularly for those who did not want to be deeply indebted.

In many cases, the newcomers also followed the 'family logic' and worked with agnatic kin – usually a father and son(s) or two or more brothers. In the first stage of such firms, they bought a second-hand cutter, working their way up until they were able to order a new vessel. Of at least thirteen newcomers in the offshore segment between 1961 and 1969, two firms began with a father and sons as co-owners, and three started with two or three brothers as co-owners. As soon as they consolidated their firms, the family dynamic of occupational inheritance applied. The father would leave at a certain stage, the brothers would cooperate until sons entered the crew and the firm would split. To give an extreme example: one of the firms that began with one boat in 1962 had three vessels by 1967, owned by a man and his four sons who were also on the crew. One of the brothers died in a fisheries accident in 1968. Subsequently, one of the cutters was sold and the two remaining brothers each continued with a cutter, setting up independent firms that would eventually disappear from the local fisheries arena. The other new family firms do still exist and belong to the most successful of the Texel fishing fleet. Among the newcomers were two firms that began as a partnership between two former crewmembers. These were not successful – the firms were dissolved after only a few years. Two former partners set up new firms with siblings, while the other two went it alone. Two brothers-in-law set up another firm, but as they disagreed on almost everything and could not get along, they quickly went their separate ways, each establishing a firm of his own. Both firms only existed for a few years. One cutter sank; the other was sold, as there were continual engine troubles. Although it would be too simplistic to say that family involvement alone determined success, it certainly was an important contributing factor.

Because of family involvement in fishing firms, ownership was rather widely distributed in the occupational community of fishermen. From 1960 until 1971, between thirty and forty per cent of the Texel fishermen co-owned a vessel. The vast majority of Texel owner-operated family firms had a single fishing boat. In some cases, fisher families with several agnatic kinsmen who could be on a crew owned more than one vessel. For example, in 1969, four firms owned two cutters and one firm owned three. Although a large percentage of the fishermen had financial stakes in fishing firms, the owners usually could not do without external labour. Deckhands often had a fisher family background. Family enterprises generally used the share system of remuneration to hire crew. Both the involvement of kindred and co-adventurers contributed importantly to the Dutch offshore fleet's success in comparison with the wage-based distant water fleet. By 1950, the former had surpassed the latter in economic importance. 'Family-owned boats, skippered and crewed by family members generally perform better than company-owned boats with hired skippers and crews. Share fishermen bearing the full catching risk will perform better than fishermen having a partly fixed or minimum wage' (Banks et al. 2001:46). However, the share system does have a number of disadvantages for crewmembers. So, let us have a closer look at the pros and cons of co-venturing as perceived by owners and non-propertied fishermen.

Co-Venturing: Sharing Risks and Revenues

The share system of remuneration – in which all crewmembers are co-adventurers and share costs, risks and rewards – is common in many fishing communities across the world. Anthropologists argue that the system elicits the crew's commitment and labour motivation, stimulates productivity, alleviates tensions, minimizes conflict, promotes cooperative behaviour and fosters a spirit of egalitarianism (Acheson 1981:278-279; Jorion 1983:143; Lummis 1985:42; McGoodwin 1990:33). Both owners and crew benefit from attempting to obtain the maximum output while keeping operating costs to a minimum. As has been remarked about British fishermen, the share convention 'undercut class attitudes in two ways, for it fostered both individualism, and interdependence across the normal boundaries of class' (Thompson et al. 1983:244). Generally, anthropologists and historians would seem to emphasize – perhaps even overexpose – the share system's positive sides. In practice, however, there are some serious cons, as I will attempt to show (see also Chapter 6).

Dutch fishermen – including those from Texel – usually also operate under a share system, with crewmembers and owner(s) signing a partnership contract (*maatschapscontract*), stating that all crewmembers receive a certain percentage of the net revenues and that owner(s) and crew have a joint responsibility and a stake in the result.[4] Accordingly, in theory, the parties should consult each other on such matters as fishing positions, fishing strategies and times of departure and return (see also van der Vlist 1970:13). The *maatschapscontract* states that the owner will provide for the vessel and the nets (the means of production), while the non-propertied crewmembers will deliver 'labour, knowledge and assiduity'. The contract is valid for a year and automatically renewed, unless one of the signatories terminates it within a set term, usually four weeks in advance. In principle, owners benefit because crewmembers share the fishing operation's risks, are dedicated and work hard to maximize the gross revenues, are motivated to limit costs and are conscientious in keeping the vessel and equipment in tip-top condition. If feasible, the crew conducts maintenance and minor repairs. If a vessel cannot sail due to repairs and if catches or prices are bad, owners do not face the problem of having to pay fixed weekly or monthly wages. Deckhands are better off because they share in the proceeds of capital. If catches and prices are good, they earn a lot, creating an incentive to put in extra effort because they will share in the rewards. Hence, under the share system labour productivity is high, but the deckhands' weekly incomes fluctuate considerably. They may reap a windfall or earn nothing at all. The share convention ties them directly to ecological and market contingencies. If operating costs surpass gross proceeds, for example if fishing has been particularly bad or inclement weather requires riding out storms at sea, they will usually be passed on to the next week's bill. If share fishermen fall ill, are on holiday or cannot work for other reasons, they do not earn a penny. The same is true during regular maintenance or when vessels are docked to be repaired or overhauled, which may take weeks or even months. It is therefore deemed wise to set something aside in good times in order to be able to weather bad times. The disadvantage for owners is that they have to bear the costs of damages and the boat's and gear's wear and tear.

The net shares come about as follows. First, operating costs are subtracted from the amount of money the fish fetches at the auction (the gross revenues). These costs include expenses for fuel, lubricants, ice, commissariat, rent of electronic equipment, fish-handling and auctioning fees, and 'mending money' (*boetgeld*), which is set aside to pay the one or two retired fishermen who work on the nets ashore. The remainder is divided between the vessel (meaning the owner or owners) and the crew according to some predetermined ratio. The division of vessel and crew shares is dependent on the type of fishery. The boat share should cover instalments, maintenance, repairs, renewal of the engine, nets, electronic and other equipment and so on. Part of it is for investments, reservations, depreciation and – in some cases – an income for the owner(s). The remainder is profit. Unlike deckhands, owners can deduct certain costs related to running the firm (for example, a car, phone bills and so on). Each crewmember receives a fixed share, occasionally varying somewhat according to seniority and position. Experienced deckhands get a full share; a hired skipper may obtain a small bonus, while a young apprentice's share is usually slightly lower than that of the other full-blown crewmembers. If an owner works aboard as skipper or deckhand, he also receives a share. In the offshore cutter fisheries, the vessel and crew share used to be fifty–fifty, and with equal shares: so for example for a crew of five, each crewmember would receive ten per cent of the net proceeds.

A fisherman born in 1928, with experience as both deckhand and owner, explains the mutual benefits:

> If a skipper-owner does well, you earn more as a deckhand. Owner and crew have the same interest in running things as smoothly as possible. Everything has to be perfect, not just the nets, but also all the equipment. If there is a problem because something busts, it will cost the crew money too, because you will make less. [...] The better the relationships aboard, the better the returns. If relationships are good, you will look after the vessel much better: you will take care that everything is in perfect shape, you will grease and oil it and you will make sure that the nets are in tip-top condition. Maintenance is part of your work, even though you do not earn a penny for it. If you do not land fish, you cannot get a share and you will not receive any benefits either, because as a crewmember you are a small self-employed entrepreneur.

Crewmembers are meticulous about their jobs. They have a stake in keeping everything in good repair. Many handle the boat and the equipment as if they owned it. Their work practice is conscientious, which is not imposed, but elicited through the share system of payment. The system stimulates a penchant for independence and self-reliance. Crewmembers accept long working hours and a relentless labour rhythm because they benefit directly. They will agree to sail in rather inclement weather, because if they refrain from so doing they will not earn a penny.

As I have alluded to above, crewing owners usually agreed on not actually cashing their full crew share but instead taking out a pre-fixed sum of money; their 'week money'. They remitted the remainder to the firm. The boat share could lead to profits, but these were usually reserved for future investments. This implied that deckhands often had higher cash incomes than propertied fishermen. A retired skipper-owner, born in 1930, remembers:

> After a week of fishing, a deckhand would sometimes take seven or eight hundred guilders home. Back then, that really was a lot of money, you know. [As a co-owner] you just had a hundred-and-fifty guilders, which was not so handsome. Then again, a wage labourer had eighty guilders, so if you had almost double that amount you were still in clover. [...] Sometimes the boat share yielded some extra money, but it could also be quite disappointing. For instance, if you lost a net, it cost you a lot of money. There have been times when we lost so many nets that we had to work a month to make up for it. Well, that was the risk of being an entrepreneur.

With their 'week money', skipper-owners sometimes had trouble making ends meet within their nuclear families. The system of a weekly allowance occasionally gave rise to tensions and conflicts. The difference between making do with a hundred-and-fifty guilders for a household of six and for one of three was quite a difference, as a skipper-owner's wife, born in 1921, elucidated: 'We had four kids and my sister had just one. So that was not entirely fair.' For larger expenditures, co-owners could take out money from the firm. This, however, would affect their share in it.

Generally, deckhands remember the fifty–fifty share convention quite favourably. Come Saturday night, they would go to the skipper-owner and receive their money in an envelope. 'There have been months when I earned what Mr Average had in a year, especially in the herring season. You have to be lucky,' related a crewmember born in 1938. An older colleague, born in 1911, assented: 'We had ten per cent [with a crew of five]: that was before the beam trawl. We earned well, although we did not get rich. In comparison with onshore labourers, we earned at least twice as much.' A deckhand, born in the previous year, who worked for the same Oosterend skipper, agreed. 'When it was transmitted on the radio what our cutter had grossed, the neighbours would tell my wife: "You can hold out your apron again"' (implying that quite a bit of money would land in it). For what is pejoratively termed 'unskilled labour', deckhands indeed made extremely good money. Incomes could fluctuate heavily, though. He continued:

> One particular week or another could make a huge difference. There was not always lard in the frying pan. We have had weeks that we earned hardly

> anything at all. In those days, you would not set sail in a force 7 or over. Therefore, you always had to set something aside: otherwise, you would get your fingers burnt.

'One week you would gross so much, the next week three times that amount of money and the third week you would earn nothing,' said a retired deckhand, born in 1929. 'You always had to calculate what you earned across the entire year.' The fishermen were generally well aware of the vicissitudes of fishing and adapted to the ups and downs.

Because of the partnership model, the micro-social relationships aboard are rather egalitarian. Apart from some specialist tasks, there is no clear division of labour. Even 'specialists' (the engineer, the cook, the 'fish hold man', the helmsman), take part in all other work. Having his own responsibilities, the skipper stands out to some extent. However, in many cases, a skipper is also a father or an elder brother and this can make his authority seem 'natural' (although with brothers this is certainly not always the case. Siblings may even be disinclined to accept each other's authority when working together on a boat). Moreover, the usually long-standing and multi-stranded relationships between crewmembers – in many cases they are consanguineous relatives, in-laws, neighbours or fellow villagers – foster a relaxed shipboard atmosphere and enhance a team spirit (see also van der Vlist 1970:13). What Byron remarks concerning Shetland fishermen also applies to Texel fisherfolk in the 1960s: 'the structure of shipboard social relations articulates in an idiom of kinship that stresses the qualities of equivalence, trust and cooperation' (1975:147). Age can also be an intervening variable in hierarchy: the social distance between a skipper and an older, experienced crewmember – who has perhaps already been crewing with the skipper's father – is usually much smaller than between a senior skipper and an apprentice deckhand. Ideally, synergy should pervade and characterize social relations aboard cutters. It is important that one can trust and depend on the other deckhands, especially in inclement weather. Being interdependent also necessitates a great measure of tolerance. Skipper-owners take meticulous care to hire men who 'fit in' with the rest of the crew. They seek to avoid tense relations at sea. If, despite these precautions, they have recruited someone who appears to be out of tune, they will not hesitate to sack him. According to an owner-operator, 'it only takes one bellyacher to spoil the mood.' A colleague confirmed this view: 'The crew constitutes a temporary family with smooth relationships. Individuals who do not fit in are usually gone quite soon.' Another experienced skipper, born in 1923, told me:

> The mood aboard has to be good. It is like a marriage, though: there is always bound to be something [that leads to friction]. You have to talk it over. That is also the reason why we never go off to our bunks right after work on deck. We always chat over a cup of coffee first.

Although he emphasized that it should be possible to discuss everything related to fishing, he also stressed that when it comes down to it, the skipper decides and the crew should accept this. He acknowledged that 'with a good team spirit, motivation and dedication are better, even if the work is ashore and [being share fishermen] they don't get paid.'

The partnership rhetoric notwithstanding, there is hierarchy: owners and crew are unequal partners. For instance, crewmembers perceive owner-operators as 'bosses' (*bazen*) and call themselves mates (*maten*) or hands (*knechten*), while owners often refer to crew as personnel (*personeel*). Ideally, however, there is an ethic of cooperation, relations are congenial and egalitarian and a skipper downplays his power. Anthropologist Fredrik Barth explains this egalitarian ethos, which is also apparent aboard Norwegian herring fishing vessels, in terms of a transaction: skippers underplay their authority in exchange for a stable, motivated and hard-working crew, often resulting in higher catches (1966:10). However, other anthropologists point out that this ethos is a façade and that Barth's emphasis on harmony as an adaptation to technological demands completely ignores tensions and conflicts (Byron 1986:99; Knutson 1991; Pascual Fernández 1999). True enough, status differences exist that may surface in subtle ways. On New Year's Day, deckhands used to go to their skipper-owner and his wife and wish them a happy New Year, but this tradition has since vanished. Sometimes the egalitarian façade is a thin veneer. A deckhand (born in 1928) maintains, 'There are skippers who think they're always in the right, even if they clearly are not.' It was a moot point and several of my interlocutors complained about it. A former deckhand, born in 1938, explains:

> The crew's say is limited. You may talk things over, but the skipper determines what will happen. Some skippers do not communicate at all and just strike off as they see fit. However, it can fly in their face if catches are bad. In that case, there is bound to be a lot of moaning and groaning. It also depends on whether it happens incidentally or structurally. If the deckhands have been involved in decision making, for example to steam for twenty-four hours to remote fishing grounds, and the trip ends up in complete failure, it is much easier to swallow.

Despite egalitarianism being the principle, in practice the skipper does indeed make the decisions (also see Chapter 6). In this regard, some of my interlocutors perceive a difference between operators who are born into the traditional fisher families and those who have experience as deckhands themselves. The former are usually less inclined to involve deckhands in decisions. Allegedly, they are also less transparent concerning the grossing and the costs. Therefore, ideal and practice are often at odds. Although in theory the share system appears to be to the advantage of crewmembers – which benefit from the owner's capital input – not all of them were particularly enthusiastic about it (see below).

The fact that deckhands can leave a vessel at short notice mitigates a skipper-owner's power, however. To some extent, labour is footloose. A retired skipper-owner, born in 1904, revealed that his son had had a joint venture with someone who did not communicate with the crewmembers. 'He was honest and a fine skipper, but he always experimented and never ever told the men what he was up to. He could not keep a crew, they were always leaving and that was the worst. He was impossible.' Often, deckhands will attempt to get a berth aboard the vessel of a good skipper – a 'catch king' or 'high-liner'. Conversely, skipper-owners not only compete for fish but also for experienced deckhands. This means that proficient skippers usually have an experienced, dexterous and highly skilled crew who excel in tasks related to, for example, nets or engines. This in turn reinforces the success of a successful enterprise, while making it difficult for less fortunate skippers to recruit the best deckhands. A retired skipper-owner said: 'The skippers sometimes pick on the men. But a good crewmember is worth his weight in gold.' There is also peer pressure to have a 'golden team' as crew. Deckhands sometimes find it unfair that less experienced, less qualified or less dexterous deckhands should receive equal shares while forcing them to compensate for inabilities. Share fishing implies that each crewmember 'has to earn his share', meaning that he has to work hard and keep up with his colleagues aboard ship. 'The labour ethos is forced upon you,' a retired deckhand (born in 1938) says. 'You get used to performing well as you will earn more. It turns into a compulsive drive.' Staying in tune with the ceaseless work rhythm is important and there is considerable peer pressure and social control to conform. Shirking may imply lower gross revenues and therefore slackers are teased and gossiped about. Pranks should indicate to them that they must not slow down the pace of work. Given the importance of good shipboard relationships, this is the most common way of putting someone in his place. Should he fail to understand such 'mild' signs he may turn into a scapegoat and eventually be socially ostracized. With shipboard congeniality jeopardized, the owner is likely to sack such a crewmember.

Young deckhands who are novices to a crew usually face ragging rituals in the guise of jokes and pranks played on them. This way, the peer pressure to perform well is brought home to them early on. It is a rite of passage that most fishermen have to undergo to learn the occupation's ropes and 'to prove their worth'. The quicker they learn to work hard and show themselves dependable colleagues, the sooner such teasing and nagging ceases. They are then accepted as full-blown crewmembers. Some skippers would not tolerate bullying, however. A sixty-seven-year-old owner-operator, who was still active as a shore skipper when I interviewed him in 1990, told me:

> If you are a newcomer, you are unfamiliar with many things. You have to develop an instinct for all sorts of things; you have to be on the alert continually. Take for example the waves. We always sense when a wave will crash

onto the deck in rough weather. We simply *know* when to move off the deck. It is dangerous, for your oilskins will make you float and you can easily be washed off the boat, or be in water up to your neck. Usually, the young lads have to learn this the hard way, and they will be a laughing-stock. The other deckhands will take it out on him. If he does not respond light-heartedly, they will keep picking on him. As a skipper, you have to back the lad. My father always did that and I did the same. I hate ragging rituals. If you help those lads, you should see how they appreciate it and in the end, they will be excellent fellows and fine deckhands, as they know they can trust you.

For a skipper, it is not entirely for moral or disinterested motives that he will stick up for young scapegoats aboard. He is well aware that all deckhands need an on-the-job training regardless of whether they have attended Fishery School or not. The more he invests in this, the more likely it is that the novice will turn into a dexterous and loyal deckhand.

For fishermen in their fifties, it can be increasingly difficult to follow suit. Consequently, younger crewmembers have to put in extra effort to make up for them. Depending on crew relations, they are often willing to do so, but only to a certain extent. After some time, the older deckhand – even if he is a co-owner – will realize or be made to understand that it is time to abandon ship. A skipper-owner, born in 1921, related the following:

Once you get older, it gets harder to catch up on your sleep aboard. You begin worrying whether the person in the wheelhouse has everything under control, the way you would do it yourself. I have heard other older fishermen say similar things. Those young lads go off to their bunk, turn over and fall asleep. However, if you have only been sleeping for half an hour or less and you have to get up again, it starts getting tough. It gets more demanding, mentally and physically. It also depends on the season. In the winter, the seafloor is firm and you can make long hauls, but in the summer, it gets weaker and the tows are much shorter. You get insufficient kip and you worry. Do not forget that the work is demanding. If you are out on deck when you are 55, you do not have a pretty life. My wife's cousin also told me about this. He said, 'They [the other deckhands] do spare you, but you want to chip in your bit.' [...] The physically demanding labour forces you to quit because your body can no longer handle the rhythm of those young lads. [...] Your knees and your back suffer from the boat's continual motions, and you have to sort the fish in a stooped position. There are all sorts of occupational diseases. The rule of thumb is that the youngest deckhand is the cook, but sometimes the crew will let the oldest one aboard be the cook, saying, 'You just prepare the coffee and the meals, and we'll do the work on deck.'

Legally, share fishermen are considered self-employed entrepreneurs, as fishing is regarded a co-venture of petty independent businessmen. This implies that deckhands are not seen as employees to whom Collective Labour Agreements apply. Therefore, they have to arrange for their own insurance and old-age pension contributions. Industrial disability is a considerable risk, so insurance has to be good. Even so, if crewmembers depend on an industrial disability allowance, they will not receive the level of income to which they have become accustomed. In the Texel and Dutch beam-trawl fishing industry fishermen are expected to retire early: that is, when they are around fifty-five years of age, and that they should have accumulated capital to stick it out until they are sixty-five, the age at which they are entitled to a state pension. Consequently, they have to save up for a rainy day and this goes for deckhands in particular. Upon their retirement, owner-operators can still obtain an income from the boat share's profits. With the improving financial results of the 1960s cutter fisheries, early retirement became feasible. In 1947, slightly less than half of the Texel fishermen were younger than forty; almost a third were older than fifty. In 1970, more than three-quarters of the Texel fishermen were younger than forty; only approximately ten per cent were older than fifty. Still, some deckhands encountered difficulties bridging the gap until they were entitled to a state pension. They had to find shore-based work as quinquagenarians or even sexagenarians. If they were skilled net-menders, they could often earn extra income by taking on such a job for an owner-operator.

The fact that share fishermen as self-employed co-adventurers were exempt from the Dutch welfare society's social security system and had to arrange for private insurance was a thorn in the side of labour unions and several political parties and politicians. For example, a 1956 Labour Party report stated, 'Today, hundreds of fishermen are still exempt from the blissful effect of social insurance laws by concluding so-called partnership contracts' (Anonymous 1956:59). As of the early 1920s, the Texel vessel owners had arranged insurances for their crew, but these were limited to accident and illness insurance and the rather modest benefits lasted six months only. In the 1950s, insurance coverage improved after the Texel fishermen discussed the issue in Co-op meetings. The Co-op arranged for collective insurance in 1954, but this did not cover pensions. A government committee in 1959 began studying the possibilities of a comprehensive social insurance and other social issues in the cutter fisheries. The discussion did not lead anywhere, however, and ended up in some bureaucratic drawer. Several years later, it was rekindled. Although the self-employed share fishermen did not desire any kind of state intervention concerning social insurance in the fisheries, the Minister of Social Affairs and Public Health, Bauke Roolvink, sought to change this situation and wanted to introduce mandatory social insurance. Christian Democrat Roolvink, who was in office from 1967 to 1971, had a union background and was a member of the Anti-Revolutionary Party, a political party that traditionally could count on many fishermen's votes. Despite lacking their support, the legislation was changed and as of 1 July 1967, a part of the share income of fishermen had to be deducted and set aside for social insurance, including old-age pension, unemployment benefits, and social costs. Owners were exempt, but the share fishermen did not agree. The sum could amount to half their gross income and cost them considerably more than private insurance. The skipper-owners, in particular, insisted on maintaining the extant mutual insurance system, as they feared that they would not be able to pay the state-induced insurance fees in lean years. They opposed state interference and were of the opinion that mandatory arrangements would 'breed a wage labourers' mentality' with the crewmembers (Visser 1950:138). Boat owners pointed out that the partnership had created a 'community of interest' between the providers of labour and capital (Kranenburg 1968:554; Dubbelaar 1980:122-123). On Texel, insurance was arranged collectively – 'a matter of community sense' – as someone commented in a Co-op meeting. They feared that under the arrangements Roolvink proposed, relations between owners and crew would become adversarial, that the crewmembers' labour ethos would suffer and that sick leave would become much more common. They did admit that the extant system had its pitfalls. Initially, beam trawling was highly dangerous and many vessels capsized. Scores of fishermen drowned (see above), often leaving the bereaved widows and orphans uninsured, and under the care of the state. This was one major reason why the government intended to alter the social security system for fishermen.

The Texel crewmembers met several times to discuss the matter. There were also talks between them and the owners. Fishermen hired by fishing companies (*reders*) usually worked under Collective Labour Agreements, receiving fixed wages plus a bonus. However, that was not what the Texel deckhands and owner-operators desired. The share system's success formula was that putting in extra effort meant that crewmembers would potentially earn more and could even reap windfalls. Most share fishermen earned considerably more than wage-labour employees. Nevertheless, in 1968 Minister Roolvink took another step towards bringing the fishermen in line with 'ordinary' employees. He intended to introduce minimum wages and a holiday allowance for deckhands. The share fishermen were outraged and intended to protest in the country's seat of government, The Hague. However, after talking to representatives of fisheries organizations, the Minister decided not to introduce the measure, but the fishermen remained discontented.

The threat of losing their self-employed status, which would have consequences for both owner-operators and crewmembers, made it abundantly clear to Texel fishermen that they had to organize more tightly to gain political power. As we have seen in previous chapters, cooperation was usually restricted to the local level because of socio-economic differentiation within the occupational community of fishermen, the concomitant specific interests and problems of its component parts, and socio-cultural differences within the island society. As they had for decades, the Oudeschild and Oosterend fishermen still maintained their own separate associations. In 1970, however, they began talks regarding arriving at a merger. They realized that they needed to join forces to be better able to cope with outside intrusion on the fishing industry that affected fishermen's lives.

Oudeschild's Progress through Unity had been languishing for some time. When the number of its members had dropped to a mere seventeen in 1968, the board – three of whose members had been serving for twenty-five years – began successfully recruiting new members. A year later, seventy-nine fishermen had joined the association. Since its inception in 1931, the Co-op had also functioned mainly as an association for Oosterend owner-operators. As we have seen, it served as an important linchpin in the local fishing fleet's offshore segment. In 1970, the Co-op represented thirty-four vessels and their owners. By then, its board felt that Co-op matters and representation of general interests – particularly in connection with the social insurance issue – were at loggerheads. In addition, as members of the national Fishermen's Union, Texel fishermen shared the same interests.

The Co-op and Progress through Unity joined forces in the sense that the Co-op continued to operate as a commercial cooperative, but no longer functioned as a fishermen's association. This task was relegated to the new association, which was officially established on 1 January 1971 and continued under the name of the Oudeschild organization: Progress through Unity. The old board members stepped down and the members elected a new board. The North Sea fleet's owner-operators (who would deliver the chair), the non-propertied deckhands, and the Wadden Sea fishermen would each have three representatives. The first chairman was Ben Daalder, a young fisherman aged twenty-eight who had been skippering a boat for seven years. Although his father was a postal worker and his ancestors did not belong to the island's fishing elite lineages, he was to become a leading face in the Texel and Dutch fishing industries. Looking back on this, he claimed it was easy: 'If you got up twice to intervene during a meeting, you were on the board. If you got up a third time, you were elected chairman.' Within a few months, about 170 members – nearly the entire occupational community – signed up to the local fishermen's association. Interestingly, about forty per cent of the members hailed from Oudeschild, the village that had suffered most from the decline of inshore fishing. Apparently, offshore skippers much appreciated the fishing knowledge and skills of Oudeschilders. Almost thirty-five per cent of the association's members came from Oosterend and Oost and about twenty-four per cent lived elsewhere on Texel. A few members were deckhands from Den Helder. With its broad representation that included deckhands, the association would turn into an important vehicle for the promotion of the members' interests and the social integration of Texel fisherfolk. Other local, regional and national fishermen's organizations were run by and for skipper-owners; deckhands could be members but were not on the board. The main subject that was initially discussed in Progress through Unity was the social insurance issue. For example, in one of the first general meetings, the Fishermen's Union's chair, Klaas Hoekstra, delivered a speech in which he pointed out that joining forces was a wise idea, against the background of growing state intervention in the fishing industry. He continued by saying that the battle to get an exemption from the social insurance system finally appeared to be successful. The disadvantage was that fishermen would not be eligible for unemployment benefits, but the advantage was that if they arranged for their own insurance system, fees could be kept relatively moderate.

Minister Roolvink finally proved to be susceptible to the fishermen's criticism and resistance. In the course of 1970, he defended a proposal in Parliament to exempt fishermen from mandatory social insurance provided that they arranged for insurance collectively. Indeed, the next year it was determined that, as previously, fishermen would be considered self-employed and could arrange for insurance mutually. This demand led to the establishment of the Social Fund for the Share Fishery (*Sociaal Fonds voor de Maatschapsvisserij*) in which all mutual insurance of fishermen was to be combined. Unlike wage labourers, fishermen were not legally entitled to state unemployment and disability benefits, health care and sick-leave allowances. As per 1 May 1971, the Texel fishermen joined the Fund and effected additional types of insurance: a health insurance and a local mutual insurance. Overall, this implied that insurance fees were considerable, detracting from the deckhands' gross incomes. In addition, the Fund did not cover unemployment benefits. If deckhands were laid off, they would end up on the dole. Still, the fishermen preferred to arrange for such matters themselves rather than being subject to what external authorities decided for them. 'Don't forget that Texel is a community that likes to do things autonomously. It is a homogeneous community, perhaps more homogeneous than [communities] elsewhere,' a skipper-owner explained. To the share fishermen, it finally appeared to be a case of all's well that ends well. However, new difficulties were afoot and, again, they were connected with the share system. This time, however, the problem came from within.

Because of the rapid capital-intensive fishing fleet's modernization, skipper-owners began discussing the share arrangement. For decades, it had been based on a fifty–fifty division between vessel and crew, but with the greater capital outlay newer vessels required, the skipper-owners deemed it reasonable that the vessel share should be increased. The issue was discussed at firm level and also in the local fishermen's association. It would lead to heated debates. In 1972, the Progress through Unity board suggested maintaining the fifty–fifty share base for vessels under 1,000 h.p., and changing the division into a fifty-five–forty-five per cent division for vessels with greater engine power, subject to negotiations between owner and crew. Crewmembers were not amused. The deckhands who were on the association's board faced a difficult position: they had to represent the crewmembers' interests, but they also had to be receptive to the owners' position and take into account the general interest. It would make for a degree of reluctance on the deckhands' part to be on the board. In order to avoid social division and labour unrest, Texel skipper-owners initially adopted a passive stance. The vast majority of vessels maintained the fifty–fifty division. The matter would stay on the agenda, and in the wake of the 1973 oil crisis discussions concerning a rearrangement of the share percentages intensified at local as well as national levels. The Fishermen's Union proposed a scheme of proportions for four categories of vessels: a fifty–fifty share division for small boats (0-399 h.p.); a fifty-two–forty-eight per cent share division for the next category (400-699 h.p.); a fifty-five–forty-five per cent arrangement for a 700-999 h.p. category and a fifty-eight–forty-two per cent arrangement for the most powerful vessels of 1,000 h.p. and over. The Texel fishermen opposed such uniformity, arguing that the shares should be fixed in agreement between skipper-owner and crew. The Progress through Unity meeting in which the issue was discussed attracted nearly a hundred fishermen. In its 1974 annual meeting, the association decided that the fifty–fifty share agreement would be maintained in the category of boats with up to 1,000 h.p. engines, while on beamers over 1,000 h.p. the shares would be fifty-five–forty-five per cent for boat and crew, respectively. This change of the share convention reflected the fact that powerful boats tended to be more capital intensive and had greater exploitation costs than smaller ones. The skipper-owners firmly believed that they would be able to explain the need to change the share convention and that the crew would accept it if they did. Even in hindsight, they legitimized changing the share convention: 'For the crewmembers, the fifty–fifty arrangement was sacred, but as skipper-owners we were confronted with ever higher exploitation costs. Initially, the deckhands protested and it gave rise to tension. When they discovered they were earning more because boats were bigger and better, all was serene again,' claimed a skipper-owner, born in 1946.

However, changing the shares did lead to crew dissatisfaction aboard some boats. The recession hit the crewmembers thrice: firstly, they earned less due to the below-average proceeds after subtraction of the fuel costs; secondly, they saw their share percentages diminished; and thirdly, on top of that their insurance fees increased. This happened while the pressure to perform well had increased along with the growing investments in and modernization of the fleet. The deckhands discussed the situation among themselves and this 'quay talk' worsened the deterioration of the relationships between skipper-owners and crewmembers. The association's crew representatives therefore asked deckhands to see them about problems on board individual vessels so that they would not need to resort to gossip, and the difficulties could be dealt with through the organization. The deckhands continued their attempts to redress the share convention to the fifty–fifty arrangement. Their efforts stood little chance. There was more to be disgruntled about. A retired deckhand, born in 1928, recalled:

> During the oil crisis, the owners received state subsidies on fuel. We never got anything at all, though. That was improper, for we also paid for the fuel

> because the costs were subtracted from the gross revenues. I told the owners that it was unfair, but they just said, 'We are not allowed to pass it on to you.' [...] The deckhands certainly did not appreciate the smaller percentages, but if there are many crewmembers, there is bound to be a lot of grumbling, although not in front of the skipper-owner. You would not tell that you were dissatisfied, because there were many people looking for a berth. Particularly if you crewed with a top skipper, you shut your mouth and said Amen. Money alleviates labour.

Another thorn in the crewmembers' side was that some skipper-owners chose to maintain the traditional share convention so that some crews fared better than others. Generally, skipper-owners who were not heavily indebted had more leeway to do so. For example, a skipper-owner told his colleagues: 'I will give my deckhands what I think they are worth. It is nobody else's business.' He clung to the traditional share convention. Another skipper-owner allegedly told his deckhands: 'As long as I am the skipper, we will stick to the fifty–fifty arrangement.' However, when his sons took over, they immediately changed the share division. The deckhands – particularly those who received lower shares, of course – deemed heterogeneity with respect to the share system unfair. Skipper-owners justified altering the shares by pointing out that investments were up and hence capital risks had increased. They needed a larger share to remain solid with the bank and to cover for their boat's depreciation. The deckhands' say was minimal. In a Co-op meeting, an Oosterend skipper-owner remarked, 'The owner stands at the edge of the abyss while deckhands are earning a reasonable income.' The mounting tensions between owners and crew during the oil crisis revealed that the share system and the partnership contract were not so perfect after all.

The Dynamics of Intensification

At local and national levels, rapid modernization and expansion characterized the share and family-firm-based fishing industry of the 1960s and early 1970s. A number of intertwined developments that mutually amplified each other brought about these processes. Technological, ecological, economic, political, social and cultural structures and contingencies all contributed to the dynamics of increases of scale and intensification of resource exploitation. Taken by themselves, these variables would insufficiently explain the course of events. For example, reference to greed or an innate penchant of fishermen to over-invest and overexploit would amount to psychological reductionism, while merely pointing to financial incentives would boil down to economic reductionism. Instead, as I hope to have shown in this chapter, it was the specific interlinkage and interaction of various factors that coalesced into a process of rapidly intensifying capitalization – 'capital stuffing' – and exploitation. Technologically, the kind of equipment available to fishermen determined to a great extent what species and quantities they could catch and under which (weather) conditions. For instance, bigger boats and sophisticated fish-finding and navigation instruments meant more days at sea and greater exploitation accuracy; more powerful engines, beam trawls and nylon nets enabled greater catching efficiency of targeted species. In addition, preservation methods and means of transportation and communication were significant concerning the markets that could be supplied. Ecologically, the variety and richness of marine living resources were important. With intensifying exploitation, aggregate catches initially increased and fish stocks – flatfish stocks in particular – did not seem to decline. Beyond a certain point, however, intensification endangered sustainability. Sole catches, for example, diminished after 1966, although they remained at a fairly high level. However, the catch per unit of effort declined. Catches were not subject to human-induced change only; climate oscillations and other natural causes also contributed to fluctuations. Economically, food supply and demand (linked with cultural food preferences), fish prices, costs of fuel and other commodities, the share system of remuneration, subsidies, taxes, the ample availability of bank loans and credit all added up to specific exploitation bills that either enabled or hindered firm continuity and fishing industry expansion. Politically, structural support measures on the one hand, and access, use and gear regulations and restrictions on the other hand impacted upon the range of choices open to fishermen. Socially and culturally, the family-firm ideal and the pervasive labour ethos contributed to specific economic orientations that stimulated growth.

Despite feeling entangled in a ruthlessly competitive game, most fishermen-owners as individuals saw no alternative but to participate in the horsepower race. Apprehending that they would lag behind if they did not follow suit, skipper-owners invested in expansion even though they were well aware that the sum total of their individual decisions could potentially be devastating. It was a drama of collective action with an outcome that no one desired but all experienced as being inevitable. The government ignored the desperate fishing industry pleas to check the collective expansion. Had the state heeded this advice, many problems could possibly have been avoided. The episode is inscribed in the fishermen's memory. 'The expansion of horsepower was the root of all evil,' one of them contended. Older and younger generations of fishermen alike frequently refer to the state's 'historical error' in apologetically explaining why overcapacity, overexploitation and other problems have subsequently assailed the fishing industry. They point out that it would have been nearly impossible for the fishermen to have voluntarily restrained themselves. This would have required collective action and mutual trust that were in fact lacking. The fishermen realized that free riding would soon undercut voluntary regulations, and lacking the means to enforce rules implied that the premium on flouting them was high because there would be no sanctions whatsoever. Mandatory engine power regulations enforced by an external agency might have worked. Then again, such restrictive measures would have stood a chance only if taken in a European Community or international context. Had the Dutch government restrained the national fishing industry in this respect without other states taking similar measures, the fishermen would certainly have protested and dodged the regulations. Of course, this is 'what if history'. An agreement on a European Common Fisheries Policy was at that time still lacking. However, it became increasingly clear that '[t]he initial aims of structural policy – expansion, modernisation and efficiency improvements – were at variance with the growing awareness of over-exploitation of many important stocks' (Symes 1997a:149).

The drive to expand converged with the family firm's logic and dynamic, where – under conditions of economic growth – hiving off into two or more firms was a common pattern. What is evident is that cooperating kinsfolk proved to be an important economic asset. Owners were prepared to curtail incomes, consumption and expenditures, which turned out to be extremely important in weathering bad times and building up the financial reserves needed to expand. With a single owner, such shock-absorbing and hoarding capacity obviously amounted to less than that of multiple owners or prospective owners who were prepared to defer gratification, as was the case with relatives. Young sons of skipper-owners were confident that they would in future receive their share of the patrimony. Based on this trustful expectation, they were content with a weekly allowance instead of a full share of the gross revenues. For this reason, cooperating with sons or brothers was more attractive than recruiting unrelated deckhands. Kin provided for the flexibility, versatility and resilience needed in an industry where uncertainty was rife. Through being economical and instilling 'fishing values' in their sons, the wives of owner-operators contributed significantly to family firm success. Therefore, social capital constituted at the same time significant economic capital. The cultural ideal of family-firm continuity and expansion could not, however, be attained by all the local actors in the fishing arena. At the micro-level of the family firm, there were contingencies related to the stage of the family cycle and the composition of the nuclear family that to a large extent determined the kind of options that were available and the course of action that could be envisaged. Some skipper-owners remained unmarried or had no offspring. Others had sons who did not aspire to skippering or were unfit for a fishing career, or only had daughters who – under the extant gender relations – rarely obtained or desired to obtain a berth aboard a boat. The fishing vessel was regarded as a male-dominated world by men and women alike.

Even owner-operators who could work with a core of agnatic relatives depended on recruiting additional deckhands to complete a crew. The share system of remuneration brought together capital and labour in their interest to minimize costs and maximize production and proceeds. As top-earning skippers tended to attract the most skilled and the most experienced deckhands and as labour was footloose, it was paramount to keep up with recruiting the best crew, creating yet another incentive to modernize. Based on the excellent financial outcomes of the late 1950s and 1960s, in more than a dozen cases it provided crewmembers with the initial capital to turn independent, several of them successfully following the family firm logic. Newcomers therefore also contributed importantly to the Texel – and Dutch – fishing industry's growth. Entry to the offshore fisheries was fairly easy, as no major restrictions were in place yet, and banks easily provided credit. The ideal of becoming one's own boss was deeply ingrained in many – but certainly not all – fishermen. Although starting from a relatively disadvantaged position compared to the established fishing lineages, many newcomers succeeded in redeeming their debts in a fairly short time and subsequently continuing a viable enterprise. The desire to run one's own firm makes it clear that, although benefiting from the capital input of owners and for all the egalitarianism allegedly characterizing shipboard relations, being a deckhand was not the favoured position of all fishermen. There still existed a considerable covert and overt hierarchy between owner-operators and crew, as became apparent when the former unilaterally altered the share convention. Nonetheless, deckhands earned substantial incomes, even after the share division changed. A considerable part of the net revenues accrued to them, and revenues soared when the beam-trawl fishery for flatfish prospered.

The catching efficiency and economic successes of beam trawling led to a focus on flatfish and catalyzed headlong expansion. A vast proportion of the Dutch North Sea fishing fleet caught sole and plaice as the main target species, although the seasonal herring fishery did not entirely lose its importance. In the early 1970s, the Dutch fishermen landed about eighty per cent of sole catches in Western Europe, and approximately forty per cent of plaice catches (Kranenburg 1977:15). This would prove to be crucial with the onset of quotas in 1975. The next chapter will deal extensively with the introduction of quota regimes in the fishing industry, their significance for fishing practices and the actions and reactions of Texel and other Dutch fishermen. Having been subject primarily to the vagaries of marine ecosystems and fish markets for centuries, fisherfolk henceforth had to cope with the additional and perhaps even more fickle forces of fishing politics. However, far from being passive recipients of top-down decisions, they would attempt to steer their own course, in turn forcing policymakers and politicians to continually respond to the fishermen's actions. The blind process of this interactive dynamic would lead to a profound complication of policymaking, burgeoning red tape and impracticable and unenforceable rules and regulations. Perhaps more than anything else, the messy fisheries policy and management measures that ensued encouraged, rather than impeded, behaviour that contributed to irresponsible exploitation.

Chapter 5

Catch Kings and Quota Busters

In the mid-1970s, several international events affected the fishing industries of Western European countries in major ways. Negotiations on the United Nations Convention on the Law of the Sea regarding the establishment of 200 nm (nautical miles) Exclusive Economic Zones were underway. It would be the largest sea-based 'enclosure' operation. The lengthy talks were expedited when the second British-Icelandic Cod War (1972-1973) reached a zenith. With the Law of the Sea agreement emerging but not yet in place, Iceland nonetheless decided to unilaterally extend and enforce its exclusive fishing zone from 50 to 200 nm, giving rise to the third Anglo-Icelandic Cod War (1975-1976). The conflict between the two NATO allies ended when they agreed on a compromise. Territorial use rights were also a major issue in negotiations regarding the accession of new member states to the European Economic Community (EEC). When Denmark, Ireland and the United Kingdom joined the EEC in 1973, it was agreed that member states would have equal access to Community waters – that is, within the national 12 nautical mile (nm) limits. At a conference in The Hague in 1976, the member states decided to extend their jurisdiction from 12 to 200 nm through the establishment of Exclusive Economic Zones, while they also determined that the 12 nm zone would be prohibited for boats with an engine power exceeding 300 h.p. and 50 gross register tonnes (GRT). A forceful European fisheries policy still lacked, but with the accession of the three new member states – all of which had important fishing industries – the development of a comprehensive Common Fisheries Policy (CFP) for the North Sea and Atlantic fisheries was agreed upon. It was scheduled to be implemented by 1983. In the interim, member states should introduce measures, which needed to be consented to by the European Commission, to maintain the state of the stocks at a sustainable level.

In 1975, the North East Atlantic Fisheries Commission (NEAFC) established total allowable catches (TACs) for several species of fish. Each state that had signed the convention preceding the commission's founding received a share based on 'historic rights'. The NEAFC allocations were not binding and at the national level, the introduction of catch restrictions was very messy indeed. The European Economic Community stepped in the next year and allocated national shares of the total allowable catch to member states. Sidestepping the NEAFC, the EEC established total allowable catches following the advice of the International Council for the Exploration of the Seas, and, from 1979 onwards the Advisory Committee on Fisheries Management. On the basis of its recommendations, the European Commission formulated a proposal that was subsequently discussed in a Council of Fisheries Ministers meeting. The Council made the final decision on the maximum output, usually following prolonged political lobbying and negotiations that predictably led to considerable amendments of the initial proposal. Administrators and officials found themselves in a contradictory situation. On the one hand, they had to issue and enforce catch limitations, while on the other they sought to safeguard national interests and socio-economic stability by negotiating total allowable catches and attempting to achieve the greatest possible share for those species that were particularly important for the national fishing industry. It is clear that conservation and the national interest make strange bedfellows. Being pragmatic and having to arrive at a consensus decision, the Ministers often simply ignored scientific advice or used its alleged ambiguity as a pretext for compromise. During lengthy and opaque in camera sessions, they 'translated' the advice by capping the limits at higher-than-proposed levels, in fact creating non-limiting conditions for their fisheries and thus minimizing potential enforcement problems (Symes 1997a:146; Daan 1997:324). Their focus was consequently on the short term, which compounded rather than solved the fisheries crisis. Although there was a political rhetoric of tackling over-fishing, the policy mainly aimed at keeping the peace among fishermen and guaranteeing their incomes. Besides these goals of conflict avoidance and socio-economic stability, the common internal market continued to be the primary European concern.

Meanwhile, negotiations on a Common Fisheries Policy were profoundly arduous, in particular because the UK and Ireland sought to hold on to the exclusive right of exploiting fish stocks in what their governments regarded as their national waters: a zone of 50 nm. Following seven years of political discussions and concessions, the member states met their deadline and initiated a Common Fisheries Policy in January 1983. It would be valid for a ten-year term. Its objectives were: conserving resources; maintaining the viability of fisheries and fishing communities; and improving economic performance (Holden 1984). There were three key management instruments under the Common Fisheries Policy: effort management (input restrictions and regulations such as mesh-size and days-at-sea restrictions); quota management (output restrictions and regulations); and fleet management (capacity adjustment). Common Fisheries Policy measures were binding for member states and individual fishermen. They usually took the form of regulations that were directly applicable without requiring the implementation of national legislation. Enforcement was carried out through inspection at ports and at sea as well as by protection vessels and aircraft. The member states arrived at an agreement concerning the allocation of total allowable catches based on the principle of relative stability, implying that national quotas were allocated according to fixed shares. Hence, the individual member states could expect to retain their positions in relation to each other because national fishing industries would equally benefit or suffer from resource fluctuations. These political processes of negotiation and compromise became deeply ingrained in EC decision making.

Under the European Community's 'principle of subsidiarity', member states had some leeway in defining their own rules and were responsible for proper enforcement of CFP measures in the waters and territories under their jurisdiction. The transition from a national to a European fisheries policy that had to be retranslated into a national policy was far from smooth. The initial vagueness and uncertainty are evident from the fact that the Dutch Ministry of Agriculture and Fisheries frankly reported above-quota landings in the second half of the 1970s. Discovering that enforcement was lacking, the fishermen took this as a sign of 'no objection' to overshoot what they believed to be targets rather than limits, and instead of curtailing production, they expanded. Europe's common market ensured that they could sell their landings. In regard to the allocation of allowances within its boundaries, the Netherlands opted for licences and divided the national quotas into individual quotas that would soon become transferable. If individual transferable quotas (ITQs) meet with the characteristics of long duration, complete specificity, transferability, exclusivity and security, they make for 'complete property rights' (Scott 2000; Caddy and Seijo 2005:69-70). Champions of the individual transferable quota system – generally neoclassical fisheries economists – claim that, theoretically, individual transferable quotas have important economic, administrative and ecological benefits:[1]

1. The system improves efficiency in that the most viable units acquire landing rights that are transferred on the market, facilitating adjustment, while economically marginal units will sell them, reducing excess capacity and balancing vessel size and landing allowances.
2. This progressive rationalization also creates incentives to reduce the costs of landing quotas by economizing on labour and capital per fish landed, preventing 'capital stuffing' and rent dissipation (that is, using more capital than required to produce a good).
3. The system facilitates access to loans for capital improvements with privatized landing rights being accepted as collateral.
4. The system leads to higher and more stable prices as it ends the frantic 'race to fish', allowing for planning a better distribution of landings across the year and optimizing marketing opportunities to maximize prices, thus ending boom-and-bust cycles and production bottlenecks in processing.
5. The system therefore augments mean earnings and profitability.
6. The system enables specialization, in that operators can focus on what they do best and have the opportunity to improve.
7. Since costs are internalized, the system creates incentives for stewardship and sustainability from self-interested motives.

8. It enhances compliance with the rules, cutting on the costs of monitoring and enforcement, and it improves cooperation with fisheries biologists, regulators and law enforcers, reducing information and administrative costs, enabling a simplification of the regulatory regime and facilitating participation in management. Privatized rights, the neoclassical economists argue, are superior to common property rights. In this chapter, we shall encounter how the system operated in practice and whether the assumed economic, administrative and ecological benefits did in fact apply.

Transformations in the regimes of access and use rights affect economic behaviour and performance: 'By allocating decision-making authority, they also determine who are the economic actors in a system and define the distribution of wealth in a society' (Libecap 1989:1). In what follows, I will detail Texel and other Dutch fishermen's responses and modes of action concerning their encapsulation into European and national fisheries management regimes. With new restraints that restricted their freedom of acting in their customary way, it was inevitable that they would 'sacrifice and suffer' (Jentoft, McCay and Wilson 1998:434). As we shall see, they attempted to steer a collision course, seeking to find the loopholes of the law and, if they thought it necessary, flouting the rules and regulations. They proved to be not passive recipients of top-down decisions, but actors who made and remade their own world. Still, their daily livelihoods were impacted in several major ways. To understand their behaviour, it is imperative to devote attention to the economic, social and cultural dynamics of the fishing industry and the ways in which it was embedded within the legal and regulatory frameworks of national and supranational politics. What ensued was an interactive process: the fishermen would respond to measures in various ways, in turn forcing the national and supranational authorities to react upon the fishermen's reactions and come up with new regulations almost in perpetuity. In the course of the process it would become apparent that fisheries policy in many respects was counterproductive. This chapter will outline the successive regulatory regimes and deal with the fishermen's perceptions of and responses to the demise of mare liberum. It will transpire that harnessing the fishermen proved to be extremely difficult and had profound political consequences. I will examine the attitudes and views of Texel fishermen towards those who have played a major role in this development: civil servants, politicians and biologists. Next, I will discuss the important role that is attributed to the skipper in the catching of fish, the status he derives from being successful, and the competitive dynamics that status ranking, reputation and rivalry give rise to. These cultural considerations are important for the comprehension of why non-compliance was rife.

Fishy Business, Flawed Policy and Faltering Enforcement

Owing to the fact that Dutch fishermen had been investing so heavily in beam trawling for flatfish since the early 1960s, they obtained over seventy per cent of the total allowable catch for sole, and nearly forty per cent of that for plaice in the initial NEAFC allocations for 1975. These species represented about two thirds of all the fish Dutch fishermen landed. The national quota for sole amounted to 9,445 metric tonnes; that for plaice was set at 47,020 metric tonnes. This would mean a drastic reduction in legal landings: forty-seven per cent and nine per cent, respectively, compared to the landings of the previous year. Herring fishing entitlements were also subject to the system of total allowable catches. By then, the North Sea herring stocks were alarmingly low and biologists called for draconian measures. The herring fishery – traditionally important for Texel family firms – would be closed as per January 1977. Subsequently, many of the large Dutch trawler companies based in Vlaardingen, Scheveningen and Katwijk went bankrupt. Some converted to beam trawling and otter trawling for cod, which led to additional competition in a segment of the fishing fleet that was already experiencing unprecedented growth. At the national level, the allocation of flatfish quotas was initially left to the fishing industry – *in casu* the statutory Fish Board, a vertically integrated corporate organization representing the interests of producers, traders and processors in negotiations with the government. This did not prove to be a success. The implementation of quota measures was due from 23 February 1975, but the Fish Board only publicized the quota measures almost three months later. By then, fishermen had already caught a considerable amount of the sole and plaice they were entitled to catch (by mid-year approximately two-thirds). In addition, the state did not monitor and police the catch limitations. The Dutch Fishermen's Union asked the Ministry to take steps, also with regard to the ban on fishing within the 12 nm zone with vessels over 50 GRT and engines over 300 h.p., and landing undersized sole and plaice. Those who disobeyed the rules did not get fined, but only received warnings. The Union protested and said that the state ought to supervise and enforce its rules. If not, many fishermen would obviously ignore them. As of 7 July, the Fish Board announced that vessels with odd and even numbers should alternate fishing every fortnight, a measure that was bluntly ignored. Catches were simply transferred at sea to a cutter with an even or an odd number, depending on which part of the fleet was allowed to land the fish. The Ministry declared that policing and inspection would be tightened and doubled the number of General Inspection Service's vessels to four. Initially, quotas were not allocated individually. The fishermen's organizations, the Minister and the Ministry's Fisheries Directorate were all in favour of individual quotas, but the Fish Board had not come up with such an arrangement. All in all, the transition to the quota system was rather confusing.

Initially, most Dutch fishermen did not oppose the introduction of quota measures. The topic was discussed at a 1974 meeting of the Dutch Fishermen's Union. All local departments – including Texel's Progress through Unity – voted in favour, provided that the state avoided mandatory decommissioning and came up with support measures (Vissersbond 1994:71). The Union was willing to cooperate so as not to compromise its leverage in negotiations with state institutions. However, an Urk fishermen's association – which had left the Fishermen's Union earlier – did fiercely oppose the quota regime and advised its members not to abide by the rules. Subsequently, more and more owner-operators from other fishing ports followed suit. Many fishermen feared that they would lose their jobs. The oil crisis and economic recession contributed to their pessimism. A plethora of new rules and regulations and heavy debts made for tremendous uncertainty and anxiety. The depreciation term of a new cutter was ten years, that of an engine five years. Bank loans had to be redeemed and interest paid, making short-term adjustment difficult. Like their compatriots in other Dutch fishing ports, Texel fishermen also feared imminent unemployment as a consequence of the quota measures. They would then end up on the dole because they were not in the state's unemployment benefit scheme. The Fisheries Directorate claimed that – given overcapacity – decommissioning a fifth of the Dutch fishing fleet would be necessary. However, none of the Texel owner-operators was inclined to decommission his vessel. In a January 1975 meeting with the Texel fishermen, the head of the Fisheries Directorate, A.C. Freling, said: 'The weak will have fewer opportunities. There is a structural rift, a watershed. Heads will roll. It is very annoying that the quota measures are introduced at a time of economic recession when many are already experiencing problems.' The owner of a brand-new beamer deemed it hard to swallow that 'someone who is up to his neck in debts cannot make up for this by pushing harder to meet his commitments'. Their willingness to work relentlessly had always been a key value for fishermen, but with output limits in place they felt they could no longer put it into practice. Freling stated that for the survivors, there would be support measures in the form of subsidies. The Texel fishermen were not at all confident that this would provide a solution.

The local fishermen's association continued to hammer home the necessity of capping engine power. Its members were also in favour of allocating quotas per horsepower category (up to 1,200 h.p.), but other associations did not share this viewpoint. What bothered the Texelians was that they continually faced opaque new rules and regulations with which they had to comply, but which were not enforced at all, and could change overnight. This obfuscated what was expected of the fishing industry. In a July 1975 board meeting of Progress through Unity, chairman Ben Daalder said that he was extremely disappointed with the extant situation:

> As it stands, none of the owners and deckhands knows what he has to abide by. At present, everyone with good intentions concerning the quota system takes his own course of action. You cannot blame them. New rules change again the very same week.

Daalder certainly had a point. A blueprint seemed to be lacking and ad hoc rulemaking made for considerable uncertainty. Tensions between compliers and non-compliers mounted and emotions ran high. The former accused the latter of leading the fishing industry into a crisis, while the latter reproached the former for being overly cooperative with incomprehensible and impracticable regulations. The association's board members had to meet more and more often to address all the problems and red tape that swamped the fishing industry.

Later that year, in October 1975, the members of Progress through Unity discussed the introduction of individual quotas in a general meeting. Petty and large fishermen appeared fundamentally divided. Daalder did his utmost to bring the factions closer together: to no avail, however. Many fishermen announced that they would not comply with the rules until they received some kind of financial compensation in return. Daalder disagreed and threatened to step down. He was loath to bring this message to the Fishermen's Union's board meeting. The Union had proposed to arrive at individual quotas for the year 1976. Daalder deemed individual quota measures 'unavoidable' because of over-fishing, although at the same time he thought that until then, the measures had been ill-informed. When in the third week of November 1975 the Dutch sole quota was exhausted and the Fish Board informed skipper-owners that it was forbidden to land sole until 1 January, the fishermen collectively ignored the ban and steamed to the fishing grounds because the state did not come up with financial compensation. Texelians followed suit, although some began pursuing round-fish (a generic term for such species as cod, hake, haddock, whiting and so on). The next week, they targeted plaice and round-fish only, in line with the Fishermen's Union's advice. However, many Texel fishermen disagreed with this advice and with the board of the local fishermen's association. They favoured a tougher stand towards the government. When a delegation of Dutch fishermen, including Daalder, met with the Minister responsible for fisheries, the latter made it crystal clear that the government did not intend to compensate the fishermen. As a consequence, the fisheries organizations, including Progress through Unity, stopped consulting with the Minister. In his annual review of the fisheries in the local newspaper, Daalder referred to the violated trust in the government. The fishermen had agreed to the quota measures on the condition that the state would provide financial compensation, but the state did not even enforce the measures.

Partly at the fisheries organizations' instigation, an individual quota system was implemented in 1976. Based on the track records of their highest catches in 1972, 1973 and 1974, the Fisheries Directorate – which took over the task of allocation from the Fish Board – assigned individual, non-transferable quotas to sole and plaice fishermen.[2] The idea underlying the measure was that allocation of exclusive allowances would lead to a responsible mode of fishing and offer fishermen the prospect of earning resource rent, maximizing their profit and increasing operational certainty. In theory, this seemed attractive. The mean sole quota allocated to Texel owner-operators amounted to approximately seventy metric tonnes per firm, but there were major variations. One of my interlocutors claimed: 'Those who had overfished big time were rewarded.' The arbitrary choice of the years on which the allocation was based meant that fishermen who had not pursued flatfish during this time or whose boats were under repair during part of it obtained relatively small quotas. Moreover, aggregate sole landings had been considerably lower in the reference years than in the late 1960s (see Chapter 4). Another bone of contention was that skipper-owners who had been targeting flatfish *and* herring – as many Texel offshore fishermen traditionally did – were allocated rather small sole and plaice quotas in comparison with those who had been specializing in the mixed flatfish fisheries. When the herring fishery was closed as per January 1977, this proved to be an important disadvantage. Right from the onset, the seeds of dissension were sown.

On Monday 10 May 1976, approximately 2,500 Dutch fishermen went to the country's seat of Parliament, The Hague, in protest. They wanted to call the government's and the Dutch people's attention to the 'perilous situation' of the fisheries due to the quota regime. Fishing industry leaders had urged local associations and their members to join the mass demonstration: 'Now you should show your solidarity and concord to accomplish the objectives we set. The decision to campaign applies to the entire cutter fleet. With no exception,' a pamphlet read. Approximately seventy-five cutters steamed to the fishing port of Scheveningen near The Hague, and many demonstrators went to The Hague in buses, including three coaches from Texel. The action committee's leader, Ben Daalder, presented a petition to the speaker of the House of Representatives. Daalder addressed the protestors and said that if the Minister did not give in, the fishermen's protest would become much tougher than a demonstration alone: 'If we are forced to go down the drain, we will sell our lives dearly.' However, the national administration was firmly embedded within the European Community and, save for extending some sympathy, could neither change the quota regime nor do much to support the fishing industry. The fishermen went home empty-handed. Yet it would not be the last time that they voiced their discontent in a mass demonstration. A few days prior to the protest action, Daalder said in an interview:

> In hindsight, we can now conclude that previous governments were wrong to leave unchecked and even support the fishermen's inclination to invest. We are left with the bitter fruits. The fishermen should have been protected against themselves. Mind you, the fishing industry even called for such

> measures. With catch restrictions in place, we have been driven to the wall. Of course, we realize that catch restrictions are beneficial. However, better measures are necessary. The state ought to come up with a decommissioning scheme and an adjustment fund. We should be helped through the difficult times ahead by means of fuel and interest subsidies (*Texelse Courant*, 7 May 1976).

Daalder rightly expected the decisions concerning the fisheries to increasingly become a matter of give and take between European Community member states. A Texel owner-operator remarked in the 1976 annual meeting of the local fishermen's association that under the extant conditions, 'one can imagine that a fisherman at a certain moment says "let's catch as much as we can and throw the rules to the winds", even though this does not solve the existing problems'.

The mood was grim and the fishermen's worries about the future were genuine. The forward linkages of primary production – auctions, fish traders and processors – feared that their business positions would deteriorate with diminishing landings. In a meeting with directors of municipal fish auctions on 23 September 1976, the head of the Fisheries Directorate, Tienstra, maintained that overcapacity had to be reduced. When the directors claimed that this would be detrimental to their businesses, he allegedly replied: 'The auctions should ensure that everything adds up on paper'. Although perhaps not intended this way, the auction directors and the fishing industry understood this to mean that illegal trade circuits would be condoned as long as the account books were in order. In a startling report on the fishing community of Urk, two journalists extensively described the murky wheeling and dealing in the fishing industry as early as 1977 (Verhey and van Westerloo 1977). They stated that local and national officials were fully aware of the illicit dealings. Infringements were manifold. They consisted of: using liners or blinders (*binnenzakken*, nets within nets) and too-small mesh sizes; stashing away catches in hidden spaces; landing the fish in the dead of night when auctions were officially closed and General Inspection Service officials were not present; discharging fish abroad in ports where inspections were known to be lenient or lacking and subsequently transporting the fish with trucks to Dutch auctions where its origin was not registered or delivering directly to dealers and processing plants; transferring fish at sea to smaller boats that discharged the fish in ports without auctions; orchestrating simultaneous landings of many beamers so that only a handful of boats could be inspected; filling boxes with much more than the forty-kilo limit; landing undersized fish; putting a tiny layer of plaice on top of sole to pretend that boxes contained plaice; filling in the unspecified term 'sea-fish' on slips of paper that marked the species contents of fish boxes; trading 'consigned' fish of unknown producers through auctions; indicating the wrong catch areas in logbooks; writing fictive vessel registration numbers on administration forms, and so on. The authorities could not condone such massive contravention of the rules and regulations. Not officially, at least. They demanded compliance, but the fishermen refused to budge.

Hence, the quota system's introduction was disruptive in regard to relationships between the fishing industry and state institutions. It also proved to be internally divisive. On Texel, it provoked disagreement between the categories of larger and petty skipper-owners and between owners and crew. At national level, it translated into tensions between the Fishermen's Union's board and owner-operators who wanted to steer a confrontational course. Deeming the Fishermen's Union overly cooperative with the state and its chairman, Klaas Hoekstra, 'too nice', several local fishermen's associations abandoned the organization, including Progress through Unity, chaired by Daalder. Daalder was on the Union's executive board for some time, but he and some other fishermen were of the opinion that a new, reorganized and professional national fishermen's organization should be established since they believed the Union to be too weak and too accommodating to successfully negotiate with the government. The matter was discussed several times at local level to see whether things were improving. However, Daalder concluded that the Union was 'a mess'. He would later recall that most skipper-owners who were on the Union's executive board were septuagenarians, some of whom had even operated sailing vessels. 'They could not keep up with the developments,' he said. To some extent it was a generational conflict, with Union executives being conservative and law-abiding citizens, whereas younger skipper-owners who wanted to expand were more akin with the contemporary fashion of protesting authority – a legacy of the 1960s' contestation of power. In a September 1977 meeting, Progress through Unity decided to withdraw its membership from the Fishermen's Union. Following heated discussions, Daalder exclaimed: 'We will always have our differences of opinion, but let's show the outside world a collective and united face.' Thirty out of the thirty-two members who were present voted in favour of abandoning the Union. The Texel fishermen said they would consider returning if the Union's structure was reorganized and its board members became more engaged in defending the interests of the owners of big beamers. A new national organization, the Federation of Fishery Associations, was established. Soon, it represented almost half of the Dutch fishing vessels – mostly powerful beamers. A few Texel fishermen remained loyal to the Union, among whom were the kin-related owners of three vessels, whereas the vast majority opted for the Federation. For years to come, discord rather than unity among and within the fishing industry's national associations would be common.

With two national voluntary associations in addition to several other branch organizations and a host of local and regional associations, the fishermen's leverage in regard to state institutions was weakened rather than reinforced. In an era in which the European Community and the Dutch state introduced various measures affecting the fishing industry, a united viewpoint of all fishermen was needed to lend effective weight to the balance. The fact that they were divided among themselves prevented them from successfully opposing the growing state intervention. Moreover, the fission profoundly complicated negotiations between the parties, because membership of either of the national associations increasingly became a matter of conviction that had deep symbolical implications for the social relations in the fishing industry. 'Union' or 'Federation' turned into something similar to 'left wing' versus 'right wing', 'Protestant' versus 'Catholic', or 'orthodox' versus 'heterodox'. For years to come, the old sore would continue to ooze. If the Union said 'yes' to a proposal, the Federation was inclined to say 'no'. Mutual opposition and suspicions made it nearly impossible to come up with a united stand, undermining the fishermen's political clout. Thus, the quota regime drove a wedge into the fisherfolk's occupational world. Once again, it was not a 'need for independence' or individualism that had brought about division. Clearly, the factional process was about the diverging interests of specific categories of fishermen that subsequently developed into groups. Cooperation and conflict were not mutually exclusive phenomena but reflected changing interdependencies (Elias 1974:xix).

The Federation's tough course of action would soon become apparent. Late in 1977 rumour had it that the EEC was going to radically reduce the Dutch sole quota and take measures to enforce compliance. The news caused upheaval among the flatfish fishermen. When discussing the issue in the local fishermen's association, a Texel skipper-owner asked rhetorically: 'What kind of image do the high and mighty who regulate everything have of fishermen, and does the fisherman's opinion count at all?' The fishermen felt cornered. The Federation, provisionally established in January 1978, immediately flexed its muscles when that very month about seventy cutters – ten of them from Texel – briefly blockaded Holland's main fishing ports. The action was coordinated from Texel. Ben Daalder said that the prime goal was to impact public opinion. The fishermen, who were of the opinion that the government came down upon them too harshly for non-compliance with the quota rules, did indeed gain some sympathy. In addition, the quotas for 1978 were still unknown and they would be established month by month. The fishermen could land a twelfth of their annual individual sole and plaice quotas per month. They considered this to be unworkable, since catches tended to fluctuate across the year, with lows in the summer. Such ad hoc measures also hindered long-term business planning. The fishermen were further dissatisfied with the Dutch government's decommissioning policy, while the British fleet was believed to be expanding at the same time. Anxiety mounted and anger reached a zenith. The General Inspection Service defused the tension by temporarily relaxing inspections. In addition, a voluntary tie-up scheme with some financial compensation for weeks that boats did not fish was introduced, but the vast majority of owner-operators preferred to sail. The fishermen felt that they had no other option than to simply ignore the rules.

Contrary to the owner-operators' initial expectations, quota allowances proved to represent substantial economic value. They had received these individual entitlements free of charge, but when they became transferable, the rights-holding owner-operators could capitalize on them. Quite soon after their introduction, a lively trade in landing rights developed; a sign that owner-operators wanted to match their entitlements with vessel capacity and not depend entirely on illegal activities. Although individual quotas would officially be transferable as of 1985, de facto such transfers came about much earlier (Smit 2001). Fishermen simply bought a boat with associated quotas and then sold the vessel without landing allowances. Fission and fusion of enterprises also enabled transfers of individual quotas. Thus, the government merely put the practice on a statutory footing. Although it stipulated that only a vessel's entire quota could be sold, Producer Organizations could purchase an entire ITQ and then resell it in parts to their members. Acquiring quota rights was facilitated because the same tax benefits applied to their purchasing costs as investment in equipment. Entitlements were in great demand and prices soared, much to the dismay of Texel skipper-owners. The quota system and the market for landing allowances did not, however, lead to a more responsible and efficient fishery that enabled fishermen to respond flexibly to changes in the market, while the burden of management and enforcement still rested with the government (van Vliet 1998a:219). Property rights were insecure, since the flatfish fishery would be closed once the national quotas for sole and plaice were exhausted. This implied that those fishermen who chose to use their individual quotas steadily across the year might stand empty-handed well before year's end. They quickly learned to fish as much as they could in as short a time as possible. The social dynamic was that successful quota busting encouraged law-abiding skippers to join the transgressors in their illegal operations for fear of being put at a disadvantage. Deeming total allowable catches to be fixed at ridiculously low levels and discovering that their protests to have them altered were in vain, many fishermen simply ignored the quota regime. With quota prices being high and monitoring and enforcement of restrictions suboptimal, to say the least, it paid to continue fishing illegally. The fishermen claimed that they would be 'stealing from their own wallet' if they refrained from doing so, since 'everyone was doing it'. Being ambivalent about state intervention and strict law enforcement, they said that the state should do a better job of regulating and policing the industry, yet they continued their illegal practices. Right from the onset, the national sole and plaice quotas were overshot. In spite of the individual quota system, the derby-style competition to catch fish was not eliminated. Neither was the horsepower race.

Following a brief intermezzo, when due to decommissioning, aggregate engine power of the Dutch – including the Texel – fishing fleet diminished, the expansion of the number of boats and engine power continued almost unabatedly from 1976 onwards (see Appendix C). Since the oil crisis, the fishing fleet had been encountering negative net results for a few years in succession and the enthusiasm for investment was low. However, as of 1976, debts were redeemed at a fast pace, ameliorating the owner-operators' credit-worthiness with banks. The solvency of Dutch cutter owners increased from 52.3 per cent in 1973 to 68.9 per cent in 1978, the first year after the oil crisis to yield net profits. The mean annual gross revenues of Dutch cutters in 1978 increased by sixteen per cent compared to the previous year. In 1978, Texel cutters grossed an aggregate of 36.4 million guilders: slightly over ten per cent of the Dutch cutter fleet's total gross proceeds. With confidence in the future restored, fishermen almost immediately began ordering new and more powerful boats, nullifying recent decommissioning efforts and increasing overcapacity. Despite renewed fishing industry requests in 1979, 1980 and 1981 to introduce an engine power limit, the government again refused to intervene on account of favouring a liberal market economy and lacking the legal means to do so. It regarded self-restraint the fisherman's individual responsibility. In conditions of fierce competition, the skipper-owners' attitude was to stay in business as long as possible. In fact, owner-operators who could or would not expand began lagging behind and were ousted from the fisheries arena. Many vessel owners therefore felt a need to expand, while deckhands who aspired to become independent and sons of owners who wanted to set up an independent firm bought second-hand vessels from those who had acquired new vessels.

Fishermen were able to obtain bank loans fairly easily. Their solvency was considerable and – in addition to boats – they could use their landing allowances as collateral. Moreover, from 1978 onwards a general investment subsidy (known as *WIR premie*) in the form of a fiscal allowance of twelve per cent or more on newly built vessels created an important incentive to invest in new fishing boats, while liquidating a firm was fiscally unwise. The state's left hand did not seem to know what the right hand did: while the Ministry of Agriculture and Fisheries was discussing how to tackle overcapacity, the Ministry of Economic Affairs came up with a generic scheme of investment subsidies to stimulate national economic growth. Fishermen could also apply, in fact exacerbating the overcapacity and overcapitalization problems. Quota restrictions on the one hand, and investment subsidies on the other, gave ambivalent messages and contradictory incentives to fisherfolk, putting them 'in a somewhat schizophrenic position' while at the same time making for uncertainty (van Vliet 1999:168; also see van der Schans 2001:425ff.). Individual fishermen blamed each other for the overcapacity problem, collectively they blamed the government for failing to intervene, and the government in turn blamed the fishing industry for investing irresponsibly (Vervaele, Ruimschotel and Widdershoven 1990:133).

The Dutch cutter fishing fleet expanded rapidly, both in terms of number of vessels and aggregate engine power (see Appendix B). Although the pace of developments on Texel was slower (Appendix C), on the island, too, there was a clear tendency towards boats becoming bigger and more powerful. Along with growing engine power, fuel bills subtracted a greater and greater share of the gross proceeds. Not only did fuel consumption increase, but fuel prices also rose sharply even though they were tax-free. In 1974, fuel costs accounted for seventeen per cent of a vessel's operating costs: six years later they represented over a quarter (Rijneveld et al. 1981:122), but owner-operators continued to lodge orders for yet more powerful vessels. Expansion was the buzz-word, to which end owner-operators continued to invest. Fears about over-fishing, which had often been expressed in the early and mid-1970s, seemed to have vanished completely. By 1980, a few cutters with over 2,000 h.p. engines were launched. The first such 'super beamer' would make her appearance in the Texel fleet four years later. The owner-operators' expectation was that more engine power would result in more pulling power, so that longer beams of twelve metres or over and more tickler chains could be applied, enhancing the catching potential for sole, which dig into the sediment. At the same time, powerful engines would reduce steaming time and enable fishing to continue in rough weather. In addition, new high-tensile fibres and more sophisticated net constructions would allow towing at higher speeds and improve gear efficiency (Banks et al. 2001:44). Navigation equipment (Loran replacing the less accurate Decca) was particularly important in the mixed flatfish fishery. It enabled mapping positions, so that a particularly abundant fishing spot could be worked rather systematically by returning to a set course until productivity declined.

Some of the money invested went into improving safety and working conditions aboard. The boats had sheltered whalebacks where the crew could process the catch, and could be completely operated from the bridge. They also had central heating, a shower and a lavatory – which had previously been regarded redundant 'luxuries'. In the early 1980s, fish-sorting installations with conveyor-belt systems and gutting machines eased the tasks of deckhands. Previously, the cod ends were emptied on deck, where the fish was mostly mixed with a lot of benthos, debris and sand. While being exposed to spray and the weather on a rolling and pitching vessel, crewmen had to sort, grade and gut the catch – a repetitive job – in a kneeling or stooped position. This quite often led to back troubles at a relatively young age. The new equipment enabled them to do the job in an upright position under the whaleback. With CCTV monitors to keep an eye on the engine room, watch alarm and closed sterns becoming more and more common, safety aboard increased.

Owner-operators also acquired more sophisticated navigation and fish-finding equipment, which was particularly important in the pelagic fisheries. When the ban on herring fishing was lifted in 1983, most Texel fishermen who were entitled to do so resumed seasonal pair trawling. It required quite an investment, as a single herring net alone cost something in the order of 63,000 guilders. Shifting from flatfish to herring and back took two weeks of adjustments to the boat and the gears, weeks in which precious fishing time was lost and nothing was earned. Entitlements were even more insecure than in flatfish fishing. Herring fishing was not based on an individual transferable quota system. For licence holders, the national herring quota uptake was initially on a 'first come, first serve' basis. The largest company-owned trawlers consequently had more opportunity to catch their share in as short a time as possible, since for family firms herring fishing was a seasonal affair that was usually alternated with mixed flatfish fisheries. Nevertheless, nine Texel firms continued their switching strategies. Three of them even had identical purpose-built stern trawlers made that specialized in the herring and round-fish fisheries. They were twenty-four metres long and had two main engines of 300 h.p. each. The owner-operators expected that if they switched off one of the engines, they would be allowed to fish within the 12 nm zone. However, much to their dismay, this turned out not to be the case because the vessels were over 70 GRT. The firms had received substantial EC subsidies to have their boats built; amounting to almost half a million guilders per unit, but it would prove to be extremely difficult to operate the vessels in a cost-effective way.

Texel's rather small inshore segment of the fishing fleet was not subject to the kind of output restrictions and other measures imposed upon the offshore boats' owner-operators. Several fishermen tried their luck in the inshore fisheries. In the early 1980s, the local fishing fleet comprised about seventeen to eighteen inshore cutters with up to 300 h.p. engines, nearly all of whose owners lived in Oudeschild. Approximately thirty men manned the boats, whereas the offshore fleet employed about 170 fishermen. In contradistinction to the state-of-the-art big beamers, these small boats were usually rather old – with the exception of a few new-built vessels. Inshore fishing provided an alternative route for fishermen who wanted to be independent but who did not have sufficient capital to acquire a beam trawler or did not desire to ply the North Sea. The inshore fleet was a hotchpotch of craft that targeted shrimp, shellfish, eel and sundry other species. Most firms also had mini-quotas for flatfish. Three or four boats specifically catered to tourists; they still took holidaymakers on shrimp-fishing trips, and with the tourist season's extension the importance of their commercial fishing diminished. Approximately ten vessels specialized in targeting shrimp. Entry to inshore fishing was not always easy, however. In the early 1980s, Wadden Sea shrimp fishermen had to have licences as the state adopted a policy of non-expansion. Several Texel shrimp fishermen initially encountered difficulties in obtaining such a licence. Even with a limited number of owner-operators, competition was fierce. There were no catch restrictions, but markets were soon saturated and shrimp prices were usually low. In the shrimp-fishing industry, it proved to be hard to arrive at mutual agreements to contain production, even after a Union of Shrimp Fishermen was established by the end of the decade. It became increasingly difficult to recruit crew for shrimp boats. Most deckhands desired a berth aboard a big beamer, where the share remuneration was soaring. Many were even willing to work as a substitute crewmember on the North Sea fleet.

In summary, the times of the Dutch offshore fishing industry's rapid expansion were characterized by reports of illegal fishing, underreporting of catches, grey and black trade circuits and inadequate policing and enforcement by the Dutch state. The monitoring of compliance with fisheries legislation was entrusted to the General Inspection Service, whose inspectors were assigned as special enforcement officers. They had full control, inspection and investigation powers. In the quota regime's early days, their land-based inspections were rather lenient – although the number of warrants they handed out would appear to indicate otherwise. Usually, however, these concerned minor offences. Moreover, the legal basis of Community law was initially unclear, and several Dutch District Courts referred fishery offences to the European Court of Justice stating prejudicial questions. It often took more than a year to obtain a ruling, and many fishermen believed this to be a sign that things would turn out right. According to the Fish Board chairman, Dick Langstraat, fishermen were of the opinion that 'having an elastic conscience offered a better future than to suffocate in one's own decency' (Subcommissie Visquoteringsregelingen 1987:196). It was a confusing time, when no one seemed to know how stringent Community regulations should be applied. In addition, there was considerable uncertainty concerning the extent to which other member states would enforce the rules. Initially believed to be born out of necessity, illicit dealings soon turned into a conventional mode of conduct. Auctions, processors and dealers colluded with rule-dodging fishermen so that illegally landed seafood could be marketed. The traders needed steady supplies throughout the year and the auctions' existence depended on a three per cent commission charge from fish sales. They feared losing customers to competitors if they abided by the law. Despite the quota regime, auctions reported record turnovers in the 1980s and invested heavily in new facilities. Even the Fish Board collaborated. It distributed landing forms that had to be completed 'truthfully', but it nonetheless demanded a 0.2 percentage charge on *all* landings, including grey ones. Consequently, many parties had stakes in continuing the illegal fishing game. However, forces from without threatened to thrust a spoke into the wheel.

Mushrooming Rules, Insecure Rights

Dutch fishermen who evaded or violated the regulations abroad faced tight surveillance. In the early 1980s, Belgian, British, German and Danish authorities caught many Dutch beam trawler skippers fishing in territorial waters or using prohibited gear devices. They were fined heavily. These events had an impact on the fishermen's public image at home. For a long time, they had been perceived as 'noble commoners', toiling to eke out a livelihood. Now they were increasingly viewed as irresponsible and reckless egotists plundering the sea's resources. Dutch beam-trawl fishermen gained the reputation of being unscrupulous brigands with an insatiable appetite for fish, blatantly and quite openly flouting the regulations. Pointing fingers was popular, self-reflection was lacking. Fishermen hailing from other member states were no saints either, but much to their amusement and relief their Dutch counterparts were in the limelight. Dutch fishing-industry leaders were convinced that this was inextricably linked with jealousy concerning the state-of-the-art Dutch beamer fleet and its economic successes (Schaap 1990:102-104). They alluded to 'protectionism'. Criticism was mounting and Dutch fishermen turned into scapegoats who allegedly scoured the North Sea illegally catching fish. Soon there was a call for stricter enforcement and punishment of those who evaded or breached the rules and regulations, especially after the introduction of the Common Fisheries Policy in 1983. The European Commission, sensing that catch registration and enforcement in the Netherlands were suboptimal, sent European inspection officers and demanded tighter control in 1984. This proved to be difficult in practice as the state's grip on the fishing industry was rather weak.

The Deputy Minister (*staatssecretaris*) who was responsible for sea and coastal fisheries from 1984 to 1986, Ad Ploeg, implemented a number of new regulations and increased monitoring and enforcement efforts. He regarded fisherfolk as industrious and generally law-abiding people, but he nonetheless warned them that they had become scapegoats in Brussels, and that the number of fisheries offences in the Netherlands surpassed that of other member states (Subcommissie Visquoteringsregelingen 1987:275). Although this fact could also be indicative of a lack of enforcement abroad, Ploeg was under pressure to take measures. As of 1985, fishermen needed licences to target species for which the European Community had established quotas. These transferable licences registered the engine capacity of a boat. Licence holders would only be allowed to increase their vessel's engine power when they garnered additional horsepower licences on the market. Such licences could only be transferred within specific fleet segments. The measure's aim was to fix aggregate engine capacity at the December 1984 level and as engine capacity licences represented market value, prices went up. The measure's weak point was that extant orders for new vessels or engines were exempt. Because word about the imminent rule had transpired, shipyards booked a torrent of orders for new cutters and new engines just before it became effective. It was the same old story of keeping up with the Joneses. To contain the overcapacity problem, Ploeg introduced a mandatory five-week tie-up scheme, but the measure was ineffective. Given the mismatch between landing allowances and catching capacity, the national quotas for several species were exhausted prematurely. The fishermen nonetheless continued to fish. Ploeg instituted a by-catch maximum of five boxes of cod for the flatfish fishing fleet. The General Inspection Service began checking vessels at landing sites more thoroughly in an attempt to do away with 'grey' and 'black' trade circuits. Tensions between inspectors and fishermen mounted. Ploeg threatened to begin Navy inspections of vessels at sea.

Fishing industry leaders deemed the stricter control policy and the wheeling and dealing of establishing total allowable catch shares at European level 'nauseating' (Borghouts 1991:67). However, despite additional measures, fishermen continued to circumvent rules, which created further tensions between them and state institutions.

Texel fishermen ardently discussed how to deal with the tightened management regime. In April 1985, Progress through Unity's chairman Bert Weijdt emphasized that a united stance was urgently needed to cope with 'the dark clouds on the horizon'. In order to successfully demand higher quotas, 'all those earning a living on the North Sea should make a front together with the dealers and processors and act as a single force. Any friction between interest groups or associations, any threat with fission from whatever group will weaken our strength'. The fishermen rightly feared that the national quotas would be exhausted early. This expectation brought about worries among the deckhands, who wanted to know what they were up against. The Texelians suspected that 'Brussels' was playing foul and that the Community bureaucrats were placing all the blame for fish stock problems on Dutch fishermen. In their view, the real problem was that individual entitlements were extremely insecure. Several Texel skipper-owners had been investing in acquiring additional landing rights, but regardless of whether their uptake had been possible or not, fishing opportunities would be closed once the national quota for a particular species of fish had been exhausted. 'Look, we have bought extra quotas at great expense, so we want to be able to land what we are entitled to. The authorities must enforce the rules,' said an Oosterend owner-operator. There was thus a serious mismatch between the individual entitlements and national obligations towards Brussels. The problem was aggravated because new big beamers continued to be launched, often with little or no landing rights. The majority of the Texel flatfish fishermen favoured group quotas, whose uptake the state should protect. Pooling individual landing rights would enable quota transfers within the group so that those who had used up their individual quotas could rent quotas from other group members who had not exhausted their entitlements. Urk fishermen had already pooled their individual quotas in a group under the aegis of their Producer Organization. Although many Texel skipper-owners expressed doubts about the effectiveness of organizing this way, they nonetheless developed plans to establish a group together with fishermen from neighbouring Den Helder. The structure of the Den Helder fishing fleet was similar to Texel's, with owners holding comparable landing rights. Five Texel skipper-owners said they would stay aloof and refuse to cooperate, among whom the local Fishermen's Union members. However, negotiations had hardly begun when well before the end of the year (1985) the uptake of plaice exceeded the national quota and the state stipulated that fishermen must refrain from landing any more plaice. The Texel skipper-owners were determined to steer a head-on course and continued fishing. So did fishermen from other communities. Despite putting up an occasional show of power, state institutions again failed to effectively enforce the ban.

Ploeg was not amused and demanded that flatfish fishermen work on fishing plans for 1986 and arrange group quotas, claiming that he would only negotiate with statutory groups. The Texel and Den Helder fishermen then resumed their talks about establishing a group. All skipper-owners now voted in favour and group regulations and by-laws were formulated. The whole matter was put on hold because Ploeg – whose efforts to contain the fishing industry seemed to have been in vain – did not return in the new coalition government following the 1986 general elections. He was blamed for failing to control the fisheries, while his senior, Minister of Agriculture and Fisheries Gerrit Braks, resumed office without any complication whatsoever. This time, however, he assumed full responsibility for the fisheries. Braks immediately had his work cut out for him. As a consequence of a drastic reduction of sole quotas in the late 1980s (see Appendix D) and faltering enforcement, illicit landings increased. A plethora of new measures, including an increase of legal mesh sizes to 80 millimetres, added to the management regime's complexity and made for more rules – rules that had loopholes and that could be circumvented or flouted. The fishing industry continually made headline news about quota overshooting, early closures, clashing fishermen, illegal fishing, court cases, and clashes with fisheries inspectors. European Community pressure on the Dutch fisheries authorities to do a better job was mounting. It reported that in spite of the catch registration system, the systematic monitoring of landings, the expanding enforcement staff and the significant number of prosecutions, the Dutch record of curtailing quota overshooting was the poorest of all member states (Vervaele, Ruimschoten and Widdershoven 1990:107).

In 1987, the controversies pertaining to the fisheries climaxed. Braks and his Ministry of Agriculture and Fisheries were critiqued for failing to enforce the law and even condoning illegal fishing. This led to an enquiry of a parliamentary subcommittee into the Ministry's role concerning over-quota landings. It concluded that the Ministry's civil servants had been passively – and in some cases even actively – involved in dodging the European Community's quota measures because they identified the fishing industry's interests with the national interest (Subcommissie Visquoteringsregelingen 1987). This attitude had not stimulated 'loyal' abiding by Community obligations. Knowledge of the illegal practices was widespread in the Ministry's Fisheries Directorate, but its civil servants turned a blind eye. Other parties were complicit. The state's condoning of over-quota landings gave fishermen the impression that they could continue their illegal practices without serious repercussions, which they then obviously did. The fishing industry – including the Fish Board – resented hardly being involved in the policymaking process. They sensed that many of the measures the state announced were little more than paper tigers, bringing about mistrust concerning the rules' efficacy. Auction managers helped fishermen to logistically and administratively get rid of illegal fish, because their existence depended on an auction levee. They feared that the fishermen would go elsewhere if they did not assist in maintaining this 'second' market. Fish processors and traders were also involved. They needed steady supplies, and with higher than legally allowed landings, prices would be low, turnover high and employment ensured. Local authorities often turned a blind eye to what was going on.

Even when fishermen had been caught fishing or landing fish illegally, judges usually sentenced them to low and tax-deductible fines. Failing to deprive them of commercial gain, such sanctions did not deter skippers from continuing illicit practices. 'The last haul for the judge' became a popular expression. The flatfish fishermen consequently considerably overshot their quotas, leading to early exhaustion of national quotas and subsequent premature closure of the fisheries. Because entitlements were insecure, a derby-style catch-as-catch-can attitude surfaced. Obviously, as individuals the fishermen had much to win by contravening the rules if they succeeded in not getting caught red-handed, or after the fact on the basis of their administration. Many were heavily indebted as a consequence of recent investments in new vessels. Mortgages and loans had to be redeemed. Overshooting their individual quotas meant that they could keep their incomes up to the mark and stay afloat economically. This was especially the case for the category of fishermen with big beamers but small quotas. There was peer pressure aboard these vessels to fish more than their owners-cum-quota-holders were entitled to. As share fishermen, deckhands had an interest and a stake in attempting to obtain revenues that were as high as possible. They would benefit just as much as owner-operators. Therefore, the partnership contract usually included a term that stated that fines would be subtracted as 'costs' from the crew share. Since illegal landings had a price-undermining effect, the fishermen had to land even more to maintain gross proceeds at the level they desired. Of course, this in turn only aggravated the problem. Until the fishery was closed, fish swamped the market, usually yielding low prices. The fishermen who had not yet fished their individual quotas suffered. With this experience of being entrapped in a prisoner's dilemma, the race for fish continued, stimulated by lenient enforcement, ample opportunities to contravene the rules, and a strong demand for flatfish abroad. For a long time, Minister Braks had stubbornly insisted that the official fisheries statistics were 'absolutely accurate', and had only admitted to the subcommittee that the Ministry had known about the illegal circuit. In his view, knowing about it did not equal condoning it (ibid.:322). After his humble mea culpa, Braks politically survived the parliamentary debate that followed publication of the subcommittee's report. The head of the Fisheries Directorate was forced to move department, but the vast majority of MPs gave Braks 'the benefit of the doubt' when he promised to adequately deal with the problems in the fishing industry. 'I am of the opinion that extant and future measures will cope with the problems', he maintained.

Indeed, a host of new regulations and restrictions were implemented as of 1987, including a days-at-sea regulation and compulsory registration and check of all landings (van Vliet 1998b:70). The days-at-sea regulation sought to adjust fishing effort to the available quotas and to prevent the need to close the fishing season early (Davidse 1998:59). A team of 120 inspectors began to systematically monitor fish landings for which specific places, times and conditions were set. Fines for violations of the rules became much stiffer, and some fishermen, traders and auction directors were even incarcerated for non-compliance with the law. As an additional technical measure, the maximum beam length for the flatfish fishery was fixed at twelve metres for vessels over 300 h.p. and to four-and-a-half metres for those under 300 h.p. At long last, Braks also announced that there would be a maximum main engine capacity of 2,000 h.p. for newly built vessels. By then, there were already scores of vessels with more powerful engines, while orders for quite a few new boats had been lodged just before the deadline. The same happened when it transpired that the general investment premium (*WIR premie*) was going to be abolished the next year. Instead of containing the overcapacity problem, the 1987 policy decisions gave an unintended boost to expanding aggregate engine power, which peaked at almost 600,000 h.p. the next year. The aggregate engine power of Texel cutters also reached its zenith in 1988, when it amounted to almost 55,000 h.p. Several of the island's beamers carried engines of over 3,000 h.p., the most powerful one having a 3,600 h.p. engine. Elsewhere in the country, beamers with stunningly powerful 4,000 to 4,500 h.p. engines began to be launched. By then, a new-built cutter required the sum of seven to seven-and-a-half million guilders. The flatfish fishermen's optimism had received a fillip in the mid-1980s, when financial results were beyond expectation, and they did not worry about operating costs. Fuel prices amounted to twenty-five cents per litre in 1988, which despite inflation was considerably less than those paid in 1980, when the price had briefly peaked at sixty-four cents. Although after 1987 the Dutch cutter fishing-fleet's size steadily decreased, the category of big beamers of 2,000 h.p. and more continued to grow. In addition, a class of so-called Euro-cutters – 300 h.p. boats built specifically for use within the 12 nm zone and so named because this vessel type was developed with EEC subsidies – quickly arose (see Appendix C).

As one of its manifold measures, the state created a quota reserve of five per cent to cover for overshooting of individual quotas and to allow others to fully use their rights. Here was the crux of many problems. Individual quota entitlements were extremely uncertain because once the national share of the total allowable catch for a species had been exhausted, the fishery would be closed regardless of whether fishermen had exercised their individual entitlements or not. This weakness proved to be particularly calamitous for round-fish fishermen. The beamers' cod by-catches consumed a large part of the national cod quota. This was detrimental to the specialized round-fish fishing fleet as it led to early closures of cod fishing. In 1986 and 1987, for example, it was closed as early as May. Although they were only allowed to land five boxes of cod per week, the beam-trawl fishermen claimed that they could not avoid catching cod and it went against the grain to throw the fish back into the sea because it would certainly die. They regarded it an example of a perverse policy. For the twenty holders of a specific cod-fishing licence (a 'cod document'), which was implemented in 1987, the fishery seemed no longer viable. As a Katwijk cod fisherman exclaimed:

> The beamers have brought shit upon us. They have ruined the fishery. Over-fishing is getting worse every year. They come up with more horsepower all the time and they only want more, bigger and ever bigger. This way the one firm brings down the other in disaster. We are about to become our own hangmen (*de Volkskrant*, 20 June 1987).

His words are symptomatic of the rifts and ruptures that had come about in the expanding fishing industry. They caused new fissions in the sector's organizations, including the Federation, which failed to adequately represent and defend the small owner-operators' interests. In November 1987, round-fish fishermen blocked the IJmuiden sea sluice gates to protest the flatfish fishermen's dodging of the by-catch measure. Minister Braks had to admit that the individual quotas 'would appear to be insufficiently secure' (*de Volkskrant*, 21 November 1987), but nonetheless the government summoned police troops to end the blockade. Early in 1987 Braks had introduced the measure that flatfish fishermen should stay ashore a number of weeks to give cod fishermen the opportunity to take up their quotas. This decision was successfully contested in the Supreme Court, and the Ministry was forced to renegotiate with the fisheries organizations. The burgeoning rules and regulations led to increasing litigation and stifled the fishermen's time-honoured strategy of gear and species switching. For example, the cod document was only allocated to fishermen whose incomes between 1984 and 1986 had depended for at least sixty-five per cent on catches of cod and haddock. The measure caused upheaval among the three Texel round-fish fishing firms that also pursued herring, because they did not meet these conditions. Another hotly contested rule was that fishermen's individual over-quota landings of the previous year would be subtracted from their current quotas.

Along with the 1987 measures, a voluntary decommissioning scheme was adopted. Owner-operators who decommissioned their vessel only returned their engine capacity licence and could sell their individual quotas to other boat owners. Vessels were not necessarily scrapped, but could be sold, so that capacity did not diminish but was often merely relocated. The option to sell individual transferable quotas at high prices was intended to facilitate exit decisions. Early in 1987, sole quotas sold for twenty guilders per kilo. This price tripled by mid-year. A 1,500 h.p. vessel had a mean individual sole quota of seventy metric tonnes; the mean value of landing allowances for sole thus increased from 1.4 to 4.2 million guilders. However, it was still unattractive for owners to have their vessels decommissioned and to leave the industry, while the prohibitively high costs of acquiring quotas made it extremely difficult to match catching capacity with entitlements. Despite a fifty per cent contribution from the European Community and a twenty-five per cent contribution from the national government, the state raked in a disproportionate amount of decommissioning money through its fiscal measures. Forty per cent of the compensation had to be paid in tax at once. In 1988, this amounted to 2,160 guilders per GRT, whereas the state's decommissioning subsidy was only 1,800 guilders (Schaap 1990:49). In order to deal with the overcapacity problem, in 1988 ten-week and eight-week mandatory tie-up schemes were applied to the big and small cutter sections of the fleet, respectively. All the old 'irrational' fishermen's motivations to continue with their occupation still applied and the incentive to decommission was meagre for those vessel owners who had few prospects of finding another job. The state's Development and Decommissioning Fund (*Ontwikkelings- en Saneringsfonds*) was good at achieving development but not at decommissioning. Moreover, there was no social scheme for non-propertied deckhands who would become unemployed due to decommissioning. By this time, slightly over 3,000 fishermen manned the Dutch cutter fleet, 500 more than in 1975.

On Texel, the occupational community boasted 200 to 220 fishermen in the 1980s. About eighty-five per cent of them worked in the offshore fishing fleet. Most big beamers were 'overstaffed'. They had at least one and often two or even three more than the legally required crew of six, so that crewmen could alternate having a week off. Given the situation of overstaffing, the expectation was that it would be difficult to find a berth for deckhands whose skippers opted to decommission their boat. Owner-operators whose vessels were decommissioned would be the beneficiaries of the decommissioning subsidies. Deckhands were not entitled to any of the money. Nor would they – as self-employed entrepreneurs – receive state unemployment benefits. If they owned property they would have to sell it before being eligible for the dole. The issue brought about considerable unrest among crewmembers, who were already dissatisfied, having just had to swallow new changes in the share percentages, which were commonly fixed at a fifty-eight per cent boat share and a forty-two per cent crew share. In a special meeting of Progress through Unity in June 1989, the deckhands and owners discussed the matter. In order to avoid fission, they arrived at a solution. The owner-operators decided that they would use part of their contribution to the Decommissioning Fund to compensate those deckhands who became unemployed and could not find work aboard another cutter. Again, the local association proved its worth.

Faced with all the restrictions coming their way and in an attempt to maintain at least some say in their businesses, the Texel owner-operators resumed their talks on the formation of a group, an issue that had been postponed for some time following the 1986 change of government (see above). By the end of 1988, fifteen Texel skipper-owners decided to establish a local group. They pooled their individual quotas, facilitating swaps, leasing and selling of quotas and quota uptake within the group. The group would only function for one year, however. It was abolished in January 1990, when Braks disallowed groups in an attempt to tackle difficulties with catch registration and quota over-fishing. Nine out of ten groups had overshot their entitlements. The Texel group was the exception to the rule – on paper at least. Some of the larger quota holders had refrained from joining the local group, having preferred to take up their individual entitlements. 'It really was a mess,' a skipper-owner told me about the group: 'It was entirely non-committal.' As soon as the group was disbanded, discussions ensued about linking individual rights and horsepower with allocations of days-at-sea. The problem was that every boat was assigned the same number of 'base days' (*basisdagen*) and that firms had to request more in order to be able to fully exercise their quota entitlements. Some smallholders successfully applied for extra days. This went against the grain for large rights holders, who suspected that the former would intentionally use these days to overshoot their quotas. Most Texel owner-operators had large quota entitlements, and therefore they preferred a system that would link up fishing time with fishing rights. However, the small quota holders vehemently opposed such an arrangement, particularly those who had vessels with powerful engines. It proved to be yet another matter that nourished disharmony in the local fisheries arena. However, on another front, success was in the pipeline. The Progress through Unity board members began negotiations with their counterparts in the Den Helder fishermen's association to take over the municipal auction in the naval town. It was privatized in 1989 when owner-operators from Den Helder and Texel as a joint venture established the Cooperative Fish Auction Den Helder/Texel. Forty-five firms chipped in 1,000 guilders each. About eighty per cent of the turnover was henceforth realized by the home fleets.

A new chairman of the local fishermen's association, Bas van der Beek, was to lead the organization into the 1990s. He was an ardent advocate of uniting forces. In his maiden speech to the association's general meeting in February 1989 he expressed his apprehension regarding the fact that each sub-sector in the fishing industry preferred to safeguard its own interests. He rightly pointed out that 'the danger is that the opponent [the state] will divide and rule in negotiations'. By repeating his message over and over again, he could at least attempt to close ranks at the local level. Progress through Unity would have to speak with one voice in submitting proposals to the national fisheries organizations and state institutions. The most pressing topic on the association's agenda was how to enable quota holders to fully exercise their entitlements; hence the idea of linking individual quotas and horsepower with days-at-sea. Most board members believed that such a measure could solve the problem of premature exhaustion of the national landing shares since 'smallholders' with big boats would not be able to fish as often as large rights holders. The entitlements of owner-operators would then be better protected, quota uptake would be ensured and fish prices would improve because of a more even distribution of landings across the year. The weaker firms – of which the island had few – could either invest to match their allowances with vessel capacity or decommission. They could sell their individual quotas to fishermen who wanted to reinforce their position. It all seemed quite simple and fair, but it took time to convince all owner-operators, particularly those who had below-average quotas and mixed fisheries firms. For reasons I will discuss later, the Texel fishermen would soon become convinced that such a management system might be just and efficacious. However, they did not dominate the national fisheries arena, and skipper-owners from several other fishing ports rejected a system of days-at-sea related to landing entitlements (Connolly 1997). They feared that such a measure would stifle their operations as many of them were smallholders.

For a part of the Texel fleet, seasonal herring fishing was traditionally important. It was combined with beam trawling for flatfish or round-fish fishing. Four big beamers and three smaller stern trawlers had herring landing allowances, but these were not individual quotas. To counter the derby race for herring, in 1989 the allocation of landing allowances was changed. Based on historical rights, the national share of the total allowable catch for herring was first divided between the company-owned offshore fleet and the fleet that was owned by a score of family firms. The share of the latter was subsequently allocated individually to rights holders following lengthy negotiations, which were supervised by Fish Board and Fisheries Directorate representatives. The allocation was based on horsepower categories and, later, gross tonnage. Every year, this resulted in an agreement on the annual quota each cutter could land. The beamers would, for example, receive 6.84 per cent of the national herring quota; the stern trawlers 2.65 per cent. In addition, there was a days-at-sea regulation. Days used for the three-month herring season were initially subtracted from days-at-sea for beaming, regardless of flatfish entitlements, which caused considerable problems for the big beamers' owner-operators. They would be assigned additional days when they could prove that it was impossible to take up their individual sole and plaice quotas. In the herring fishery, too, quotas and sea days usually fluctuated. For instance, a firm operating a big beamer had landing rights for more than 2.5 million kg in 1987 and 1.1 million kg three years later, while the assigned days-at-sea declined from ninety-two to forty. With few competing dealers and processors – most of them had not survived the herring ban – herring was not auctioned but sold directly to the processing firms, at prices negotiated in advance. It proved to be difficult to reach agreements, not only regarding prices – which were extremely low – but also concerning an even distribution of herring landings across the year. In 1989, two beamers that pair trawled could land a maximum of 6,000 boxes of herring per week. With a prefixed price of eighteen guilders per box, the gross revenues would amount to 54,000 guilders per vessel per week at most. This was hardly deemed cost-effective, as substantially more could be earned in the mixed flatfish fisheries. Skipper-owners nonetheless continued the herring hunt to retain their rights, hoping that the herring market would recover in the future. However, this was not the case (see Chapter 6 and de Jonge 2005). Texel's offshore fishermen would therefore increasingly come to depend on flatfish.

However, the difficulties in the flatfish fishing industry also seemed to multiply. Despite all the measures that had been taken, reports about Dutch fishermen who evaded or breached the law again figured prominently in news bulletins in 1988. Politicians and state officials obviously considered non-compliance a serious offence, while fishermen regarded it as a survival strategy. They increasingly felt entangled in a net of red tape. In October 1988, the Zeeland port of Vlissingen became the scene of occasional rows and riots. It was the home-port of Vlissingen and Arnemuiden beamers, which represented ten per cent of the national catching capacity but only four per cent of landing entitlements. This discrepancy indicates why these fishermen strongly depended on black and grey fish landings to stay in business. They clashed with General Inspection Service officials who attempted to check for illegal landings and police trying to protect them while doing their jobs. The skirmishes came to a head when an inspection team was awaited by furious fishermen and their friends and families – including women and children. They set fire to an Inspection Service van and the inspectors were forced to make a swift retreat. The atmosphere had turned ugly. In subsequent weeks, platoons of special police troops were deployed at the dockside when the score of big beamers from Vlissingen and Arnemuiden discharged their catches. The fishermen from these fishing ports in the southern part of the Netherlands were branded as 'professional criminals'. Their compatriots from other communities disapproved of the events as they rightly feared they would also be stigmatized. Texel fishermen firmly condemned the incidents: 'In a constitutional state, it is intolerable to express discontent in this manner,' said Progress through Unity chairman, Bas van der Beek. 'The events have damaged the reputation of fishermen, even though the vast majority dissociate themselves from these actions' (*Texelse Courant*, 25 October 1988). Ben Daalder, who was about to step down as chairman of the corporatist fishing industry organization (*Visserijschap*), pointed out in several interviews that the state ought to tackle skipper-owners who had few or no landing allowances for species to which quota restrictions applied. The two national fishermen's associations also believed that decommissioning measures should target the owner-operators with insufficient entitlements to run a firm within the limits of the law.

Mutual trust in the national fishing industry had completely withered away. Fishermen from different communities accused each other of being the culprits. For example, Texel owner-operators firmly believed that their Urk and Den Helder colleagues were tampering with the days-at-sea registration. They were also convinced that some officials were leaking information to the Urk fishermen, so that they could avoid being caught in the act when they illegally discharged their fish. A lack of trust thwarted cooperation and was an obstacle to the emergence of mutual agreements on collaboration. Solidarity was also undermined. When asked to contribute financially to a decommissioning fund for round-fish fishermen, the Texel skipper-owners responded by saying that they had already to discard precious cod so that round-fish fishermen could take up their quota; they were not inclined to come up with money as well. For the Dutch government, the fisheries sector was a terrible headache. Its reflex was to further tighten regulations and to come up with new legislation. Most of the Fisheries Department's leading staff had a background in law. Their legalistic approach led to a stacking of rule upon rule, in an attempt to grapple with the problem of non-compliance. They erroneously ignored the question of whether the torrent of regulations were comprehensible, equivocal, efficacious and enforceable. This was clearly not the case and the process was a clear example of a bureaucratic catch-22: complexity increased, coordination diminished and rules were contradictory. In addition, enforcement was difficult. In 1989, no less than 1,100 warrants were issued, more than forty per cent of which had to do with inaccurate catch registration. Despite this show of force, flatfish landings continued to exceed individual and national quotas, and as a consequence of his failure to contain this problem and his alleged misinformation of Parliament, Braks was forced to step down on 19 September 1990.

Anxiety, Encapsulation and Ambiguity

When Braks had to leave office, I was in the midst of conducting field research on Texel. The local fishermen were not pleased with the situation, arguing that some of the Ministry's civil servants should have been kicked out. They feared that monitoring and policing would be tightened even further. After Braks's resignation and with the growing risk of heavy fines for non-compliance, many were anxious that their situation would deteriorate as a consequence of draconian measures that would almost certainly descend upon the fisheries sector. Braks's successor, Christian Democrat Piet Bukman, would not want to run the risk of being censured by the House of Representatives for incompetence in harnessing the fishermen. In addition, the fishermen comprehended that they had lost control over their individual businesses and the fishing industry as a whole. Moreover, prices were far from optimal due to grey and black landings and the premature closure of flatfish fisheries that invariably followed the national quotas' exhaustion. Flatfish catches were excellent in 1990, but with a saturated market fishermen had to land even more to keep their earnings up to the mark, but doing so meant exceeding their entitlements. Therefore, most Texel fishermen became more and more inclined to invest in landing rights to match their vessel's capacity. Early on, Texelians had been advocates of individual quotas, whose uptake should be ensured and protected, and a system of days-at-sea linked with landing rights. They loathed the fact that big beamers with small entitlements had enjoyed unrestricted access to the fishing grounds. However, the strategy of acquiring additional quotas would only be fruitful if they were secure, which was still not the case. (I will return to this in the last section of this chapter.)

Texel fishermen had certainly not been sea-green incorruptible. They were involved in illegal fishing and fish landing activities perhaps as much as their counterparts from most other fishing ports. Many got away with it without being caught. It did create new worries though, as a retired Oosterend skipper-owner (born in 1921) explained:

> Look, when you had a nice catch in the old days, you landed and sold it. Currently, you wonder how to get rid of the fish. You have to sell the fish, but there are quotas. It is mentally demanding. Everything used to be free, but the freedom is gone.

The expression 'to get rid of' ('*wegwerken*') was a commonly used euphemism for illegal landings. Occasionally, the owner-operators would have to appear in court for overshooting their individual quotas. They were also privy to the way the Den Helder auction dealt with 'consigned' fish, which could not be traced by inspection officers. The 'second market' – which in addition to 'consigned' fish included sole and plaice landed under the code name of fictive 'sea fish' to which no catch restrictions applied – provided a lucrative channel for Texel and other fishermen's over-quota landings. In addition, black landings were fairly common.

As in other fishing communities, many could not resist fishing within the 12 nm zone with big beamers, and the use of blinders was also widespread. Their utilization was believed to be necessary to catch marketable sole. Sole larger than the legal minimum size can get through eighty-millimetre mesh; hence the temptation to use blinders or smaller-meshed nets. Several skipper-owners were vigorously opposed to utilizing them. 'Blinders are detestable. You'll catch more than you're entitled to land and you'll land smaller fish. You have to get rid of those fish and it will undermine prices,' said a shore skipper. In several meetings I attended, the leaders of local and national fishermen's associations called on the owner-operators to behave responsibly and ban blinders. Fishermen's Union leader Johan Nooitgedagt begged on one such occasion in February 1990:

> Don't use those things! How can we arrive at compromises with politicians when the next day they are used again? We just cannot explain that to politicians or in Brussels. You'll burn your own fingers! Don't do it. It has to stop, or else we won't be able to achieve anything at all any more.

In 1990 meetings of the occupational community of Texel fishermen there were occasional pleas and promises to ban the blinders. This seemed to be a discursive reassurance that '"we" don't indulge in using illegal gear devices'. The fishermen acknowledged that should it become evident that many were nonetheless still using them, the self-discipline would evanesce. Later that year the issue was brought up in a meeting again by an angry young skipper-owner: 'I'm so furious... There are perhaps five, six or seven [skippers] who use blinders. I thought Texelians wouldn't be so stupid, but shit, they are.' When a few months later a skipper-owner in a meeting confessed to using blinders 'because I also want to take up my quotas', a colleague burst out in anger and shouted: 'That's how we go down the drain. We can't even stand by agreements on Texel. They should put a GIS [General Inspection Service] inspector on every cutter!' Later, an owner-operator and his father told me that they didn't blame the culprit: they themselves also used blinders and they were convinced that the outraged skipper was doing exactly the same. Pointing fingers and looking for scapegoats had become a popular pastime.

Several Texel skipper-owners admitted that a number of government measures made sense, if they could only be enforced equally. In their view, here was the crux of many problems. In an interview with a skipper-owner and his wife from Oosterend, both born in the early 1930s, she said:

> You can never strike agreements because no one will stick to them. They will always think they own the fish. In the days when there was little policing, everyone messed with the rules, and those whose messed with them were rewarded. If you abided by the rules, you burned your own fingers.

Her husband nodded, adding:

> Well, some rules are incomprehensible and you have to run a firm. Look, if you have huge quotas and you will get the same number of days-at-sea as firms with half your quotas, that cannot be right, can it? They will have to land their catch illegally. [...] If the fisheries were entirely 'clean', in the sense that everyone could be certain about what he could catch, the fishermen would not have a problem with that. You should be absolutely sure that the fish is not snatched away illegally. That's the whole point: You can never be sure!

The differences of opinion and the differences between words and actions – some owner-operators said they opposed illegal fishing but in fact indulged in it – had an impact on the mutual relationships between local fishermen. Tensions rose, and this was also felt in the wider community. The skipper-owner's wife quoted above explained that even in church this was the case:

> If there are certain fishermen who get rid of their fish, and they are on the Church Council, this will breed ill feeling. Every year we have a thanksgiving for the harvest, a shared meal of bread and fried fish. This year, my daughter did not want to go. She said: 'They can eat their own fish.' They sit up front in the church, but still get rid of fish.

The vast majority of my interlocutors deemed capping engine power an example of a sound measure. 'Without it, we would certainly have ousted each other from the fishing industry,' an owner-operator voiced the dominant opinion, adding:

> Fishermen are ambivalent about state intervention. They always think about their short-term self-interest, not the long-term general fishing-industry interest. There's a change of mentality now. Most Texel skipper-owners have invested in landing rights. Currently, if there's an early closure in sole fishing, they'll say: '*They* [that is, others] have caught *our* quota.' But that's nonsense. Texelians also have messed with the rules; I won't close my eyes to that.

These views are indicative of a fundamental ambiguity. The Texel fishermen deemed regulations necessary, but they would not abide by them unless the state systematically monitored and enforced them. Despite this stance, the older ones nostalgically recalled the days of 'much liberty and few rules'. They were of the opinion that the tighter measures and enforcement stifled their autonomy. A few examples I obtained during my first stint of fieldwork illustrate this.

In the shed where he is mending his nets, an Oosterend beam-trawl fisherman (born in 1946) contemplates the transformations in the fishing industry. 'It used to be a fine occupation, an excellent occupation... It still is, but with all the rules and regulations it has become less and less attractive.' He cherishes memories of the days when fishermen could still freely roam the North Sea in pursuit of prey. Although he is only forty-four years old, he rarely joins the crew of the beam trawler he co-owns. He prefers operating as a 'shore skipper', arranging all sorts of businesses for the family firm while staying ashore. Recently, his twenty-two-year-old son became the other part-owner and skipper. 'These young lads know how to deal with the tension of inspections,' he says. 'They have grown up with them, whereas we were used to freedom, and then one day we suddenly were confronted with all sorts of regulations and inspections.' He explains that he had a bad time learning to deal with officialdom and tightened enforcement:

> Some time after the introduction of quotas I had a nervous breakdown. After every week at sea I needed a week to recuperate. I couldn't stand the inspection officers. When you are dead tired returning from a week's trip at sea and they are awaiting you ashore, you get pretty fucked up. Don't forget

> that if you haul the nets for the last time at, say, one o'clock Friday morning, it is still hours of work before you're back in port. You're completely knackered and then a guy like that comes aboard. Regardless of whether he fines you or not, he will receive his pay-check anyway.

With a tinge of embarrassment, he continues:

> We were in port once and while I was busy repairing a burst pipe in the engine room, an inspection officer shows up saying: 'Skipper, I will have to fine you.' I was half tucked under the pipe with a huge pipe wrench in my hand and I got so angry that I nearly bashed his head in. I've never seen anyone get away so quickly! Of course, that was completely wrong and in hindsight I became aware of this. ... I apologized to the man later on, but I also told him that he shouldn't have come into the engine room just like that. ... I just couldn't stand it anymore and I stayed home on sick leave for some time.

With the state attempting to wield its authority, Texel fishermen felt that inspectors encroached on what they regarded as their private space. To fishermen, a vessel is not merely a material object; it is a piece of equipment that they perceive as having special characteristics and vagaries, a body and a soul. In other words, boats are saturated with symbolic meaning. Fishermen can wax lyrical about the qualities and aesthetics of their vessels and develop an emotional bond with their cutters. They depend on their boats for their survival, and many have encountered situations in which this was unequivocally brought home to them. In many fisherfolk homes, paintings and photographs of vessels that have been or still are family property decorate the walls. These vessel genealogies often take as prominent a place as family photographs, and show the analogous importance of age. Owners in particular connect their own life history with launches of new boats and they remember all the important events in the vessel's span of life: when a new engine was installed, when major alterations were made, when and where bumper catches were taken, who were the crew at particular points in time and so on. Many owners give their vessels names referring to their wife, father or mother, a grandfather or grandmother, a brother or a sister, a daughter or a son. The tradition reflects the significance of family ties in fishing identities. The photo shoots of the various stages of the construction of a new boat resembles that of a wedding. There is a baptizing and a naming ceremony for a new vessel, and a minister or a priest says a prayer for the well-being of boat and crew and usually donates a ship's bible. On its maiden voyage, there is a rite of passage when the exchanging of the shipyard's flag with the owner's marks the official transition into new hands. A new cutter arrives in a blaze of glory and its inauguration is celebrated with a party on board. Scores of people are invited to admire all the boat's qualities and to wish 'her' and her owners well. A family member or a long-time crewman addresses the owner(s), usually going into the family firm's history and stressing the importance of its continuity. Afterwards, there will often be a lavish dinner for invitees. The local newspaper extensively covers the festive event, presents a detailed description of the boat's qualities and publishes several photographs. Many fishermen – owner-operators *and* crew – take profound pride in their vessel and experience her as an extension of themselves. Boats are a referent of social status and self-respect and a dominant symbol of identity. The accommodation and the bridge are carpeted and fishermen only enter these domains on stocking-feet to keep things neat and clean. That is the reason why the uninvited boarding of a vessel by strangers can lead to outrage: it is experienced as an intrusion of the private – deeply symbolic – domain and even the self. It partly explains why inspection officers were usually not really welcome aboard.

Inspectors are renowned for their nit-picking of insignificant infringements. Such actions have a psychologically stressful effect on fishermen, especially if they perceive hefty fines to be unfair. For instance, the beam-trawl fishermen resent the subjective nature of measuring the mesh size with hand-held wedges. Depending on the inspector, more or less force can be used to push the wedge into the mesh and this may lead to sizes either being deemed correct or too small. The stricter regime does indeed have an emotional impact upon most fishermen, particularly the older ones. The owner of one of Texel's most powerful beam trawlers, a man from Oudeschild, born in 1940, contends:

> When the quota regime was implemented, and enforcement got tighter, we had a tough time. Sometimes I didn't go to the harbour for weeks. People my age couldn't handle it. I don't like having to mess with the regulations. When we had caught too much fish I would almost pee my pants upon arriving at the fish auction. Those young guys don't give a damn though. They have no problem with landing sole or cod illegally; they have grown up with it. It's not right, though. It makes me nervous just thinking about it.

When conducting fieldwork in 1990, I frequently heard such lamentations from the generation of skipper-owners who had had first-hand experience of the introduction of quotas, the burgeoning national and supranational regulations of the fishing industry, and the increasingly strict enforcement of rules and laws. In this connection, they often refer to 'before' and 'after' and this perceived watershed colours their occupational world. They complain that 'the heydays are over'. Younger skipper-owners and deckhands do not perceive such a watershed. Despite tight regulatory regimes and policing, they appreciate their métier, even though the measures have had a strong impact on the daily lives of fisherfolk and fishing communities. The regime change has also put a mental strain on family life, as the sixty-year-old wife of an owner-operator who had recently sold his cutter makes clear:

> Just before he [her husband] sold the firm, I told him: 'If you go on with this mess, I'll leave you.' I mean, you get to a stage in your life that you can no longer handle it mentally. I mean, all the fuss and bother and situations, you just can't cope. It backfires on the family. Although I have never done the administration and things like that, I have always felt committed to the firm. But at a certain moment I thought as far as I'm concerned you just blow up. I hated it at some stage, that's the point it had got to. The continual moaning on this or that new measure... Occasionally, I talk to fishermen's wives and listening to them I think to myself, I'm so glad I got rid of all that, honestly. They return home and they are all nerves and misery.

Leaving aside the more general problems fishermen have with regulations and their enforcement, these interview fragments point to the following. Firstly, there would seem to be a difference between the older and younger generation with respect to non-compliance and 'cheating'. The younger generation of fishermen is supposedly less nervous about it than their older counterparts. The tensions and mental strains also affect family life. Young fishermen began going to sea when quota regimes were already firmly in place and so they are used to having to deal with them. Consequently, they do not continually compare the 'present situation' with the 'era of unlimited freedom'. The second point concerns the perception that the work aboard is extremely tough and tiring. Beam-trawl fishermen are at sea for four or five consecutive days, and they work continual shifts of for example two hours on, one hour off. The work rhythm is determined by the nets' shooting and hauling and the gutting of the fish. In between there is time for coffee and a nap, but once the bell rings to signal that the gear will be hauled again, the fishermen have to be on deck in minutes. The only 'breaks' from this monotonous and exhausting routine are the meals. Fatigue makes it difficult to strike the right tone when dealing with officials. Thirdly, owner-operators often retire at an early age; they still have an income from the ship's share. It is not unusual for them to become 'shore skippers' when they are still in their forties. Their role is to manage the firm from ashore, and only when the need arises to occasionally accompany the crew. Although they like the occupation, they also perceive it as vigorous. Fourthly, there is the pride skipper-owners take in their vessel, a pride that appears to be put down by the growing criticism fishermen meet. As independent entrepreneurs they have attempted to continually modernize their firm, and for a long time the Dutch state and the European Community supported them in doing so. Currently, however, many people find fault with their behaviour. Lastly, it is clear that 'outsiders' who interfere with the fishing industry –including inspection officers, regulators and biologists – are viewed quite critically. Fishermen regard them as people who meddle with their business without running any risk for their own livelihoods, while fishermen themselves bear all the risks of their entrepreneurship and are relentless when it comes to the hours and the energy they put into their work. 'Civil servants,' says one vessel owner, 'try to make your life miserable, but they earn a living whatever they do. We have to make sure everything is running smoothly and that we take care of our family'. On the other hand, many fishermen think that the officials' mentality as 'wage labourers' gives them an advantage. 'Those Inspection Service lads work office hours, but at night they are in bed with their spouse. So you have to land your fish when they are sound asleep.' Fishermen thus resist the stifling of their perceived independence, which – in their self-image – is one of the occupation's core values. They have the feeling of being boxed in from all sides and losing their autonomy in decision making. Still, there is more to it than a straightforward integration conflict of resistance to encapsulation. Some rules and regulations they find at best simply silly and at worst perverse. It goes against the grain for each and every fisherman to return prime fish they have caught to the sea, particularly if they cannot survive, as is the case with cod. It is in flat contradiction to common-sense behaviour. Just abiding by administrative orders that lack legitimacy and lead to wasteful practices seems to make little sense.

The resentment towards onshore technocracy also extends to scientists. Although not all fishermen oppose catching effort limitations, many are aggravated by the fluctuations in the biologists' 'guesstimates' and the usually short notice on which quota measures for the upcoming year are published (see also Chapter 6). Fishermen deem the biologists' advice to be politically coloured. Their mutual relationships have the additional dimension of battles over competence. Fishermen say that the biologists always carry out research in fixed spots and do not listen to the fishermen's opinions, whereas the former follow the fish or switch positions if catches are low. They often have a clue or a hunch where the fish might be. On the quayside, one of them points out that the fishing industry has to cope with more and more highly educated managers, who know nothing about the fisheries, earn high salaries but do more harm than good to the fishing industry. What is at stake here is a feeling of self-esteem and an awareness of the value of the fishermen's own knowledge and expertise concerning fishing and the seascape. The gist of opinions is that biologists may have a university degree, but they should not think that their theoretical knowledge is more valuable than the fishermen's practical knowledge. Fishermen are out at sea most of the week and work around the clock. They live and breathe fishing and consider themselves to be experts par excellence in fishing matters. It is precisely in the domain of knowledge and expertise that some of the occupation's core values are situated. To understand the fishermen's responses to the state's tightened enforcement of management regimes, it is necessary to dwell a bit longer on the way fishermen perceive their occupation. Fishing is not a nine-to-five job, but permeates all aspects of life. It is as much a way of life as it is a way of making a living. It is an existential matter, an important marker of identity and a cherished lifestyle to be continued by successors. This goes especially for skipper-owners. In the following section, I will deal with success, competition, communication strategies and the changing views concerning prestige and reputation. The perceptions and the concomitant modes of behaviour in these domains can tell us a lot about what makes fishermen tick and why they appear to be such ardent opponents of regulations and restrictions or – more accurately – failed enforcement.

Pursuing the Big Haul: Rivalry, Reputation and Respect

The feeling that they are able to operate independently is an important aspect of the fishermen's mental world. For a long time this outlook on life has been accompanied by rivalry. Fishing used to be – and to some extent still is – highly competitive. 'Beating the others' is part of the fishing game. Successful fishermen enjoyed tremendous prestige among fisherfolk and in their communities. Skippers and crews judged and ranked each other on the basis of their ability to land as much fish as possible. They paid close attention to the latest innovations in boats and gear that would give them an edge over their rivals. The number of boxes filled with prime species of fish a skipper could put ashore was not merely a measure of his economic success, but also of his self-esteem and prestige. It earned him social recognition. In this respect, in the first instance kilos, not revenues, seemed to count. 'My father used to say: I fish to catch fish, not to earn money,' an owner-operator (born in 1963) told me. However, with the limitation of landing rights and stronger enforcement, fishermen lost the liberty to land as much as they could. They have had to learn how to control their penchant for competitive rivalry. If a skipper has long been evaluated on the size of his catch, this no longer applies, much to the disappointment of the fishermen themselves. Nonetheless, it still is prestigious to be the week's 'top grosser' (*topbesommer*) and fishermen – skippers and crews alike – always compare how they have performed by the end of the fishing week. They say that 'there are fishermen and there are catchers'. Some fishermen consistently land more fish than others and there is fierce competition as to who will be the 'catch king'. However, catches fluctuate from week to week, on occasion quite sharply.

In my interviews with Texel fishermen, the fact that one wants to catch more than 'the neighbours' was often alluded to. 'If they [the other local skippers] envy you, you're doing fine,' according to an owner-operator. A colleague maintains: 'At sea we begrudge each other everything.' Within the local fishing fleet, the fishermen keep a sharp eye on each other. A former owner-operator explains: 'At sea, there's competition. You're rivals there, because a fisherman is a hunter. When you land your fish and there are plenty of boxes at the quayside, you're proud. When the catch has been bad, you have a lousy weekend.' A locally renowned skipper, born in 1922, intimates: 'When you're in the auction with half of what your neighbour lands, it just cannot be. You're so ashamed; you feel you could sink into the ground.' Many Texel fishermen have told me about how important it is to them to outdo their colleagues-cum-competitors. They regard their fishing prowess as 'a matter of honour', a phrase I heard used time and again in this context. In this sense, there is definitely a local fishing *arena*. Even – or perhaps particularly – between co-owning brothers who each skipper their own boat, there is competition as to who will land the bigger catch by the end of the fishing trip. The occupation's competitive nature is important for the fishermen's appreciation of the job but also leads to envy. Success is in short supply, as by its very nature it is always measured in relative terms. Catches and incomes may be satisfactory or even good, but this is not what counts in the social ranking of skippers and crews. The vast majority of the Texel fishermen are convinced that success is linked to craftsmanship and dexterity not fortune. 'Sure, you have to be lucky, but you can't always be fortunate, so skill and expertise are primary. On the other hand, as they say "without luck no one fares well",' says an owner-operator, born in 1941. While many anthropologists report that among fishermen across the world luck is an important emic concept to explain differential success, this is certainly not so on Texel, although most local skippers acknowledge that chance is a contributing factor.[3] All of them have experienced weeks in which catchability proved to be poor, but a single 'lucky haul' could partly make up for this.

Occupational communities of fishermen are commonly characterized by fierce rivalry and competition between skipper-crew units and, at the same time, cooperation, loyalty and egalitarianism. For social and moral reasons, fishermen must strike a balance between these seemingly opposed stances.[4] They might be viewed as reluctant rivals or cooperating competitors. The key issue is that there are conflicting incentives between the social and the individual. As an individual, no one can accumulate sufficient knowledge to capture the quantities of fish that are potentially available. Sharing information is necessary, but this social characteristic conflicts with the individual interest of catching fish (Wilson 1990). So let us explore how Texel fishermen attempt to solve this dilemma. Communication with others is a continual game of poker. All skippers have their familiar and favourite fishing grounds that they will attempt to keep secret. It is only with certain categories of colleagues that they share information concerning catches and niches where there is plenty of fish. These categories usually include people from local communities and especially kinsfolk and friends. To some extent, the time-honoured differences between Oudeschild and Oosterend surface in the small networks of cooperation, although these often also coincide with kin networks, making it difficult to determine which one is the more decisive factor. 'We always cooperate in small clubs, swapping data. You assist each other and it works out fine,' says a skipper-owner, born in 1941. Some agree on certain secret codes that indicate how one is doing. Exchanging knowledge about prime fishing locations is strictly based on reciprocity. Those who do so scramble the channel to avoid others from listening in. Even in this case, however, the information that is volunteered is often limited and not entirely accurate. For example, only the approximate not the exact position of good catches will be given and a difference of half a mile can make a substantial difference in yields. Alternatively, one can volunteer information only towards the end of the week or underestimate catches to some extent. When joining a crew for a week, I was able to observe these communication strategies. The radio was on all the time and I could hear voices utter incomprehensible phrases when news was exchanged with 'friendly' skippers. There was also the continual squeaking of scramblers when cutter skippers who were fishing nearby exchanged information. When 'Biem' called the skipper of our boat, one of the deckhands intimated that as a partner the man was useless. 'He lies and cheats. He wants you to think that he's caught zero fish, but come Friday he'll still land a nice number of boxes.' Knowledge of micro-niches is precious cognitive capital that is exchanged only as part of a tit-for-tat strategy. Temporarily monopolizing a niche may give important benefits by the end of the week. Often, a few good hauls determine above-average landings. Hence the tendency to claim rich fishing positions for oneself and to be reluctant to share the information (see also Byron 1975:152 153). With improved communication, the potential to exchange information has increased, yet owner-operators want to make sure that it is restricted to those who reciprocate. A virtual 'arms race' has therefore been going on 'in means to keep confidential information secret and means to get access to it' (Banks et al. 2001:43).

In general, fishermen regret the fact that today competition is fiercer than previously, in their perception at least. Perhaps somewhat nostalgically, they recall that in 'the old days' there used to be more cooperation and relations were more amicable. With scramblers distorting communication, they are of the opinion that something of the traditional congeniality has been lost. At the same time, however, they all use scramblers and deem it necessary to participate in this competitive game. To a considerable extent, they exaggerate about the willingness to cooperate and exchange information in the past. With the limited electronic means to communicate and the scant opportunities for others to listen in, it may seem to them that earlier on distortion of information was lacking. However, in actual fact this was certainly not the case. A skipper-owner, born in 1915, vividly remembered that competition was perhaps even fiercer in the days of sail: 'When the weather was foul you always waited until someone dared to go. When someone did, all the others with some guts followed suit. If you didn't participate in that game, you were always at the bottom end.' It is not just secrecy concerning good catches that is part and parcel of competition: deceit can also be involved. Therefore, fishermen have to learn to interpret other fishermen's behaviour and this skill even has to become a second nature. It involves knowing where other boats are fishing and for how long, whose distorted voices you hear on the radio, whether a skipper is prone to underreport or exaggerate catches, where a vessel was before its lights were switched off, what pattern of tracks other beamers are trawling according to the radar. It is a *habitus* that fishermen cannot learn at Fishery School; they can only acquire it through everyday action and years and years of experience (see also Pálsson and Helgason 1998). This also goes for deckhands, who are in the bridge as watchmen. Strict boat solidarity is required of them in that they are expected not to leak precious catch information to other crews.

Not all specific types of fishery are equally competitive. With herring being a migratory species that travels in shoals, there are advantages in closely communicating and cooperating with each other during specific stages of the hunt. Although skippers may have a hunch where the herring might be, they can never be certain about this. Consequently, they have to search the North Sea in pursuit of prey. Foraging until they strike upon a school of herring costs time and energy. To save on these expenses, it pays to cooperate closely. Once a skipper hits upon a good herring shoal, he and his partner-skipper will shoot the pair trawl. If the catch is rewarding, he will immediately inform other pair-trawl skippers. By the time they arrive at this particular position, the first-comers will have caught their share. It usually takes only a few short tows to fill up the fish holds of two cutters. The skippers then have to market the catch. Since the herring will be elsewhere by the time they return to the fishing grounds, monopolizing the information is of no use whatsoever. In addition, the fishermen have struck an agreement with processing firms about prices and maximum weekly landings (see above). In the herring-fishing season, skippers will therefore refrain from scrambling their radios and keep each other informed about prime catches. This conduct is based on an ethic of reciprocity. This does not mean that secrecy and deceit are lacking: quite the contrary. In herring pair trawl fishing, too, information-sharing and cooperation are restricted to certain clubs, there are leaders and followers, and landing a bumper catch in as short a time as possible is considered an honour. Most skippers thoroughly enjoy the 'herring hunt' thrill. One of them leaves skippering in the beam-trawl fishery entirely to his son, but always assumes responsibility in the wheelhouse during the three-month herring season. However, following the herring ban (1977-1982), the number of Dutch skipper-owners who combine beam trawling and pair trawling for herring is small – most of them hailing from Texel – and this facilitates close communication and cooperation. To keep relationships in good repair, the Texel skippers who partner in pair trawling usually also exchange information in beam-trawl fishing.

To be at the top of the skippers' hierarchy demands a considerable amount of nerve, stamina, skill and knowledge. This was particularly true in the era when sophisticated navigation and fish-finding equipment was lacking. For example, several skippers pointed out that the navigational aid Decca was highly unreliable at night: 'If you knew exactly where you were, you could sometimes have fabulous catches without others being able to trace you,' related one of them. 'Today, with GPS [Global Positioning System], you just tell the watchman what track to trawl and you can go off to your bunk to sleep. Back then, you would hardly leave the wheelhouse. I had trouble handing over the responsibility.' Although state-of-the-art technology is important, it still requires more than electronic devices to find the fish. Being able to 'read' and understand 'natural signs' continues to be of great significance. Skippers of beam trawlers are continually building up their expertise. Fishing patterns comprise alternating exploration and exploitation stages, and foraging decisions involve tradeoffs between perceived costs and benefits (Rijnsdorp et al. 2000a). On the basis of their previous experiences, expectations and hunches, skippers have to decide where and how they will trawl to have a rich catch and minimize the risk of damages to nets and vessel. Their decisions depend on the behaviour of other fishermen, the crew's commitment, the location of shipping lanes, the occurrence of physical obstacles such as wrecks, oil rigs and cables, the sediment's contours and the seabed's suitability for beam trawling, and, of course, the target species' abundance. Once they have located a concentration of fish, skippers will intensively trawl such a niche. They attempt to stay close to previous fishing positions, but also try to avoid fishing exactly the same position (see also Piet et al. 2000). With modern navigation equipment such as track plotters and autopilot, the skippers will tow the gear on 'a line' (that is, a track) for, say, two hours, then haul, stop, turn, shoot and trawl a parallel line in the reverse direction at a small distance, repeating the operation several times in succession. This way, they often cover a patch of a few square nautical miles. When the catch rate decreases significantly, the skipper will return to the search mode or steam to other, more remote fishing grounds where he expects to find sole or plaice. However, steaming requires the gears to be brought aboard and means fishing time is lost, while fuel costs continue to be added to the exploitation bill. Therefore, skippers usually avoid criss-cross courses and going to and fro.

Whereas some skippers would be content to 'haul, shoot and scrape' the entire week for average gross revenues, others are less quickly satisfied. 'I would sometimes continue fishing on a Saturday,' says a high-lining skipper, born in 1959. 'It might lead to tensions with the crew, but you try to catch as much as possible.' A retired owner-operator, born in 1928, contends that pluck is important:

> There are certain locations where you can expect a rich catch. When you fish there, you should not be afraid of occasionally losing a net. You have to learn by bitter experience. Now Willem and Pieter know how to do it, but Aris (Pieter's brother) doesn't.[5] ... If you're a good skipper, you dare to go to places that other skippers tend to avoid.

Nerve in itself is not enough. Some skippers go out by themselves a lot, fish near wrecks or in places others circumvent, but they are prone to damage. Building up a fund of knowledge, skill and experience and developing proficiency as a skipper are costly investments in time and money. A reputed skipper-owner ('Willem' in the quote above) intimated that when he began skippering in his early twenties, the gears sometimes got carried away because he took risks in positions he was unfamiliar with. The crew became disgruntled. Through sheer trial-and-error and being perseverant, he eventually succeeded in collecting information that enabled him to fish in risky environments and gross more than most of his competitors, ultimately to the crew's satisfaction. 'I just *made* my reckonings [*bestekken*],' he said, adding: 'I've had colossal gross revenues, but have also had empty fuel tanks and hardly any fish in the hold.' Such risk-taking to seek a bumper catch is certainly not every skipper's choice, given its low probability. Many opt to be on the safe side and go for the higher probability of a reasonable catch in particular spots (see also Pálsson and Durrenberger 1982). At the same time, they attempt to avoid the shame of landing the least fish. The crewmen usually respect the skipper's decisions, but if catches are poor he has to bear the brunt of their overt or covert criticism. In their eyes, the skipper is a somewhat ambiguous persona who can make or break their week. 'We will be the asses again this week,' said a deckhand in my presence, while throwing an angry look at the wheelhouse. Catches were initially poor during that particular fishing trip. However, if things go well, it is also the skipper on whom crewmembers bestow respect and symbolic power. All of them prefer a berth aboard a vessel skippered by a high-liner, although social chemistry is also extremely important. Serving under a successful but grumpy skipper would certainly not be everybody's choice. In general, success tends to be self-reinforcing. Outstanding skippers can afford bigger and better boats with more powerful engines, have easier access to bank loans based on their past performance, and are able to recruit the most experienced, dexterous and skilled deckhands. Since the outcome of fishing depends to a large extent on teamwork, this social nature of production probably contributes as importantly to success as do the material factors of production.[6]

Working one's way up in the pecking order is not easy. Following a successful skipper may be a strategy, but not a strategy that always works out well. Low proceeds and considerable costs because of damaged nets and lost fishing time is disappointing in itself, but in addition a skipper who is 'stalking' is not highly respected. Successful skippers will attempt to shake off such stalkers. A skipper's knowledge of the fishing grounds and the behaviour of fish and other fishermen is insufficient to be ranked a proficient skipper. Knowledge of markets and market prices is also important. Skippers have to decide where and when to land their catch and these usually hunch-based decisions can make for considerable differences. Still, it is their comprehension of the marine ecosystem and the behaviour of fish that is valued most. In the course of his career, a skipper accumulates a wealth of knowledge concerning specific fishing areas. In a trial-and-error process, he develops a cognitive map of good fishing locations, the results of hauls in particular seasons and under particular conditions, the position of wrecks and treacherous rocks, currents, depths, breeding habits and migration patterns of fish, the behaviour of other fishermen and so on. Therefore, skippers treasure their personal logbooks and private seacharts, which contain a detailed record of the outcomes of their stratagems. In order to prevent others from acquiring this knowledge all too easily, these logbooks are kept in safe-deposits. The skipper's job is demanding. They have extremely long working hours, both because other crewmembers are not sufficiently competent to judge the continual flow of information, and because nearly every skipper has to obtain the information on which his own catch depends (see also van der Vlist 1970:102). Indeed, some skippers hardly leave the wheelhouse, save for inspecting what a tow has yielded. 'When others were in their bunks, I would always have a look at what was in the nets,' comments a retired skipper. The size of the fish, the kind of by-catches and 'trash' provides experienced skippers with valuable information. Through years and years of building up know-how and experience, skippers get to know certain fishing locations like the backs of their hands. However, it is difficult to monopolize knowledge for great lengths of time. Eventually, others get to know about these locations. Secrecy tends to be short-lived because such information will eventually percolate through social networks. Clubs are small, but their boundaries are permeable and sooner or later information will 'leak out'. It is therefore important to join social occasions such as birthday parties, weddings, meetings and quayside gatherings in order to keep up to date. Conversations often focus on fishing and news is exchanged about the behaviour of other fishermen, good fishing locations, new electronic instruments, particular gear devices, and so on.

As a consequence of the increasingly tight regulation of the fishing industry, the quota regimes, and the stricter supervision of compliance with the rules, the fishermen's view of differential success has changed to a considerable extent. The perception concerning who deserves respect changes with alternating modes of production (Pálsson 1991:163). It is no longer landing the most boxes of fish that necessarily leads to respect. 'When fishing was free, there was a healthy rivalry, but today the dominant question is: "How did they get the fish?"', a skipper-owner indicates. A colleague adds: 'Who is messing around? That's the issue now. There's no longer a yardstick to measure who are the good skippers and crews.' During my first spell of fieldwork on Texel, relationships among fishermen were growing increasingly tense as a consequence of differences of opinion regarding compliance with the law and suspicions that certain fishermen still evaded or violated particular rules. Such violations included, for example, illegal fishing within the 12 nm zone; registering and operating vessels as Euro-cutters while in fact their engine power exceeded the required limit; using nets with too small mesh sizes or using blinders; messing with landing rights through sales in the grey circuit; and keeping part of the landings out of the books by selling them on the 'black' market. During a fishermen meeting I attended, one fisherman angrily accused others of fishing within the 12 nm zone when it had been agreed at the local level not to do so. The mutual mistrust increased as a consequence of the tighter management regime. An accusing finger was no longer pointed exclusively in the direction of fishermen from other communities

– as used to be the case – but also in the direction of Texel compatriots. Those who admitted to using illegal gear provisions said they did so because others also did. The problem was aggravated because one could not even be sure that local fishermen refrained from using blinders, let alone fishermen from other Dutch fishing ports and other European countries.

At the same time, I was able to observe the changing mood regarding the regulatory regime. In early 1990, many Texel fishermen still defended and legitimized illegal practices by saying that the plethora of management measures stifled their operations, while sole and plaice were believed to be so abundant that they 'practically jumped on deck' and 'shouldn't die from old age', as a sturdy young skipper-owner phrased it. Relationships among fishermen became increasingly tense as a consequence of differences of opinion concerning compliance with the law and suspicions that some fishermen still evaded or violated certain rules. One of the recurring discussions pertained to the illegal use of blinders. Some skipper-owners advocated the view that 'we ought to get rid of those things', while others maintained that they could not do without them because too many legal-sized soles would escape their nets, especially in rough weather. By the end of the year, the dominant opinion was that restrictions were necessary and should be complied with (which obviously does not mean that all fishermen actually complied with all rules and regulations). Another hotly debated topic was linking days-at-sea allocations to quota rights and horsepower. Since most Texel skipper-owners had been investing in acquiring additional landing entitlements, the majority was much in favour of such a measure. They believed that it could effectively tackle the problem of quota overshooting by those fishermen who had insufficient landing rights to be at sea throughout the year. However, Minister Braks perceived no advantages in linking days-at-sea with individual quotas, and his successor left the issue to rest because he was brooding on something entirely different. As we shall see in the next chapter, a profound change of affairs would have a soothing influence on the Texel and Dutch fishing industries.

Vexing Problems, Daunting Tasks

The developments described above have thoroughly transformed the métier of Texel and other Dutch beam-trawl fishermen. Since 1975 they have been restricted in their operations more and more through national and supranational input and output regulations that came about in a rather piecemeal manner. The fishermen felt increasingly suffocated by burgeoning rules and regulations about which they had not been consulted and upon which they had little or no influence. Believing that the extant national fishermen's organization was too lukewarm to oppose the fisheries authorities' measures, factional strife resulted in fission in the Fishermen's Union. The newly established Federation was more militant but accomplished little in terms of redressing the regulatory regime. On the contrary, the fishermen got entangled in ever more red tape. This is not exceptional. Throughout the western world fisheries management is usually characterized by top-down modes of policy design and implementation that involve centralized, hierarchical, command-and-control decision making and monitoring to make up for market imperfections and ecological problems (Symes 1997a, 1997b; Dubbink and van Vliet 1996, 1997). The Common Fisheries Policy is a prime example: measures affecting the fishing industry are determined in Brussels with little or no involvement of fishermen and their organizations. Such top-down policymaking often leads to poor transparency and high information, monitoring and enforcement costs, as well as exasperation and a lack of legitimacy and compliance on the part of fishermen (McCay 1995:16). With burgeoning rules, management turns increasingly complex but less effective because the quality of information available to regulators deteriorates while at the same time reducing the fishermen's incentive to abide by the rules and providing them with a rationalization for violations (see also Healy and Hennessey 1998:116-117). State intervention under the European fisheries regime, writes Symes, 'has left fishermen confused, alienated and embittered' (2000: n.p.). All this is evident in the Dutch flatfish fishing sector, but other European fishermen have also had to cope with this. The fishing industries in each member-state attempted to deal with the difficulties in their own particular ways.

The Netherlands was among the first countries to implement individual transferable quotas. This happened *de jure* in 1985, although de facto the practice of trading individual flatfish quotas developed in the late 1970s, quite soon after the establishment of total allowable catches and quota regimes. Neoclassical economists generally advocate such a regime as a potentially effective and efficient mode of managing fisheries. Expansive firms can acquire fishing allowances from marginal firms that are eventually ousted from the fisheries arena. Secure property rights are supposed to increase efficiency, and to make (self)-enforcement easier and less expensive. Theoretically, the system has important economic, administrative and ecological benefits. In practice several adverse consequences have been observed. Firstly, quota busting and cheating still occur if monitoring and enforcement are suboptimal, requiring large public expenditures to coerce fishermen into compliance. Secondly, the system leads to irresponsible fishing practices (high-grading, discarding) for purposes of maximizing the economic value of entitlements. Thirdly, the 'race to fish' continues unabated if landing rights are insecure and national quotas are prematurely exhausted. Fourthly, productive capacity is consolidated or even expanded. Fifthly, the inducement to provide information – a public good – remains weak because there is an incentive to free-ride and underreport. Lastly, the commodification of quotas does not simplify the administrative regime (see Davis 1996; Grafton 1996; Shotton 2000). All of these unforeseen and undesired outcomes have been evident in the Dutch flatfish fishing industry. In the real world of multi-species fisheries, the ideal-typical maximizer apparently behaves less rationally than predicted.

Furthermore, sociologists and anthropologists point to equity problems and the social costs of introducing property-rights regimes.[7] The initial allocation usually amounts to a giveaway of public resources, privileging the generation of fishermen who receive allowances to a certain percentage of total allowable catches free of charge over future generations who have to pay for them. If based on historical rights ('grandfathering'), it rewards operators who took disproportionate shares of the fish stocks and thus contributed most to overexploitation (which is generally the reason for the implementation of quotas). This windfall typically accrues to the boat owner while co-venturing crewmen do not share. It thus brings about a shift in the power balance between the rights-holding owner-operator or company and the crew, widening the social distance between them, and it may even result in capitalist relations of production if the share system of remuneration is replaced by a system of flat wages. The assignment of allowances creates further inequities in that it benefits some (the 'haves') while disadvantaging or excluding others (the 'have-nots'), for instance part-time and seasonal fishermen. The system creates relations of dependency in that some operators come to depend on large rights-holders – including non-operating 'slipper skippers' (see the next chapter) – to lease rights. It may also lead to the phenomenon of 'agency capture', in that government agencies tend to serve the interests of the most tightly organized groups, with large rights-holders having the best opportunities and resources to increase their bargaining power. The system virtually blocks social mobility in that crewmen and newcomers have few opportunities to acquire allowances and as capital is substituted for labour, it leads to diminishing employment in conditions in which crewmen are usually not liable to receive redundancy payments. The accumulation of rights leads to concentration, capitalization and industrialization. Rights are often accumulated in particular communities and the 'rights drain' makes for regional concentration of entitlements that may be to the detriment of fisheries-dependent communities and areas. This in turn may thwart locally important knowledge and skills and more generally a traditional culture of fishing in families and communities. Once in place, the system is rigid, consolidating vested interests, while it is difficult to revoke since property rights are solid and the costs of reimbursing rights holders through buy-back programmes would be prohibitive. Although neoclassical economists are aware of some of the detrimental social effects (Grafton 1996:12-14), they seem to accept them as regrettable but inevitable side effects of improvements in efficiency and economic rationalization.

The Dutch experience shows that following the introduction of individual transferable quotas and encompassing regulatory regimes, a mass of problems prevailed. Despite being prime stakeholders, Dutch fishermen felt twice removed from the policymaking and implementation process. It came about in the Brussels and The Hague bureaucracies, with little or no grassroots involvement or consultation. The measures that were imposed upon the national and local fishing industries often held little water with the fishermen, while in turn *their* ideas, suggestions and solutions were usually bluntly ignored by policymakers and regulators. The paternalistic authorities firmly believed that they were acting on behalf of the fishing industry, without so much as listening to, let alone involving, its practitioners. Feeling disempowered, the fisherfolk took their own course of action. Marginalized fishermen maintained that those who had accumulated a large share of the landing rights had done so through initial illegal fishing, investing the perks of their practices in additional quota entitlements. Those who bought or leased quotas countered the accusation by saying that they had the right kind of entrepreneurial and managerial spirit and that it was those who did not have sufficient landing allowances who indulged in illegal fishing. The problems pertaining to enforcement and compliance led to huge transaction costs. The crux of the matter was that rights proved to be extremely insecure and thus did not meet with an important precondition of the neoclassical economic school of thought. The professed benefits require that rights are exclusive and durable and that fishermen comply with the rules: 'If fishers who are not quota-holders are able to fish with impunity, the quota becomes valueless as a meaningful property right' (Grafton 1996:8). Regardless of individual entitlements, exhaustion of national quotas led to early closure of the fisheries. Individual overshooting of entitlements consequently led to sanctions that impacted the collective of fishermen, undermining their trust in external authorities, since rights could not be exercised. Under these conditions, a derby race for fish was encouraged rather than mitigated. Moreover, investment in individual transferable quotas was expensive, while long-term certainty was lacking because European and national fisheries policies might change and quota cuts implied that entitlements diminished in absolute quantities (although they could of course also increase).

Rather than as vehicles to enhance secure rights, Texel fishermen therefore perceived external authorities as a source of anxiety. Increasing state intervention also stymied the fishermen's freedom at sea and brought about more paperwork, leading to less job satisfaction. In addition, tensions mounted due to policing and enforcement and the public image of fishermen deteriorated because their activities were seen as environmentally damaging and they were perceived as notorious law-breakers and reckless egotists, leading to a declining status of fishing as an occupation. Furthermore, mutual mistrust and suspicions increased as bumper catches were believed to be caught illegally. Catch kings turned into quota barons, placing entrepreneurial skills higher than fishing skills and affecting the system of prestige and status ranking. Crew loyalty declined, since investments in quotas meant lower percentages in the share system of remuneration. This in turn led to a declining interest in becoming a fisherman, while at the same time it became more difficult to maintain family firms. However, fishermen – owner-operators in particular – do not easily decide to leave the fishing industry. Their occupation is a way of life that gives them an identity. They highly value continuity of family firms and all of them have experienced good times as well as bad times, and they are inclined to attempt to weather adverse times. In general, Texel fishermen felt they had lost control over how to run their individual enterprises, but they tried to muddle through. The quota regime undermined what they regarded to be the backbone of share fishing: their labour ethos of working hard to catch as much as possible. Initially, they continued to do so, but with tightened policing and enforcement they often collided with law enforcers. The system of entitlements also undermined the culturally important system of status ranking, since the yardstick to measure catch kings – their 'conspicuous productivity' (Byron 1980:228) – was believed to be corrupted by illegal behaviour. In short, the regulatory regime's new realities profoundly changed the traditional meaning of the fisherman's métier.

Confrontation and litigation rather than consultation and compromise were the dominant political processes in the Dutch fishing industry of the 1980s and early 1990s. With increasing regulation the number of violations rose, leading to more warrants, fines and court cases. Judges were often lenient or deemed evidence of infringements of the law insufficient and acquitted suspects. If not, fishermen's lawyers often successfully appealed against District Court judgments in a Court of Appeal, the Supreme Court or succeeded in having cases referred to the European Court of Justice for prejudicial questions. This stalling tactic gained the fishermen precious fishing time. In theory, the Common Fisheries Policy provided a blueprint for fisheries management. In practice, a host of ad hoc measures had to be taken as a consequence of stakeholder resistance and litigation. The skipper-owners also filed law suits against the national fisheries Ministry – often successfully. The fishermen regarded the burgeoning bureaucracy a nuisance, and it did not lead to improving compliance. Quite on the contrary: the more petty and equivocal rules and regulations, the greater the number of violations. Abiding by the law was difficult with imperfect, ambiguous and contradictory legislation, and rules deemed unworkable or unfair. Although the chances of being caught increased with tighter monitoring and enforcement, many fishermen were inclined to continue taking risks. First and foremost, their labour ethos of working hard to achieve success was deeply ingrained. It constituted the backbone of 'being a fisherman' and top-down rules that challenged its very foundations could not change this overnight. As long as the benefits of flouting the rules outdid the costs of penalties, the incentive to fish illegally remained high. In addition, fishermen were still actively looking for loopholes. The state's legalistic approach proved to be counterproductive. Its reflex was to come up with more instead of better and more acceptable regulations. The problem of rule beating seemed endemic. All in all, therefore, the Dutch fisheries bureaucrats and regulators faced vexing problems and daunting tasks. They were squeezed between the national Parliament's and the European Commission's demands and the prying eyes of European politicians and fisheries inspectors on the one hand, and the national flatfish fishing industry's resistance against and dodging of tighter management regimes on the other. Having been reputed as law-abiding citizens for a long time, fishermen seemed to have developed a penchant for civil disobedience.

The stricter management of the fishing industry was not entirely without result. Instead of continuing to fish illegally, many owner-operators began investing in quotas, hoping that certainty in regard to landing allowances would increase and that additional fishing entitlements could keep their revenues up to the mark. This inclination led to a greater balance between catching capacity and landing rights and as a consequence of stricter enforcement, demand and prices for quota entitlements rose sharply, making them an additional production factor. Prices of quota rights continued to be extremely high from 1988 to 1992. Quotas were even traded above the net present value of future returns from fishing, suggesting that owner-operators wanted to remain in business despite the high costs of matching vessel capacity with landing allowances (van Vliet 1998b:70). The commodification of individual quotas brought about a measure of concentration of entitlements in the hands of fewer owner-operators. Thus, some could expand while others were ousted from the fishing industry. By 1992, the number of Dutch cutters had declined to 472 (see Appendix C), crewed by 2,195 fishermen. While acquiring landing rights led to economic mobility, at the same time it blocked upward social mobility. With the capital needed to purchase a vessel and quota entitlements, it became hardly feasible for deckhands to become independent, especially if they did not belong to a family of owner-operators. The initial investment would be prohibitively high and probably not profitable, which was sufficient reason for financial institutions to refrain from providing loans (Davidse 1997a:9). Newcomers were therefore effectively barred from entering the flatfish fishing industry, leading to a consolidation of vested interests with extant ownership being in a relatively comfortable position. However, even with landing opportunities becoming somewhat more in line with catching capacity, monitoring and enforcement by external authorities remained highly necessary.

At the same time, the Dutch government was looking to rid itself of the increasingly heavy burden of implementing and enforcing the rules and regulations pertaining to the fishing industry. The costs of devising and instituting a watertight inspection system were simply unaffordable. Landing inspections relaxed after a while. The initial target was to monitor eighty per cent of all landings, but this was eventually reduced to thirty per cent. According to on-the-ground inspection officers, fishermen always landed about twenty per cent above what they jotted down in their logbooks. If they got caught they would not get a warrant; the only thing that happened was that the twenty per cent would be added to their logbook. The easing of inspections brought back peace and quiet in the fisheries sector. In addition, an extremely large increase of the sole quota in 1990 (see Appendix D) was accommodating enough to make rule-breaking more or less unnecessary. It helped to considerably calm down the flatfish fishermen's discontent with the European and national fishery measures and led to greater compliance with quota regulations. Still, despite the eighty-five per cent increase, the race for sole continued. Sole fishing was banned on 19 November 1990 because the national share of the total allowable catch was exhausted. The owner-operators had been expecting an early closure and in order to get their portion had been fishing quite intensively during the preceding weeks. It was a standard example of a self-fulfilling prophesy: 'if men define situations as real, they are real in their consequences' (Thomas and Thomas 1928:572). Soon illegal trade circuits functioned again (van der Kroon 1994:270-271). In 1991, it was estimated that illegal landings still amounted to twenty per cent. At the 1992 annual meeting of the Dutch Fishermen's Union, Johan Nooitgedagt said in no uncertain terms that catch restrictions were being massively circumvented again and that this ought to stop. Bona fide fishermen suffered from the fraudulent practices of compatriots and, breaking a taboo, Nooitgedagt openly called for the former to act against the latter.

It was abundantly clear that something needed to be done. Yet more restrictive rules and tighter policing and enforcement were not an option. The allocation of public funds to do so had already gone beyond limits that politicians deemed acceptable. 'We have reached the end of our possibilities', Minister Piet Bukman admitted. The situation that had emerged was a classical example of reducing and socially de-contextualizing the conduct of fishermen by focusing on the rational individual. Another policy would have to bring change. Mutual confidence had to be restored and the legitimacy of fisheries policy regained. As we shall see in more detail shortly, state agencies and fisheries representatives alike became convinced that the top-down command-and-control style of fisheries management – which involved considerable risks of non-compliance and therefore extremely high transaction costs – should be replaced by something entirely different. In the next chapter, I will scrutinize how the external authorities attempted to tackle the difficulties and how the fishing industry responded. It will transpire that the days of the catch kings and quota busters were numbered. At last the government and state institutions realized that the solution was to be found in legitimacy, not more red tape and more policing. The task was formidable. The new regime would require nothing less than a leap of faith from fishermen on the one hand and politicians and regulators on the other. In superimposing its rules, the European Union – so renamed in 1992 – would tremendously complicate the mental transition.

Figure 13. Texel shrimp fishermen, 1930s (courtesy of Klaas Tuinder).

Figure 14. Oosterend family crew, 1940s (courtesy of Wim Ellen).

Figure 15. Sorting fish in front of the Oudeschild auction, 1950 (photograph Jan de Waal, courtesy of *Texelse Courant*).

Figure 16. Texel herring pair trawlers in the port of Calais, 1958 (courtesy of Sam van der Slikke).

Figure 17. Texel cutters caught in the ice, 1963 (courtesy of Sam van der Slikke).

Figure 18. Texel beam trawlers in their home port, 1965 (courtesy of Sam van der Slikke).

Figure 19. Backbreaking work on the deck of the TX 5, 1975 (courtesy of Jan Ellen).

Figure 20. Three generations of Oudeschild fishermen mending a net, 1970s (courtesy of Fup Krijnen).

Figure 21. Emptying the cod end aboard the TX 37, 1990 (photograph by the author).

Figure 22. Crewmembers of the TX 38 enjoying supper, 2006 (photograph by the author).

Figure 23. The TX 19 *Elisabeth Christina* fishing, 2006 (photograph by the author).

Chapter 6

Commissioned Cooperation: Plentiful and Lean Years

The previous chapter showed how the fishermen's discontents and their contraventions of the rules and regulations combined into a tense situation that politicians and bureaucrats found difficult to handle. Fishermen tended to ignore and flout the rules, prompting a response from state institutions to force them to comply. This state coercion was unsuccessful. Despite huge monitoring and policing efforts and expenditure, compliance problems continued to exist. At face value, it was a typical integration conflict in which the fishermen attempted to hold on to their perceived autonomy, which was being sapped because they became increasingly entangled in burgeoning regulatory regimes. While the state ignored their calls for change, the fishermen in turn simply ignored the state's management regimes. The rift between them widened and deepened. The real problem was that the fishermen felt the rules ought to make sense, should apply to each and all of them and should be enforced equally. Those who had obtained quota entitlements wanted to exercise what they perceived to be their rights (instead of privileges or allowances). This made for considerable tension within the fishing industry, particularly between large rights holders and skipper-owners who landed more than their fair share.

Having become convinced that additional top-down decisions would only aggravate the fishermen's opposition, and would hence be expensive and ineffective, in 1993 the Dutch authorities finally attempted to involve flatfish fishermen in fisheries governance. Using their discretion of arranging the specifics of management structures within the Common Fisheries Policy framework, they delegated considerable responsibility to the fishing industry for quota management, self-regulation and self-enforcement. Group management of individual transferable quotas is a key feature of this governance system. The new regime has proved to be a turning point in the Dutch fishing industry's recent history. In a sense, it amalgamates theories of neoclassical economists, who proffer that individual transferable quotas are efficient because the strongest and more efficient units of production survive economically, and theories of institutional economists, sociologists and anthropologists concerning a proper, participatory and effective fisheries management system in which consultation, delegation and legitimacy are core elements. The Netherlands was the first

EU member state to work with individual transferable quotas, co-management groups, and days-at-sea restrictions.

The predominant viewpoint of many neoclassical fisheries economists and regulators posits a need to encourage economies of scale and reduce inefficiency. This puts a premium on those units that are able to capitalize on investments, while marginalizing small-scale units that cannot afford the rising costs of new harvesting technologies and investments in acquiring landing rights. Economists and regulators regard the latter units as redundant; they believe them to be inefficient 'excess capacity'. This ideological framing of a particular category of owner-operators – which amounts to blaming the victim – has profound consequences. Comparatively speaking, the petty production units are precisely the ones that contribute least to the problem of over-fishing, yet they are the ones that usually suffer most from new technologies and new regulatory regimes. Owner-operators with the highest levels of capital, education, knowledge and skill are in the best positions to adjust to and adopt technological innovation. With individual quotas being transferable, they tend to accumulate rights. In Chapter 5, we have already encountered some of the adverse consequences of this regime, which regards individual actors as atomistic and one-dimensional profit maximizers, and to some extent creates them. Neoclassical economists consider transformations in the social relations of production, and the concomitant creation of social inequalities and changes in the socio-cultural fabric of fishing communities, to be regrettable but necessary payoffs in order to achieve economic and administrative efficiency in the fisheries.

The really difficult question is whether the pain is worth the gain. Arguably, any regulatory system creates winners and losers. Therefore, with every shift in the regime, political bargaining and political contracting will be inevitable. The disadvantaged are likely to demand modifications that, when enacted, create still more drawbacks – a 'paradox of fairness' (Healey and Hennessey 1998). The powerful actors with stakes in the fisheries arena will seek continuity and protection of their interests. Such consolidation may not necessarily be detrimental to the fishing industry or fisheries governance: the political participation of established stakeholders might ensure that a degree of legitimacy and effectiveness of the management system emerges (Jentoft and McCay 1995:241). In contradistinction to neoclassical economists, anthropologists and sociologists proposed another avenue of tackling the prevailing problems in the fisheries: co-management, a concept that began emerging in the social science fisheries literature in the late 1980s (see, for example, Jentoft 1989; Pinkerton 1989).

Co-management involves consultation of stakeholders by external authorities concerning management decisions and implementations, and delegation of certain management responsibilities to user groups. Decision making pertaining to and governance of particular resources are shared to a greater or lesser extent between state agencies and interest groups exploiting these resources (see, for example, McCay 1995:14,

2000:210; Nielsen and Vedsmand 1999:20; Singleton 2000:6; Schreiber 2001:377). The basic premise is that 'when users obtain more management responsibility in functional terms, they are likely to behave more responsibly in moral terms' (Jentoft, McCay and Wilson 1998:427). Advocates of co-management strongly believe that such hybrid power-sharing and participation of user groups promote a sense of belonging and involvement, in turn fostering communication, collaboration, cooperation and compliance (Bennett 1999:10). It was precisely this dimension that was taken into account when in 1991, under the aegis of the Dutch Fish Board, negotiations were begun between fishing industry representatives and state officials to develop a management scheme in which certain responsibilities were devolved to producers, traders and processors. The hope in the Dutch situation was that the less legalistic approach of fisheries governance would leave more discretion to the fishermen and fishing firms 'to adapt their conduct to "the spirit of ... public policy"' (Dubbink and van Vliet 1997:183). Assuming greater responsibility to manage one's own affairs would supposedly give a boost to the legitimacy of government measures and augment compliance to its rules and regulations (ibid.:184). The primary objective was to reduce the potential for conflict.

The fisheries management system that emerged in the Netherlands seems to have combined 'the best of both worlds': individual transferable quotas embedded within a co-management regime. The present chapter addresses the initial experiences with this governance system, focusing especially on the views of fishermen, biologists and state representatives regarding its efficacy and its unforeseen and unintended consequences. Special attention will also be devoted to the ways in which, and the extent to which, beam-trawl fishermen were indeed able to participate in fisheries policy and management, and their perceptions concerning the current governance system's benefits and pitfalls. Increasingly, they faced new opposition. If they regained some of their autonomy under the new management regime, they were also up against formidable resistance from environmentalists and public opinion. Beam trawling became increasingly controversial because of the alleged damaging effects on benthos, forcing the fishermen to defend and legitimize their mode of operation. The new regime also impinged upon the practice and perception of their occupation in several major ways. More than ever before, the fishermen's topics of conversation focus on what the seats of political power have in store for them. To understand their outlook on their métier, it is necessary to dwell on some of the management scheme's particulars.

Grassroots Involvement: Co-Management and Compliance

In 1991, the government inaugurated a committee to investigate the possibilities of a system of mandatory decommissioning. Another committee – presided over by former Prime Minister Barend Biesheuvel – had to look into ways of devolving management tasks to the fishing industry (Stuurgroep Biesheuvel 1992). This committee comprised two working groups, representing the producers and the traders and processors, respectively. Fish Board executive Dick Langstraat chaired both working groups. The government and the House of Representatives threatened that, should the fisheries sector fail to cooperate with the Biesheuvel committee's suggestions, a harsh compulsory decommissioning scheme, limiting the engine power of each vessel, would be inevitable. This approach simultaneously pushed and pulled fishermen representatives to the negotiation table. A time of reconciliation ensued. The government and the fishing industry began talks on the establishment of co-management groups that would be closely associated with Producer Organizations (POs). The major difference compared with the groups that had briefly existed previously (see Chapter 5) was that the state now envisioned much tighter group regulations and monitoring. A new head of the Ministry's Fisheries Directorate, Peter Draaisma, made it abundantly clear that the fishing industry would suffer if it refrained from accepting the co-management system. 'The state has lost confidence in the fishing sector. You are at a dangerous point,' he admonished the participants at the annual national fisheries meeting in 1992. The national fishermen's organizations' leaders wanted to start with 'a clean slate' and visited the fishing communities in order to convince fishermen that they had few alternatives but to join a group.

It was indeed decided that fishermen should organize into co-management groups, dubbed 'Biesheuvel groups' after the chairman of the committee that advised on the new policy. The state aimed at increasing fishermen's responsibility and social control through devolution of specific management responsibilities, and returning some of the decision-making authority to owner-operators. The idea was that fishermen's involvement in policy and management would lead to greater legitimacy, and in its wake to increased compliance with the rules and regulations and cooperation with the administration. In turn, this should enhance exploitation of marine living resources in an ecologically and economically viable way. In the early 1990s, there was growing pressure for the integration of fisheries concerns and marine environmental management. This harmonization policy of balancing economic and ecological interests was laid down in the 1993 white paper *Vissen naar evenwicht* (Balanced Fisheries), a document embodying fisheries policy until 2003 (Anonymous 1993). It reflected the fact that the environment had taken pride of place on the political agenda. Henceforth, the prime goal of Dutch fisheries policy was to achieve a responsible way of fishing and a sustainable exploitation of fish stocks. Fishermen therefore had to take into account other functions of the seascape, especially its value as a nature area. If it proved to be impossible to reconcile fishing effort and natural values, priority should go to the ecosystem. In addition to seeking a balance between economic and ecological interests, the white paper aimed at giving responsibility to the Dutch fishery sector through co-management and new forms of cooperation. It explicated that further intensification of state enforcement and policing would be financially and politically unfeasible (Anonymous 1993:5-6). It sought to implement a policy that would fit within the Common Fisheries Policy and at the same time enhance the social and political feasibility of the regulatory regime with less, rather than more, government. More generally, cutting back on state expenditures was the political fashion of the day. Essentially, the Biesheuvel system of public–private management was a compromise between the long-term interests of nature and the short-term interests of fishing enterprises. It was dominated 'by a desire to "keep the social peace" within the limits set by scientific research and the public debate' (Salz 1997). It reflected the then predominant Dutch modus operandi of solving economic, social and political problems and conflicts through consultation and compromise, a political practice that had been conspicuously failing in fisheries matters since the quota regime's introduction.

The threat of mandatory decommissioning helped convince the owner-operators of beam trawlers to participate. Eight co-management groups, each comprising between twenty and ninety vessels, were established. Ninety-seven per cent of the skipper-owners joined a co-management group, even though initially several were reluctant to cooperate. The high percentage came about due to Parliament's threat of taking coercive structural measures should the percentage remain below seventy-five per cent, leaving the fishermen little choice but to join. There were also positive incentives. Group members were entitled to ten per cent extra days-at-sea in comparison with non-members and the latter could only trade quotas during the last two months of the year. The allocation of sea days was based on landing rights, fulfilling a long-term wish of large rights holders. Moreover, the state announced tight monitoring of non-members. This 'carrot and sticks' approach was successful. The management groups' aim was twofold: firstly, to arrive at an effective and efficient system of quota compliance that would be supported by the fishermen; and secondly, to improve economic performance within the quota restrictions. With the derby race for fish, skipper-owners tended to fish quite intensively at the beginning of the year so they would not be left with unused landing allowances should the fishery be closed prematurely. As supply would be high, this tendency drove prices down in the early part of the year. Agreeing fishing plans within a group would make for a steady supply of fish and hence higher prices across the year. The Fish Board supervised and coordinated the groups. If necessary, it harmonized their regulations, assisted in secretarial work and intermediated between owner-operators and the government. It organized meetings with group delegates to discuss problems and performance, and it advised the government and its agencies. The Fish Board thus acted – and still acts – as an intermediary organization between co-management groups and between groups and the state.

Internally, a board of directors – which supervises group management and has certain rights and obligations – heads each group, considerably reducing state involvement and enforcement costs. A chairman without any stakes in the fishing industry leads the board; the other board members are producers elected by group members. Group members must sign an agreement committing them to abide by the group's rules, remain in the group for the entire year, and provide logbook and auction data to the group and the General Inspection Service. The board controls and manages the quotas of individual members at the group level, arranges for quota transactions and days-at-sea transfers within a group and warns individual fishermen when their quota uptake has reached eighty per cent. It also sees to it that fishermen do not speculate with individual transferable quota shares and that redistribution takes place according to need rather than to the highest bidder (van der Schans 2001:364-367). Hence, to some extent, the board is governing a socially embedded, moral economy type of exchange. Individual fishermen bring their quotas and days-at-sea entitlements into a group but remain proprietors, and they are responsible for establishing annual fishing plans to achieve a better distribution of days-at-sea and quota uptake across the year. The Fish Board must approve the fishing plans and determines how to allocate the group quotas and how and when they will be caught. Members may buy, sell, lease, rent or exchange individual quotas throughout the year and the state has lifted several restrictions on quota transfers. For example, landing rights are now divisible. These measures provide for greater flexibility to smooth out surpluses or shortages and to respond to unexpected events and contribute to higher price levels. It is mandatory for group members to sell landings through one of the eleven designated Dutch auctions to ensure that the quantity and price of fish can be effectively controlled and that adequate information is available (van Vliet 1998b:71; van der Burg 2000:48).[1]

The Biesheuvel co-management regime largely hinges on the idea and practice of social control and peer pressure. What is important in this respect is that the co-management groups are relatively homogeneous, since membership is mostly arranged according to the type of vessel and gear used, the species sought, the region from which fishermen hail and membership of one of the two national fishermen's organizations (Hoefnagel and Smit 1997b:163). In addition, there is considerable overlap between specific co-management groups, Producer Organizations and local fishermen's associations. For this reason, the owner-operators are socially well integrated. The group board can prosecute members who exceed their quotas under private law, while also ensuring that group members who are unable to take up their quotas receive compensation. Fines are hefty, outweighing any gains an offender might have from non-compliance. It is possible to expel members who fail to comply with group rules; their fishing opportunity may be limited or the number of days-at-sea to which they are entitled reduced. All members suffer if the group exhausts its quotas early because the state can then prematurely close the group's fishing opportunity. Moreover, groups have to apply for government recognition each year, and it may withhold approval if the group has previously exceeded its quotas. The underlying idea is that fishermen will indeed report on offenders from their own group. Given the fact that an individual firm's transgression disadvantages the entire group, in theory, the incentive for reporting on offenders is high.

Reporting may be at variance with community values though, and fishermen regard it as 'tale-telling' or even 'treason' (Hoefnagel and Smit 1997a:17, 1997b:164; van der Schans 2001:345, 371). For example, fishermen complain about colleagues using illegal net provisions exceeding horsepower limitations or fishing within the 12 nm zone, if they think they will fail to exhaust their own individual quotas before the end of the year, but they do not consider it their duty to report specific cases to the group board or the General Inspection Service. There is, therefore, a social code preventing reporting on infringements of the rules – perhaps with the exception of days-at-sea regulations, since fishermen dislike seeing colleagues who have used up their days going out to sea. Because they are acquainted with offenders, or are even their relatives, group board directors occasionally feel exposed to social pressure not to mete out fines (Dubbink and van Vliet 1997:198; van der Schans 2001:351-352). Reporting on colleagues, friends or relatives to the authorities jeopardizes community cohesiveness and fragile power relations. Group board members therefore have to strike a balance between the interests of the group as a whole and its individual members. Given these limitations, some researchers criticize the co-management system for being inadequate in its self-enforcement aspect, requiring a statutory system of penalties and procedures (Berg 1999a, 1999b). However, this view entirely ignores the power of informal control mechanisms. Although they may avoid officially reporting on offenders, the owner-operators keep a keen eye on the conduct of their fellow group members and may resort to gossip and social ostracism. The effect of such social control, which also works indirectly via fishermen's wives and the rest of the shore community, boils down to peer monitoring and self-regulation (see also Connolly 1997). Not reporting on offenders is the unwritten rule, but subtle – or not so subtle – peer pressure often makes fellow group members conform.

This is also clear in the case of the co-management group established by the Texel owner-operators. During discussions leading to the group's formation, they openly expressed their doubts as to whether the new management system would work. Although they were not particularly enthusiastic, not a single skipper-owner voted against the idea. The Texel co-management group began with twenty-six cutters, and was soon joined by more. Its board, chaired by the ferry company's chief executive, was keen on disseminating moral rules regarding abiding by the regulations. It proved difficult indeed to fine group members who transgressed group rules. They often responded quite emotionally and sometimes refused to pay. Nevertheless, mutual social control was certainly efficacious. The group exerted more social pressure on the owner-operators than imposing a penalty would have done. If the group board called a culprit to account, this amounted to being shamed in front of colleagues. Such reprimands were not uncommon, for instance when a member had overshot his entitlements or ignored the obligation of auctioning catches, but soon they grew increasingly rare as the co-management group's benefits became apparent and the members internalized group norms. A simple warning often sufficed to end any conduct that might breach group rules. However, the board members believed that being too lenient was not an option, as this would sap their authority. At the same time, they did not consider it to be their task to be overly strict, as this would undermine the support for and the legitimacy of the group. They were consequently treading on eggshells. One of the board members said: 'You have to be careful, for the members might dig in their heels. You impinge upon their identity and dignity. However, if it works well, everyone is satisfied, but you should not behave like police officers.'

There were also some teething troubles. In the co-management system's early stages, the Texel fishermen discovered that other Biesheuvel groups included 'ghost boats'. These were small craft with one to twenty-five h.p. engines that were listed on the Fishery Register. The co-management group received the full number of days-at-sea for them, which it subsequently reallocated to the big vessels. The Texel group had no such boats and its members deemed it unfair that owner-operators from other fishing ports should have more fishing opportunities. In addition, the allocation of ten per cent extra days-at-sea applied to flatfish fishing, not herring fishing. As Texel had quite a few firms that combined flatfish and herring fishing, this led to further disadvantages. The Fisheries Directorate refrained from intervening as it regarded the issue a matter to be solved by the group board. Once the board had tackled the problem, the owner-operators began fully supporting the co-management system. In general, the Texel co-management group worked well. It created a strong community of interest. The board informed members when their quota and days-at-sea uptake surpassed a certain level and it facilitated quota swaps and leases within the group and with other groups. It also acquired cod and whiting quotas, part of which members could rent to cover by-catches. Nearly all Texel owner-operators eventually joined, while a few 'outsiders' – herring fishing firms from Scheveningen – were also members for some time. Only two Texel owners opted for membership in a Fishermen's Union co-management group.

With the Texel co-management group established, the skipper-owners deemed it expedient to establish a Texel Producer Organization, which they did in 1994. Until then, they had been members of one of the two extant Producer Organizations in the Dutch fishing industry, namely PO West. The islanders had been dissatisfied with the manner in which it functioned for several years and their influence on its internal affairs was minimal. Hiving off provided an opportunity to do things in the way the Texel owner-operators desired, and to link the co-management group's affairs closely with those of the PO. With these new voluntary associations, the Texel fishermen were over-organized rather than under-organized. Of course, they maintained the local fishermen's association Progress through Unity, the Fishery Cooperative and – with their Den Helder neighbours – the Cooperative Fish Auction Den Helder/Texel. In 1996, the Texelians established yet another association that exclusively dealt with collective vessel insurance. Running all these organizations was (and is) demanding on the small local occupational community of fishermen, and on the owner-operators in particular, although there were (and are) some overlapping directorates. The voluntary associations enhanced democracy, functioned as linchpins for the integration of fishermen and proved their worth in negotiations with external agencies impinging upon the (local) fishing industry. The owner-operators also benefited economically through the Co-op and the auction. By the end of the century, the Co-op's annual turnover amounted to between thirteen and fourteen million guilders and it provided jobs for nine employees. The Cooperative Fish Auction's annual turnover fluctuated between 89 and 107 million guilders in the 1990s. A completely new auction opened its doors in 1995. On peak days, it employed about a hundred fish sorters (fifty per cent of them women). Despite being members of the cooperative auction, a few Texel owner-operators continued to discharge catches in IJmuiden, as they had established good connections with a trader there. Once more, the occupational community's leaders exerted social control to conform to predominant group behaviour and land catches in Den Helder. In addition to all these local organizations, the Texel firms were members of national fisheries associations; about two-thirds were members of the Federation and a third were members of the Union, which had three co-management groups under its wing.

The Texel – and other Dutch – owner-operators of beam trawlers generally appreciate the co-governance system. It gives them a say in the management of the group and their own firm. It also increases their flexibility because they can transfer quotas and days-at-sea. In addition, it creates the certainty that they will catch and land their individual quotas, and do so when they expect landings to be economically most rewarding. Furthermore, it reduces the likelihood of dodging the rules and regulations by making the effects of overshooting quotas and other transgressions by a single skipper-owner felt on all the other group members. Moreover, it increases transparency through the system of mandatory auctioning in the Netherlands (Davidse 1997b:270). The co-management regime has thus enhanced resilience, security of rights, certainty, transparency, responsibility, cooperation and trust. It has also taken a lot of work out of the hands of individual owner-operators, as the group board is pivotal in quota transfers and takes care of the financial settlement. Perhaps most importantly of all, the system has put an end to the race for fish and has brought peace of mind. The fishermen value the current stability in the sector and the regained control concerning their day-to-day operations. As a Texel owner-operator remarks:

> The Biesheuvel groups work fine. The black market is gone. I purchased extra landing rights, like many others. If we were to avoid the auction, it would be detrimental to the prices and we would harm ourselves. If, on the other hand, everyone sticks to the rules, prices are much better and you do not have to go to court.

Indeed, one of the bonuses that attracted fishermen into the Biesheuvel system was that on average prices, and therefore economic results and profitability, would improve. The system allows for a better distribution of flatfish landings across the year with positive effects on market prices. Mandatory auctioning also leads to higher prices, since price-undermining illegal landings are now a thing of the past (van der Schans 2001:344).

In general, the beam-trawl fleet's economic performance was rather good in the 1990s. The new governance regime's outcomes proved to be generally beneficial and this led to a willingness to accept and work within the quota rules (Davidse 1998:66). The co-management regime reportedly brought about complete compliance with quota regulations, a drastic reduction of other offences and therefore administrative and political stability in the fishing industry (van Vliet 1998b:72). Consequently, enforcement officers met with significantly less resistance than previously and could do their job under less inimical circumstances. They did not need to board vessels and could monitor catches in the auction. This alleviated tensions, as fishermen perceived shipboard inspections as an invasion of their private domain (see Chapter 5). It also led to a substantial reduction of control and enforcement costs (van der Schans 2001:358, 371ff.). From having a reputation within the European Union of being completely unable to enforce quota regulations, the Netherlands was 'being widely regarded as a model of good landings enforcement, even if individual violations still occur' (Valatin 2000:300). Symes regards the Biesheuvel system as coming 'close to providing a template for group management' (1997b:113).

Having reviewed the governance system in 1996, the government decided to continue it, mainly because the problems with exceeding the national quotas had ended and the number of violations of the rules had decreased spectacularly. Thus, both owner-operators and the authorities seem to be satisfied with the outcome of the co-management process (Hoefnagel and Smit 1997b:175). Having experienced several benefits of the co-management system, the fishermen are generally content with the way it functions and feel relieved that the 'wild west' era of rule beating and quota busting is over. Other factors that have undoubtedly contributed to the Biesheuvel system's success are that the flatfish sector is relatively homogeneous and that it is entitled to a large chunk of the European total allowable catches for sole and plaice. The fish-processing industry and fish traders are also happy with the devolved management regime, since they can be sure of a steady supply of fish throughout the year. The Dutch beam-trawl fishermen even take pride in 'their' co-governance mode of operation. They boast about their 'good behaviour' and point to the fishing industries of other European Union member states where, in their opinion, illicit practices and lack of enforcement continue to be the order of the day. Generally, the fishing industry, the state, and the researchers herald the Biesheuvel system as a successful example of fisheries governance.[2]

To be sure, the Biesheuvel regime has not solved all the fishing industry's problems. Violations of rules still occur, albeit to a much lesser extent than prior to its implementation. The majority of the Dutch fishermen argue for stricter monitoring and enforcement to stop these illicit practices, so that they can be certain that gear and engine-power regulations are complied with more. They are of the opinion that rules and regulations ought to apply equally to all owner-operators. Therefore, they deem state coercion necessary, even though they are ambivalent about specific rules they view as 'unworkable', 'bureaucratic' or simply 'silly'. More importantly, many owner-operators feel that state institutions do not take them seriously enough, consult them insufficiently, and that their involvement and participation in fisheries matters are haphazard. In their view, groups and Producer Organizations should have more influence on the policy-making process (Hoefnagel and Smit 1997b:172). They contend that the state should listen to them and heed their advice. Although the Biesheuvel regime has delegated considerable responsibility to fishermen for quota management, government control with respect to the fishing industry as a whole is still great. In effect, the Fisheries Directorate 'determines the conditions under which the groups are allowed to manage their own affairs' (van Vliet 1998b:72). In addition, the government remains strongly involved in enforcement through its General Inspection Service (Berg 1999b:152). Indeed, participatory governance is extremely limited and primarily directed at group quota management. Through their national organizations, the Fish Board, Producer Organizations and national and local voluntary associations, fishermen attempt to exert some influence on the policy and management process, but in general the national organizations' rank-and-file members are not usually very pleased with what their leaders have been able to accomplish in their interests. The fact that there are two such organizations is generally perceived to be an obstacle in gaining political influence, especially now that the fishing industry is becoming smaller and smaller. The organizations' leaders maintain that the distance between them and the Ministry is considerable because they see new faces continually, and fishermen state that they do not have a say at all and are not listened to (Hoefnagel, Visser and de Vos 2004:36-37, 41, 44, 46). However, at the individual level, owner-operators of beam trawlers have had some leeway to steer their own course.

Rights Accumulators, Slipper Skippers and Quota Hoppers

Under the system of individual transferable quotas, many Dutch skipper-owners invested in landing rights (see also Chapter 5). Several did so with alacrity. The utilitarian logic of fishermen as economic actors resulted in a gradual concentration of fishing allowances in the hands of fewer skipper-owners. Those who were in a position to acquire additional entitlements from owner-operators exiting the Dutch fishing industry did so. The concentration trend slackened somewhat after 1994. There was less trade in quotas. The number of titleholders remained stable and quota prices continued to be high. However, to reduce the number of both vessels and quota holders, the state introduced the measure that entitlements were transferable only to vessels that already had landing rights for that specific species. As of the late 1990s, decommissioning of vessels again led to some concentration, since quota entitlements changed hands or were registered on one vessel instead of two or more boats. By then, the owners of big beamers of 1,500 h.p. and over possessed about eighty per cent of the national flatfish quotas (Davidse 1999, 2000). In 2002, due to low quotas and high fuel prices, the owners of eight big beamers applied for the decommissioning programme. It was the first time in all the years of fleet reductions that vessels of this size were decommissioned. More decommissioning rounds would follow (see below). They facilitated exit decisions, since boat owners were able to collect compensation and sell their vessel and quotas. With more landing rights entering the market, prices fell to a considerable extent, making it easier for those who hung on to accumulate additional entitlements. Investments in quotas partly absorbed the depreciation for new vessels. Consequently, the fishing fleet began ageing. In 1990, forty-three per cent of the cutters were less than ten years old, while fifteen years later this applied to less than a fifth. Currently, the launch of a new cutter is rare. With older boats, major breakdowns – for instance engine failures – become more likely. In crisis conditions, this could mean the end of many a weaker firm.

The vast majority of Texel skipper-owners also opted for investment in additional landing entitlements. They hoped that doing so would provide certainty about fishing opportunities. They therefore ardently advocated strict enforcement of quota regimes, this being in their direct economic interest. At the same time, Texel owner-operators doubted and feared the European authorities' trustworthiness. In their view, the certainty of exercising their rights was – and still is – lacking and they felt they could not depend on future policies. For this reason, some skipper-owners were reluctant to invest considerable sums of money in landing rights. They preferred renting rather than buying quotas, fearing Europe's unpredictable fisheries policy. It gave them the additional benefit of being able to wait for the right moment for leasing quotas: when they began running out of fishing opportunities and/or when prices were at a low. Often, this also proved to be a gamble. However, most of those who hung on have heavily invested in additional quota entitlements. Generally, they believe that only those who tenaciously cling on will economically survive in the end. With fewer units on the Fishery Register, there will be some room to breathe. With their relatively small fishing fleet, comprising seventeen North Sea cutters (August 2006) that represent only five per cent of the North Sea cutter fleet, Texel skipper-owners hold approximately ten per cent of the national plaice quotas and about twelve per cent of the national sole quotas. There is a catch, however. The relative shares may be secure; the actual quantities of fish fishermen are allowed to land are not, because they depend on the annual establishment of total allowable catches. Fishermen therefore deem wildly fluctuating quotas a serious threat to their economic existence.

Dramatic cuts of the plaice quotas in 1995 and 1996 (see Appendix D) caused turmoil in the flatfish fishery sector. The fact that plaice prices were on average seventeen per cent higher than in previous years due to the diminished supply alleviated some of the pressure. Usually, rising market prices for sole and plaice because of scarcity compensate for lower quotas. The fishermen are keenly aware of the fact that there is a ceiling to this compensation. Cheaper species of fish or pork, beef and poultry may provide alternatives that lessen demand for flatfish. In addition, quota cuts may be out of kilter with their actual experience. Abundant plaice catches in 1996 made for speedy quota uptake in the first few months of the year, even though effort was at a low. Many beamers remained in port in order not to exhaust quotas too soon. Sole catches were also extremely good. The fishermen argued that the North Sea was teeming with flatfish and that the biologists' advice to cut back on total allowable catches had been completely mistaken. They demanded an interim increase of the quotas, to no avail. Conversely, quotas may be too high. The very same year that the catchability of flatfish was beyond expectation, the European Commission took a drastic measure: mid-year, it reduced the herring TAC by half. This meant that Dutch herring fishermen faced sudden thirty to forty per cent reductions of their landing entitlements. Nineteen Dutch family firms still pair trawled for herring. They were outraged and protested. Seven of them hailed from Texel, where herring fishing in combination with beam trawling was still important in the annual fisheries cycle. The fishermen believed the measure to be illegal because no compensation had been offered. Understandably, they deemed it unfair that the European Union could subtract from their individual quotas, part of which they had acquired at great expense. Such ad-interim adjustments also thwarted their initial fishing plans. Despite being rights holders, there are no constitutional guarantees against changes in the system. This was a decisive moment for the co-management system's resilience, for chances of non-compliance were considerable. However, the Dutch flatfish and herring fishermen perforce and grudgingly adapted to the circumstances, although the herring-cum-flatfish fishermen subsequently began selling their herring entitlements. Soon after the millennium, not a single Texel owner-operator had any herring quota left. The offshore fishermen became increasingly dependent on flatfish entitlements, and there were more demerits.

Although related to individual transferable quotas, and not specifically to the Biesheuvel regime per se, the co-management groups tend to look after the interests of their *present* members who are likely to resist any change that would diminish the value of their property rights. They do not hesitate to go to court to seek compensation if new measures threaten to reduce the value of their assets. The state automatically renews their entitlements and as long as they retain their value, the rights holders will remain satisfied. Due to the owners' 'campaigning skills and ability to mount legal challenges' (Valatin 2000:306), there is little room for change. To become a group member, a fisherman must already possess an individual transferable quota, a fishing licence and a vessel. Starting a firm from scratch is nearly impossible, since outsiders cannot obtain a licence and quotas unless they buy a firm. The value of entitlements makes the costs prohibitively high. Therefore, aspiring newcomers are effectively barred from entry to the fishing industry, as access boundaries prove to be insuperable, while the initial introduction of individual quotas has unintentionally created a 'millionaires' club'.

The common pattern of fissions of family firms is also nearly impossible. Dividing entitlements would jeopardize maintaining an economically viable firm. Even continuing a family firm is extremely difficult because high prices have to be paid (Dubbink, van der Schans and van Vliet 1994:33; Hoefnagel 1996:68). The value of quotas usually exceeds the vessel's value. It used to be possible to bequeath these allowances, but when tax inspectors started to take into account their value, succession duties and other taxes rose phenomenally. The same applied when a retiring co-owner transferred rights to his agnates. Even if a brother or another co-owning relative was prepared to sell his share at a reasonable price, the tax collector still demanded the percentage of the estimated value, not the percentage of the actual price paid. Consequently, it became increasingly difficult to continue a family firm. Since the sense of continuity linking generations of fishermen is so pervasive, this situation has created exasperation, as the traditional pattern of succession and inheritance is no longer a matter of course. Many owner-operators have changed the juridical form of their firms into limited liability companies, in order, among other things, to make succession of ownership easier (Davidse and de Wilde 2001:33). Special tax arrangements have meanwhile made succession from father to son easier. It is now possible to bequeath rights without duties. However, if a co-owning brother who does not have a successor wants to pass on tangible and intangible assets to a nephew, there is still a heavy tax burden. In several instances, it has forced siblings to continue operating a firm together much longer than they would have done previously. Heirs who inherit quotas face the need to take out loans to buy out those heirs who do not fish. Thus, a new generation of skipper-owners has to produce at higher costs than the previous one (Davidse 1997b:107, 217). The organizational character and legal structure of the family firm have also changed in other respects. Along with the increasing value of assets in fishing firms, most spouses have opted for a marriage settlement instead of community of property, as used to be commonplace until the 1980s. If married in community of property, a spouse is liable to an equal portion of the joint property. If her husband predeceases or divorces her, she can demand her share of the assets tied up in the firm, making it extremely difficult to continue it (Hoefnagel 1996:68).

Whereas management decisions used to be relatively simple and were made with short time horizons, with the introduction of individual transferable quotas the fishermen increasingly need the knowledge and skills of external specialists who can advise them on administrative, legal and fiscal matters related to landing rights. Acquiring additional quotas also means that skipper-owners have to adopt a long-term perspective, since it may take several years to get returns on these investments. On the other hand, several so-called 'sofa fishermen' or 'slipper skippers' did manage to live comfortably by leasing out their entitlements without ever fishing. If owner-operators sold or decommissioned their boat, they could officially hold on to their quota entitlements for a two-year term. However, many circumvented this rule by keeping their vessel without actually operating it or by transferring landing allowances to tiny boats listed on the Fishery Register. They then leased the rights to other fishermen, cashing the money as a kind of old-age pension. Young skipper-owners had to buy or lease quotas at disproportionate prices, and so the younger generation was being pitted against the older. Many argued that beneficiaries of quota entitlements who were not actively fishing should give up their rights in favour of active owner-operators. Several loopholes enabled slipper skippers to continue their practice, however. They operated according to the letter, but not the spirit of the law. To tackle the problem of slipper skippers, the 2005 decommissioning details stipulated that quota entitlements should be either used on new vessels or sold within a three-year term.

Many flatfish fishermen who could not buy or lease quotas to match vessel capacity were ousted from the business. There was an escape route, however. Along with the tightened enforcement of quota restrictions and the soaring prices of landing entitlements, many Dutch owner-operators began developing an appetite for foreign fishing opportunities, especially upon discovering that some European countries did not take up their total allowable catch shares of sole and plaice and had a national rather than an individual quota system. Enabled by the European Union rules of freedom of establishment and free movement of capital, scores of flatfish fishermen bought vessels and purchased licences and fishing entitlements or re-registered abroad. Since 1990, approximately a hundred Dutch-owned flagships have reregistered in other European countries, predominantly in the United Kingdom, Germany and Belgium.[3] The re-flagged vessels fish against other countries' shares of total allowable catches. The number of

Texel quota and flag hoppers has been small, however. During the past three decades, only four firms re-flagged vessels or acquired foreign fishing rights. Today, one Texel family firm operates a Scottish-registered twin-rig vessel in addition to a Texel-registered beamer, while in the course of my second stint of fieldwork another firm, which also operates two vessels, bought a German-flagged Euro-cutter after decommissioning a big beamer. Flying a foreign flag has not been popular with Texel skipper-owners, who prefer accumulating quotas. Urk owner-operators, in particular, have used the re-flagging strategy to acquire sufficient landing entitlements to match a vessel's capacity or set up a son with a business of his own. Over half of the Dutch units that fly flags of convenience originate in this fishing town. Thus, re-flagging and 'quota hopping' has enlarged the room to manoeuvre in a situation of national limitations (Hoefnagel 1998:82; Davidse and de Wilde 2001:8). In 1998, the re-flagged fleet accounted for approximately twenty per cent of Dutch North Sea flatfish fishing in terms of vessel number, engine power, crew, fishing effort and gross revenues (Davidse 2000).

Their compatriots at home do not always view quota hoppers favourably. The reason is that some fishermen had their vessels decommissioned and were compensated in part with financial contributions levied from the fishing industry. With this money, they bought a fishing licence abroad and in so doing remained competitors fishing for the same species in the same European pond. Some owner-operators opine that there are too many foreign-flagged units that specifically target plaice, making for pressure on the resource. Understandably, fishermen in the host countries were not amused by the 'invasion' of Dutch interests either. In the United Kingdom, especially, there was fierce opposition and resentment against the re-flagged fleet. Dutch vessels fishing under UK registry landed over sixty per cent of plaice and approximately eighteen per cent of sole from the UK quotas in the Netherlands. The vessels remained Dutch-owned, skippered and crewed, and in addition to landing their fish at Dutch ports, they usually bought fuel and fishing equipment at home. Potential economic benefits were consequently lost to British regions. Forgetting that it was their compatriots who had voluntarily sold their vessels and licences for handsome prices in the first place, UK fishermen accused quota hoppers of undermining their national fisheries' interests and depriving them of their 'birthright'. When the Grimsby-based North Sea Fishermen's Organization – a Producer Organization with a small majority of re-flagged Dutch vessel owners – sought to join the national Association of Fish Producer Organizations, heated discussions ensued and several POs resigned their membership. Confronted with the fishermen's hostile response and the political mobilization of their national organizations, the UK government started supporting them and sought ways of curbing quota hopping and subjecting re-flagged vessel owners to a criterion of nationality. The measure unequivocally aimed at protecting national interests. However, referring to the 1957 Treaty of Rome, the European Court of Justice overruled the measure, although it legally allowed a 'real economic link' demand between the host state and the re-flagged vessels. From January 1999 onward, Dutch owner-operators who re-flagged their boats had to meet several criteria to prove that they lived up to the 'economic link' requirements. These included rules pertaining to landing part of the catch in Britain, employing UK residents, and purchasing goods and services in the UK. In fulfilment of these requirements, Dutch quota hoppers returned a part of the quotas they had not used in the North Sea to British Producer Organizations in 1999. Following the British example, the Belgian government also introduced 'real economic link' requirements (Lequesne 2000). At least half the crew should hold Belgian passports and the vessels should land a minimum of fifty per cent of the catches at auctions in Belgium. In 2005, the Flemish government sought ways of confining quota entitlements to Belgian nationals (read Flemish owner-operators) – deliberately risking a reprimand by the European Court of Justice.

The quota-hopping controversies cast light on the inherent contradictions of European Union policies. The point of allocating quotas according to the principle of relative stability is to give member states their 'fair' share based on an implicit criterion of nationality. Quota hopping undermines this system, and leads to resentment and mistrust. However, quota hopping is legal under EU law. Consequently, there is 'the territorial logic of an economic sector' (Lequesne 2000:779), embedded in and protected by the nation state, while at the same time there is the contradictory process of building a European common market and the – unofficial – cross-member-state trade in quotas (see also Lequesne 2004). As economic actors, the quota-hopping fishermen did little else than use European rules of law to bypass protective national policies in order to extend their fishing capacity and maximize their individual profit (Lequesne 2000:780). The same tensions between national and supranational interests surface in conflicts between fishermen from different member states concerning fishing in territorial waters. These have occasionally given rise to fierce confrontations. For instance, French and Belgian inshore fishermen hate to see Dutch Euro-cutters fishing inside the 12 nm zone, even though their skippers act in conformity with European rules and regulations. Conversely, Dutch inshore beam-trawl fishermen sometimes clash with Danish set-net fishermen fishing for sole in the Dutch 12 nm zone – a mobile versus stationary gear conflict that has strong undertones of national territorialism, based on the belief that geographic proximity and historical use lend a right to preferential access. The myriad rules and regulations are often inconsistent or contradictory because they depart from conflicting objectives, for example (subsidized) structural growth and conservation policies. So, let us shift our attention to Brussels.

Square Pegs and Round Holes: In the Throes of 'Brussels'

National fisheries regulations must adjust to the Common Fisheries Policy. However, Dutch fishermen are less than satisfied with some of its aspects and its demerits for the Dutch co-management regime, primarily uncertainties caused by sudden policy changes. Although fishermen generally appreciate the national governance system, discontentment with bureaucrats and regulators continues to be widespread, as they are the messengers and intermediaries of what is concocted in the European Union's Brussels headquarters. Dutch fishermen are in favour of fewer, and more workable and uniform regulations within the Common Fisheries Policy framework that should be monitored and enforced in equal measure across all member states: the so-called 'level playing field' in the European Union rhetoric. What owner-operators desire is assurance: the certainty that rules apply to each and all equally and are enforced in equal measure. As it stands, they have the feeling that they are the 'most virtuous pupil in school', in that they feel they have turned into the most law-abiding fishermen and – at the same time – the ones most targeted by the authorities in the European Union, while fishermen abroad are cut considerably more slack.[4] As important stakeholders, the Dutch flatfish fishermen also feel utterly disempowered concerning the manner in which their landing allowances come about.

Each year in December, the European Council of Fisheries Ministers indulges in the lengthy political ritual of establishing total allowable catches for the upcoming year. It usually is an inscrutable payoff between biological assessments and the horse-trading of national fisheries and other economic interests. Dutch flatfish fishermen anxiously await the outcome. In addition to uncertainties about catches and market prices, they have to cope with the incertitude of how much fish they will be entitled to land and how many sea days the Dutch fisheries regulators will allocate. Substantial fluctuations of national quotas and sea days make it extremely difficult to develop long-term investment plans and lead to a sceptical view regarding the fisheries' future and the role of biologists in assessing fish stocks. Uncertainty and inaccuracy characterize most stock assessments (Sinclair 1996). As in the 1980s, the flatfish fishermen continue to be sceptical about and frustrated by the biologists' guesstimates and the short notice at which quota measures are given, while they are often disappointed with what the Dutch Fisheries Minister has accomplished in their interests at the negotiation table. From one year to the next, the sole and plaice quotas often vary considerably (see Appendix D). It is easy to imagine that this gives rise to a state of apprehension. In the fishermen's view, such fluctuations now largely determine the outcome of their enterprise, instead of the hard work, stamina, nerve, knowledge, prowess and dexterity that used to be the core values of their métier (see Chapter 5). Today, nightmares about catastrophic bureaucratic caboodle probably match their dreams of bumper catches. Brussels currently pervades their thoughts and conversations. Regulations, and shifts in them, largely determine work. Since the practice of fishing is pervasive and existential, they have wider implications for experiences, social relationships and knowledge.

The fixing of total allowable catches often leads to communication problems between managers and scientists on the one hand and fishermen on the other. For example, I often heard fishermen exclaiming that the North Sea was teeming with plaice and sole, and that biologists just used the wrong methods to assess flatfish stocks. It annoys them that biologists will never apologize for getting it wrong. The owner-operators prefer multi-species, multi-annual and more stable total allowable catches that would provide for long-term planning instead of being continually yo-yoed. 'This would bring more peace and stability and we would get rid of the annual circus,' Fishermen's Union leader Johan Nooitgedagt maintains, referring to the mid-December haggling about and issuing of total allowable catches. Uncertainties pertaining to stock levels hinder not only biological management but also economic management at the industry and individual firm level. Fishermen deem annual quota allocations volatile and utterly unsuitable to the running of a business. Reductions in allowable catches potentially endanger the economic viability of firms – which happens for example when a firm's sole quota is around thirty metric tonnes (van Wijk and de Wilde 2004:63). Given the dominance of the biologists' input in the EU policymaking process, the socio-economic consequences of the occasionally heavily fluctuating total allowable catch levels are usually insufficiently included in the implementation process. To circumvent the problem of receiving the same percentage of the total allowable catch for sole and plaice, a consequence of the Common Fisheries Policy's principle of relative stability, Dutch flatfish fishermen and their organizations also prefer freely transferable quotas across member states. This would give them the opportunity to buy or lease sole and plaice quotas abroad. However, this is not legally allowed, although in fact quota hopping and re-flagging practices amount to much the same. Lastly, they are in favour of having some role in fixing total allowable catch levels. This is linked to the fact that 'fishermen often do not trust the assessments of crisis on which state action is based, and also lack faith in the effectiveness of the measures taken' (Bavinck and Hoefnagel 2004:41). The data biologists come up with often hold little water with fishermen. The latter usually refer to their 'experiential knowledge' with respect to stock levels and have little confidence in the biological analysis of fish stocks because they do not match with what the fishermen believe to be right. 'They're behind a computer screen all day, while we are men of practice who are at sea day and night,' says a Texel owner-operator. Power issues are inherent in knowledge claims and biologists would seem to have the greater authority and credibility in the ranks of fisheries policymakers and regulators. The fishermen's epistemology of practice is usually bluntly ignored.

Biologists, on their part, distrust the Dutch flatfish fisheries governance system. They maintain that its efficacy is mainly because fishermen simply cannot take up the national quotas for sole and plaice. In some years, the exhaustion percentage has indeed been below a hundred per cent. Some 'cynically suggest that compliance and the political-administrative rest are bought off with far too lenient TACs' (Dubbink and van Vliet 1997:192). In their reports to the International Council for the Exploration of the Seas, biologists adhere to the objective of what they perceive as rational exploitation: maximum sustainable yields, instead of the minimal biologically acceptable level of exploitation that was the state's point of departure. In their view, this would require a drastic reduction of fishing effort. The biologists felt frustrated by the national states' attempts to obtain the maximum for fishermen in negotiations on total allowable catches and quota allocations, which would potentially lead to irresponsible exploitation levels. Fisheries biologist Ad Corten expressly mentioned the Dutch co-management policy, 'in which the objective of rational exploitation was explicitly abandoned... The new policy would aim ... merely at maintaining stocks above a minimum biologically acceptable level ... Ministers and administrators increasingly consider quotas as amounts of "paper fish", which should be large enough to avoid any inconvenience to the national industry' (1996:5). In addition, biologists contend that the individual transferable quota system leads to the wasteful practice of high grading and discarding of low-value fish or species for which quotas are lacking or exhausted. In the 1990s, biologists and fishermen strongly contested each other's views in the weekly *Visserijnieuws* (Fishery News) and occasionally owner-operators have stopped cooperating with the Netherlands Institute for Fisheries Research's biologists.

Earlier, fishermen were more accommodative. For example, in 1989 the Dutch fishing industry fully cooperated in establishing the Plaice Box, an area of about 40,000 square kilometres of shallow coastal waters extending across the coasts of the Netherlands, Germany and Denmark. To protect important plaice nursery areas, fishing with vessels exceeding 300 h.p. was seasonally restricted there. As of 1995, it was banned altogether. However, despite a ninety-five per cent fishing-effort reduction relative to the original level, yield, recruitment and spawning stock biomass in the Plaice Box have decreased (Pastoors, Rijnsdorp and van Beek 2000:1021; Grift et al. 2004:6). When the measure proved to be a failure, the fishermen were very disappointed that the authorities refrained from repealing it. In an attempt to counter eco-political claims to establish more marine protected areas, beam-trawl fishermen often allude to the Plaice Box example. They contend that the seabed needs to be 'ploughed' with tickler chains. In their view, sole and plaice stocks will thrive when they do so: 'Where people fish you catch fish.' A Texel skipper-owner, born in 1914, says, 'We used to fish the grounds off Terschelling. We call it The Ridge. The grounds are hard there and you never caught sole. Currently, with tickler chains, you always catch sole there.' Beam trawling, the fishermen claim, enhances the growth and reproduction of flatfish. Indeed, empirical evidence substantiates this 'fishermen's lore': flatfish feed on crushed benthic organisms in the trawl path made by another trawler a short time earlier (de Groot 1984:180; Rijnsdorp and van Leeuwen 1996:1199). Fishermen say that using too heavy chains should be avoided and that there should be intervals between tows to allow time for the sole and plaice to return. They therefore keep away from positions that other vessels have trawled a few minutes earlier.

Marine biologists and ecologists would seem to agree that beam trawling is not necessarily detrimental to flatfish stock reproduction per se, but they argue that it does lead to a loss of biodiversity. They claim that the beam's shoes and tickler chains destroy seabed life (benthos) and that only opportunistic species thrive (de Groot and Lindeboom 1994; Lindeboom and de Groot 1998). Beam trawling would therefore create 'fishing fields' (*visakkers*). They further deem it a wasteful and bulldozer-like fishing method, since tickler chains allegedly penetrate as deep as ten centimetres into the seabed and 'plough marks' remain for many years. Therefore, marine biologists and ecologists – and their close allies the environmentalists – have attempted to reduce beam-trawling effort in the North Sea. In 1991, the Texel-based Netherlands Institute for Sea Research proposed introducing a 10,000 square-km no-take marine protected area in the North Sea in addition to the 'boxes' that the European Union had designated earlier (Bergman et al. 1991). The Dutch Fishermen's Union responded furiously, stating that fishing was already prohibited in vast zones and that the socio-economic consequences of additional closures would be devastating (Vissersbond 1991). Fishermen apprehend that once no-take zones have been established they will be there permanently, whatever the ecological effects. They point out that offshore windmill parks, land reclamations, sand and gravel extraction, chemical dumping sites, communication and energy cables, gas and oil pipes and platforms, marine reserves and shipping lanes already take up much of the space, in some cases in their favourite fishing areas. Ecologists and environmental NGOs, persistently promoting their viewpoint, nonetheless continued to emphasize the need for area closures. Yet area closures may be counterproductive in a literal sense.

Being ousted from the 12 nm zone and the 'boxes', bigger beamers faced more steaming time and thus higher fuel costs. Locations outside the Plaice Box and the 12 nm zone were subsequently heavily fished, particularly in the winter when sole are less scattered than in the summer. This congestion led to 'crowding externalities' as it resulted 'in the productivity of each trawl being reduced as the ground has already been partially "worked" by another boat' (Pascoe, Andersen and de Wilde 2001:202). Thus, closed areas have the disadvantage that owner-operators relocate fishing effort to other areas, intensifying exploitation there. The growing importance of the Dutch fishing fleet's Euro-cutter section is also an unforeseen consequence of this element of fisheries policy. Many vessels have been upgraded to 300 h.p., while others have been purpose-built. Several owners whose vessels were more powerful, but whose quotas were relatively small, decided to downscale their business and to use their quotas for a Euro-cutter. In 1983, Euro-cutters made up thirteen per cent of the fishing fleet, in 1998 thirty-five per cent, and currently forty-six per cent. A considerable part of these Euro-cutters fished sole and plaice as the main target species. During the same span of time, mid-size vessels (301-1,500 h.p.) almost disappeared (see Appendix C). Big beam trawlers of over 1,500 h.p. constituted almost fourteen per cent of the cutter fleet in 1983, and more than thirty-seven per cent in 1998, afterwards gradually declining to twenty-nine per cent at present. Consequently, the fishing fleet's structure became skewed towards beamers of around 300 h.p. and 2,000 h.p., the maximum engine capacities legally allowed for the two categories of vessels.

The environmentalists' opinion regarding beam trawling can be captured in two words: beaming stinks. In their view, it is an energy-consuming, indiscriminate and ecologically damaging fishing method. The fishermen are keenly aware of the ecologists' and environmentalists' criticism. They are of the opinion that information is often twisted. 'If tickler chains really went down ten centimetres into the seafloor, we would need 10,000 h.p. engines in order to tow the gear,' a Texel skipper-owner explained, 'otherwise you wouldn't move an inch.' 'And look at those chains: you can see that the wear is on one side only,' he continued, implying that the chains stay on top of the seabed. Unsurprisingly, fisherfolk resent the 'green mafia' and ecologists, whom they suspect to be political activists instead of 'value-free' scientists. They have good reasons for being suspicious. The fishermen's public image has gradually changed from hard-working people earning an honest livelihood to plunderers of the sea's resources who do not shy away from using ecologically damaging and illegal fishing methods. They are believed to pillage and rob. The pressure is on them and environmental organizations who lobby national and supranational governments increasingly have public opinion on their side. With the media exposure environmentalists and ecologists usually get, it does not even matter if their information is inaccurate. For instance, the Netherlands Institute for Sea Research ecologist Han Lindeboom (1995:593) claimed that beam trawlers were fishing the entire North Sea intensively – once to twice a year – and that for every kilo of marketable sole, five kilos of fish were discarded that would die and a kilo of benthos was killed. His alarming view was widely publicized. Subsequent detailed research by fisheries biologists showed that a large part of the by-catch was marketed and that the discards survival rate had not been taken into account. The Dutch flatfish fleet covered only forty per cent of the North Sea and seventy per cent of beam trawling took place in a fifth of this area (Rijnsdorp et al. 1998; Piet et al. 2000). Beam trawling was highly patchy. Seventy per cent of the fished area was visited less than once a year, and less than thirty per cent was swept more than twice a year. In conclusion, 'Individual vessels tend to concentrate their effort in a relatively small part of the potentially available area' (Rijnsdorp et al. 1998:409). This should not come as a surprise. Apart from avoiding rocks, rigs, wrecks and the like (see Chapter 5), skippers know from experience that not all positions are equally productive. The detailed information that countered the claim that the entire North Sea was fished did not receive significant media attention: the media would appear to be interested in headlines, not nuances. Nonetheless, in interviews and Op-Ed articles Lindeboom continued arguing that beam trawlers fished every inch of the North Sea and that closed areas were necessary (Lindeboom 2000). In claiming that the fisheries disturbed the seabed 1,000 times more than sand and gravel extraction and a 100,000 times more than gas and oil exploitation, he directly blamed fishermen for inflicting irreversible damage on the marine ecosystem.

The fishing industry is thus increasingly subject to the interrogation of ecologists and environmentalists who have influenced public opinion and manoeuvred themselves into the position of stakeholders in fishing politics. Greenpeace activists protested the European fisheries policy in general and beam trawling in particular through provocative actions at sea. The Dutch Fish Board responded to the plea for area closures in the North Sea by launching a brochure entitled *Fishing on a Square Inch*, pointing out that fishermen 'are finding themselves boxed in on all sides' (Productschap Vis 2004). The flatfish fishermen are meanwhile up against formidable forces. From once being ardent supporters of the fishing industry's modernization, European institutions began heeding the advice of biologists, ecologists, environmentalists and the public at large. A case in point is the inshore cockle fishing industry. The small Dutch cockle sector had its own co-management regime that worked rather well (Steins 1999:125ff.). Despite several self-limiting measures, cockle fishing with suction dredges met with vehement opposition from and litigation by environmental organizations. Environmentalists and biologists referred to 'plundering of conquered territory', 'systematic destruction', 'scene of battle', 'robbery', 'catastrophic ecological disaster', and – referring to seabirds that were allegedly deprived of food – 'slaughter', 'starvation', and so on (van Ginkel 2007a:136). Mechanical cockle fishing in the Wadden Sea was banned in 2004 following a Supreme Court decision based on a European Court of Justice ruling. A large company based in Harlingen owned two Texel-registered cockle cutters. Two brothers, their father and a non-related crewmember, all from Texel, operated the boat and subsequently became unemployed. Environmentalists and fishermen cast each other as the villain. In the Texel fishing communities, there is little sympathy for environmentalists. The fishermen consider them to be a threat to their livelihoods. They feel that, whatever they do, they are bound to get it in the neck from environmentalists and ecologists. In addition, Brussels is stifling their operations. Whereas European institutions had initially stimulated the fishing fleet's growth – leading to overcapacity – they began taking measures to contain its size and capacity.

In addition to the uncertainties that went along with determining total allowable catch levels, a system that continues to thwart the fishermen's comprehension and acceptance of broader policy and management frameworks, they experienced contradictions between this aspect of the Common Fisheries Policy and its other backbone, the Multi-Annual Guidance Programme (MAGP), which was introduced in 1983. Before 1992, its aim was largely to contain fleet capacity and prevent effort from expanding, but subsequently its goal changed to reducing aggregate engine power and gross tonnage to ensure that fleet size and catching effort were commensurate with fishing stocks (Pascoe, Andersen and de Wilde 2001:192). In the Netherlands, the preference has been for days-at-sea regulations and voluntary decommissioning schemes. These were socially more acceptable than a mandatory reduction of fleet capacity. The administration left to the owner-operators' discretion whether they preferred to fish with powerful vessels or not, as long as they did not overshoot their quotas, engine power limits and – since 1998 – gross register tonnage restrictions. Fully implementing MAGP targets, for example by lowering the number of days fishermen were allowed to be at sea, would jeopardize the uptake of individual fishing rights and national quotas and this could sap the Biesheuvel system and enforcement of legal landings (Valatin 2000:301). The centralized EU target of reducing capacity was increasingly at odds with the decentralized quota management responsibilities (Davidse 2000). As Fish Board chairman Dick Langstraat related, 'the MAGP straitjacket threatens to undermine the fishermen's support for our co-management system' (Langstraat 1998:12). In 1998, cuts on the number of days-at-sea led to commotion. First, members of the co-management groups no longer received ten per cent extra sea days compared with non-members. Second, the MAGP reductions were much larger than expected, in some cases amounting to a twenty-five per cent cut.

According to Johan Nooitgedagt, it was nothing less than 'a disaster'. His colleague Ben Daalder, who had succeeded Klaas Kramer as chairman of the Federation in 1997 (and van der Beek as chairman of Progress through Unity a year earlier), was of the same opinion. Fishing industry leaders argued that it was agreed under the co-management arrangement that fishermen should be allowed to take their rightful share of the national quotas, and that this right should prevail over the days-at-sea regulations. In late 1999, this led to tense relations with the Deputy Minister of Fisheries, Geke Faber, who wanted to cut sea days drastically to meet MAGP targets so that the Netherlands would not miss EU structural funds for the fisheries. The fishermen were outraged and felt punished for good behaviour. Faber was not very popular with the fishermen and gained the reputation of being a weak negotiator in Brussels. When it transpired in 2000 that senior fisheries officials proposed in an internal confidential report to close several important fishing areas so that Faber could 'score' politically with environmentalists and parliamentarians, the fishermen responded furiously, demanding that the civil servant responsible should be sacked. They deemed it a 'straight declaration of war on the fishing industry', as Ben Daalder said in an interview (*Trouw*, 13 July 2000). The responsible Minister reprimanded Daalder, but this could not prevent the fishermen from being completely fed up with her. The worst was yet to come, however.

> Thursday, 1 March 2001. Early in the morning, angry Dutch flatfish fishermen in a concerted protest action block a number of important waterways, including the ones leading to the ports of Amsterdam and Rotterdam. They are extremely dissatisfied with the sudden and panic-stricken cod-recovery measures of European Union Fisheries Commissioner Franz Fischler, which consist of a temporary cod-fishing moratorium and the closure for the duration of ten weeks of fishing areas in the southern and north-eastern parts of the North Sea as of 14 February. These are important sole and plaice-fishing grounds and the flatfish fishermen's by-catches of cod amount to less than three per cent. They are of the opinion that they are blamed and have to bleed for problems caused by UK, Irish and Danish cod fishermen. In the Netherlands, the number of fishermen specializing in cod fishing is only small. The vast majority of Dutch cod fishermen were ousted from the fishing industry in the 1980s and 1990s. What aggravates the fishermen is the fact that on top of the closure, sole and plaice quotas for 2001 are reduced considerably, while at €0.23 per litre, fuel prices are rather high. With nearby fishing areas closed, the fishermen are forced to steam further from home, increasing fuel consumption and expenses. It also means that precious fishing time is lost. A week before the blockade, fishing industry leaders negotiate with Faber to obtain financial compensation. Much to the organizations' leaders' chagrin, the Deputy Minister does not give in. She refuses to call on the European Commission for matching funds in spite of the fact that the House of Representatives moved a motion to do so. Ben Daalder is outraged: 'This has yielded zilch and zero. Terrible! Dead loss! It remains to be seen whether we can control the fleet now. Faber has turned her back on the fishermen. I am so angry with her.' Daalder fears that the closure will lead to a concentration of fishing in areas where it is still permitted. 'They will turn into graveyards,' he says, referring to the likelihood that catching effort will greatly intensify in the accessible areas. Faber leaves the fishing industry empty-handed again on Wednesday afternoon 28 February. Whilst on the fishing grounds, the fishermen learn the news about the breakdown of negotiations. Their mood is grim. They contact each other, decide on the blockade and begin steaming to the target ports with more than a hundred big beamers – including nearly all Texel-registered vessels. By eight o'clock on Thursday morning they effectively seal off the ports, paralyzing shipping. Other cutters leave the fishing grounds and return to their homeports in solidarity. It is the day's headline news. The fishermen regard the EU measure as a raid on the flatfish-fishing industry and point out that their by-catches of cod are minimal. Lacking Faber's support, they fear for their livelihoods and they are outraged that no one has ever consulted them. Dutch Members of Parliament and fisheries biologists also contest the EU measure. The politicians think that the mea-

> sure will hit the Dutch fishing industry disproportionately hard, whereas the biologists doubt the efficacy of the closure that comes after the cod spawning-season, is not targeted at the main cod-fishing grounds and in their view – shared by the fishermen – will merely lead to a concentration of fishing effort elsewhere in the North Sea. Even Dutch environmental organizations regard the measure as a useless flexing of muscles. Earlier on, the fishermen had proposed to voluntarily refrain from sailing for four weeks, but this tie-up proposal was rejected. With the closure pending and Faber refusing to compensate the fishermen, they have 'spontaneously' decided on the wildcat action that is subsequently backed by the national fishermen's organizations. Later that day, the fishermen dissolve the blockade. Under threat of being held accountable for the damages and having to pay recognizance, they capitulate and return to sea.

Soaring fuel prices compounded the cod-recovery plan and its ten-week tie-up scheme. Although by the end of the year economic returns proved less disastrous than the owner-operators had feared, the crews of big beamers faced a twenty per cent gross income reduction in 2001. Rising social security fees, which were linked with scale problems due to the declining number of fishermen (see below), meant diminishing net incomes. Much to the bewilderment and chagrin of Dutch fishermen, the European Union *increased* the 2002 total allowable catch for cod by two per cent. Understandably, they deemed supranational policies simply incomprehensible. Another instance is that the European Commission in 2001 decided that the system of mandatory auctioning, which the co-governance groups had agreed upon, was incompatible with European rules pertaining to the common market. The Commission argued that it deprived fishermen of the opportunity to freely land their catches at the port and auction of their choice. However, mandatory auctioning had greatly facilitated transparency and monitoring in the particular Dutch situation.

There were more demerits to Europe's fickle fisheries policy. In the wake of the cod-recovery programme, the EU implemented the days-at-sea system – which the Netherlands, as the only member state, had been using for quite some time to restrict effort – in the European Union as a whole. Subsequently, however, it severed the link between landing rights and sea days, which had negative consequences for the Dutch owner-operators who had previously matched vessel capacity with entitlements. Generic reductions of precious fishing time obstructed the opportunities of complete quota uptake for large rights holders. By the end of 2002, Fischler planned a forty per cent cut of days-at-sea for the Dutch flatfish fleet. The national fishermen's associations deemed this 'utterly unacceptable'. 'It really would deal the deathblow to the Dutch fishermen,' Fishermen's Union figurehead Johan Nooitgedagt exclaimed to a news reporter (*NRC Handelsblad*, 29 November 2002). By that time, Dutch fishermen hardly caught cod at all – not even as a by-catch. The expensive acquisition of additional entitlements to match vessel capacity – the preferred strategy of Texel owner-operators – proved to be insecure because rights holders had insufficient time to catch what they were entitled to land. 'What do landing rights mean if you cannot catch your quotas because there is insufficient time in which to do so?' they asked rhetorically. They loathed the fact that skipper-owners from other fishing ports who had refrained from acquiring quotas to match their vessel capacity could stay in business. 'It is just not fair. We have invested millions but we're the laughing stock of those who still indulge in illegal practices,' was the islanders' dominant opinion. The only Texel fishing firm that still specialized in round-fish fishing faced dramatic cuts of cod and whiting quotas for 2003. 'Three years ago,' related one of its co-owners, 'we still had a cod quota of 160 tonnes, but last year we only had 53 tonnes left and we have to hand in 45 per cent of that this year. Our whiting quota will also be reduced by 60 per cent.' Although the family firm had been investing heavily to accumulate additional landing entitlements, this was too much to cope with. Later that year, the four owner-operators – three siblings and the son of one of the brothers – decided to decommission their boat. 'There's simply no way you can run a business under these conditions.' Although less dramatic, quota reductions also profoundly affected what flatfish fishermen could land. An owner-operator related, 'In 1990 we were still entitled to land 650 tonnes of plaice and 193 tonnes of sole. This year [2006], it amounts to only 195 tonnes of plaice and 130 tonnes of sole, even though we acquired additional landing rights in the meantime.' It is just a sample of why, in the Netherlands, fishermen's support for the Common Fisheries Policy is lacking, intervening as it does with the rather well-adapted Dutch co-management regime that fishermen have come to consider as legitimate and just. They consider Brussels to be utterly unreliable.

One may even wonder whether there have been any enthusiasts at all for the Common Fisheries Policy. Its record of accomplishment has been excruciatingly poor, as Europe's fishermen readily acknowledge (Rossiter and Stead 2003:282; Connolly 1997). However, many ecologists, biologists, economists and social scientists also consider it a complete failure, albeit for quite different reasons (Cooper 1999; Symes 1997a:152; Gray and Hatchard 2003; Daw and Gray 2005; Rijnsdorp, Daan and Dekker 2006; Rijnsdorp et al. 2007). The CFP had myriad perverse effects. From an ecological viewpoint, the system of total allowable catches and quotas was counterproductive. It created more problems than it solved and the entire management system could easily collapse like a house of cards (Daan 1997:335). Although landing figures – which are often inaccurate – ended up in statistics, they did not represent actual catches. The negotiators did not digest the discrepancies as input for deliberations about new allocations. Increased enforcement led to high grading and, especially in the mixed fisheries, to discarding of over-quota catches and catches of species for which landing rights were lacking (Daan 2000; Pascoe, Andersen and de Wilde 2001:192). As we have seen, such wasteful practices did not sit well with the fishermen either. Economically, the CFP is said to be 'a scandalous waste of public resources' (Valatin 2001:11) in that resource rents have been dissipated or capitalized in individual quotas. Politically, it amounted to obscure interstate bargains. The policy is also a social disaster in that it protects vested interests while potential newcomers discover that the barriers of entering the fishing industry are insurmountable. Although partly designed to maintain the social peace in the European fishing industry through the principle of relative stability, its inconsistencies have led to many a conflict and nationally protective response. At the member-state level, the Common Fisheries Policy and its translation into a national regulatory framework has been socially divisive, creating wealth inequalities and impinging upon some of the core cultural values of fishermen, who now have to behave as calculating businessmen in an environment of extreme uncertainty and rigidity.

Whatever its merits in other domains of life, Brussels did a very poor job in the field of fisheries. Even European Union politicians and administrators began expressing doubts about the Common Fisheries Policy. It finally dawned upon them that they were trying to fit square pegs into round holes. The European Commission Green Paper on the CFP's future (European Commission 2001a) admitted that much had gone wrong and that more stakeholders should be involved in the policymaking process. To this end, it proposed to establish several regional advisory committees and devolve certain management responsibilities. It also called upon the national fishing industries and individual operators to submit their views on potential change. The Texel fishermen took this call extremely seriously. They believed that they would finally have the opportunity to have some input at the European level. They immediately set up a working group of seven young owner-operators, who enthusiastically drafted a proposal. After several rounds of talks, they produced a document. A Progress through Unity meeting discussed and accepted it in late July 2001. It contained a list of points elaborating the measures that the Texel skipper-owners would like the European Union to implement. They sent the document to the Fish Board, so that it could be used as input for discussion of the Dutch viewpoints on the CFP's future. It was the only document submitted to the Fish Board. The Texel proposal therefore constituted the basis of deliberations during a meeting with fishing industry representatives, held under the aegis of the Fish Board on 1 September 2001. With a few minor amendments, the Fish Board submitted the Texel response to the Green Paper to the European bureaucracy, officially as a national fishing industry response. The Texel owner-operators anxiously awaited the Common Fisheries Policy's reform, expecting to see at least some of their ideas acknowledged.

In the 'Roadmap' on the Common Fisheries Policy's reform (European Commission 2002), the European Commission openly admitted the policy's failure, a confession that amounted to a sort of covert mea culpa. The debate on its future had clearly revealed 'the shortcomings and internal systemic weaknesses of the Common Fisheries Policy, such as poor enforcement, the lack of a multi-annual management perspective, fleet over-capacity and insufficient stakeholder involvement' (ibid.:4). The lack of stakeholder participation and the unequal control and enforcement arrangements had undermined support for the policy and sapped its credibility and legitimacy. The debate also showed that 'the current policy is incapable of reversing the increasing threats to important fish stocks and of providing economic sustainability to the fisheries sector' (ibid.). The Commission believed that a future policy would succeed only 'if environmental, economic and social sustainability are put at the heart of its objectives and if good governance principles such as openness, participation, accountability, effectiveness and coherence are fully implemented' (ibid.). These principles were Brussels', new mantra following publication of a White Paper on governance (European Commission 2001b). The Commission proposed a host of new measures that should improve the situation. Predictably, the Texel fishermen found not one of their proposals incorporated. Their efforts of formulating ideas proved to have been a waste of time and energy. Deeply disappointed, the skipper-owners had their prejudices confirmed: Brussels does not care. The exalted Brussels rhetoric of openness, participation, accountability, effectiveness and coherence did not sit well with the fishermen, who continued to be confronted with measures that did not make sense to them.

The Texel fishing fleet's inshore segment also had its share of incomprehensible measures. Although inshore fishermen were not directly affected by the Common Fisheries Policy, European policies did impinge upon their métier. The only category of skipper-owners who did not seem to suffer a crisis were the operators of the three boats that take tourists on shrimp-fishing and seal-watching trips. Full-time shrimpers were less well off. Six Texel shrimp-fishing firms sold their boats and licences in the years following my first stint of fieldwork. With fierce competition and no catch limitations, shrimp prices were so low that owner-operators were hardly able to run a cost-effective enterprise. In the early 1990s, a major step ahead was made when Dutch, German and Danish Producer Organizations of shrimp fishermen agreed to a voluntary scheme of catch restrictions and minimum prices to regulate the market. Hereafter, their situation improved considerably. However, the Dutch antitrust authority deemed the system to be incompatible with EU common market regulations and in January 2003 heavily fined shrimp traders and Producer Organizations for price setting. The fines amounted to almost fourteen million euros. Dutch shrimp-fishing industry representatives argued that the owner-operators had acted in line with the Common Fisheries Policy and with PO responsibilities concerning market regulations. In addition, the scheme avoided shrimps having to be dispensed of for fishmeal reduction. In protest, shrimp fishermen blocked entry to the port of Lauwersoog. Following litigation, the fine was reduced to slightly over six million euros and then to approximately 2.2 million euros. The case is still pending, but it clearly brings to light the contradictions of European policies that may be interpreted differently. The intervention by the antitrust agency internally divided the shrimp sector. Shrimp prices plummeted due to increased landings, but consumers have not benefited from lower retail prices at all. Even extremely sensible self-imposed measures, as in the shrimp case, became a target for bureaucratic intervention.

European interference also undercut an important aspect of the Biesheuvel co-management regime. At the 2003 annual meeting of Progress through Unity, chairman and national fishing industry leader Ben Daalder pointed out that due to the cod-recovery plan, Dutch beam-trawl fishermen had been allocated fewer days-at-sea, even though their cod by-catches were low. Under the Dutch co-management regime, the allocation of days-at-sea was based on individual landing rights, while the European Union allocated sea days based on fishing gear, regardless of entitlements. For fishermen using beam trawls, fishing time was severely restricted. Daalder feared that this would sap the legitimacy of and support for the reformed Common Fisheries Policy. He made a dramatic plea to the European Commission:

> European Commission, for once please listen to the fishermen's informed arguments. We face measures that are at loggerheads with practice and that confront the fisherman with the impossible task of living up to measures such as the one-net rule,[5] an impracticable measure that increasingly widens the distance between the policymaker and the fisherman. It is a generic measure that puts all European fishermen on the same footing, as if all European fishermen work and fish in a similar way. This does not stimulate a factual dialogue to really tackle the problems. ... European Commission, why don't you listen to the fishermen's arguments?

Instead of bridging the gap, the Texel fishermen felt that the European Commission and the European policymakers were widening it. In subsequent years, Daalder frequently repeated his message that quotas should prevail over days-at-sea. Whereas in the late 1980s and the early 1990s there was a mismatch between fishing capacity and landing allowances, with the European Union days-at-sea measures a serious mismatch between landing rights and fishing days surfaced. It just does not make sense to the owner-operators to have a two-tiered, yet internally contradictory system of controlling effort. This inconsistency undermines the system's credibility, because it fails to meet the commitment that rights holders 'may take the full extent of their allocated quotas at times of their choosing' (Copes 2000:6). Moreover, apart from thwarting large rights holders from exercising their rights, there are certain negative consequences linked with reducing the number of sea days. Owner-operators tend to cut steaming time to allow for more fishing time and hence there is a further congestion of fleets and intensification of exploitation of particular locations just outside the 12 nm zone where young flatfish concentrate. High fuel prices exacerbate the problem, which is an additional reason for diminishing steaming time and avoiding the risk of fishing difficult grounds.

The manner in which the European Commission communicates is awkward and does not seem to be in line with the 'good governance principles' it so ardently claims to have embraced. Take for example the following. By mid-December 2005, the Texel owner-operators were rather concerned about the quantities of sole and the number of days-at-sea that the EU would set for the next year. There was a sigh of relief when the Council of Ministers announced its decisions. These included relatively small declines of the national quotas for sole and plaice and, initially at least, a number of sea days that seemed acceptable (although it would take months to establish the exact number, which fell five days short of what the skipper-owners had expected). Therefore, the tension that had been building up relaxed: the fishermen thought they would be left in peace – or at least relatively in peace – until the following December. However, in January, the European Commission launched a plan that the fishermen regarded as a disaster for the Dutch flatfish fishing industry. It proposed that sole and plaice TACs would be fifteen per cent up or down each year to arrive at a recovery of the stocks and to achieve maximum sustainable yields (European Commission 2006). In itself, this would have been disastrous enough for the Texel owner-operators, but what particularly aggravated them was the message's timing: just a fortnight after they had learned what the 2006 quotas would be. Moreover, they deemed the announcement premature. They believed that scientific evidence proving the need for such a drastic measure was lacking. It takes no more than a smattering of empathy to comprehend why fishermen resent everything that sprouts from Brussels.

Despite vehement contestations of European Union regulations that they consider useless, unfair or unacceptable, Dutch flatfish fishermen have become increasingly aware of ecological problems. This is partly linked with the opposition they encounter from environmentalists, but more so with the public image of fishermen as the archetypical 'Very Hungry Caterpillar', as Ben Daalder expressed it. We have already seen that they were prepared to restrict fishing effort to contribute to the recovery of cod stocks – even though their catches of cod were insignificant. The EU cod-recovery measures crosscut this proposal. In 2002, the owner-operators voluntarily agreed to restrict plaice fishing during the spawning season. The same year they held a symposium and discussed among themselves the question of how to arrive at sustainable fisheries, leading to the establishment of a Task Force Sustainable North Sea Fisheries, which has recently presented its ambitious recommendations (Task Force 2006). Moreover, of late owner-operators have begun to seek alternatives for the energy-consuming and controversial beam trawls. The efforts of finding alternatives are indicative of an altered state of mind. This also applies to the issue of self-monitoring. When the Dutch co-management system was up for review in 2003, the evaluation committee could not duly agree on a final report. The reason was that an extension of group responsibilities to monitor engine power and the use of illegal net provisions was deemed unacceptable as long as a level playing field in the EU was lacking. Dutch fishermen felt that in other member states monitoring and enforcement were suboptimal, to say the least. Nonetheless, the report appeared in 2005 and fishing industry leaders signed an agreement with the fisheries Minister, Cees Veerman, stating that the fisheries sector would fully comply with the rules pertaining to engine power and gear by disseminating norms and values through the groups and tackling and correcting each other (Stuurgroep Nijpels 2005). Although it was the Shipping Inspectorate that subsequently thoroughly checked, adjusted and sealed engines, the measure boiled down to fishermen becoming their own policemen through their co-management groups. In return for the fishing industry's cooperative stand, the Minister promised to cut the number of rules and regulations (which was in line with an EU intention of simplifying the CFP; see European Commission 2005). This show of mutual goodwill notwithstanding, the fishermen, or their leaders at any rate, believe that this is a nice gesture that will eventually amount to little or nothing: fisheries policy is, and will be, largely determined in Brussels. They are also rather cynical about all the good intentions that the European Commission announced in its 'Roadmap'. *Their* reality is that they feel they are being ousted from the North Sea.

Generally, fishermen continue to be dissatisfied with stock management measures and regulatory regimes. Their involvement through the co-governance system does not encompass policy design. Decisions are still top-down, as shown by the European Commission's recent proposal to introduce a plaice and sole management plan. In a press release, the Commission claimed to have extensively consulted stakeholders (meaning the North Sea Regional Advisory Council), but the fishermen do not think their voice has been represented. Even the Dutch fisheries Minister, Cees Veerman, was unhappy with the announcement as he felt that it would possibly frustrate the owner-operators' willingness to fully cooperate with the ongoing restructuring of the fishing industry. This does not mean that fishermen – as in the past – obstinately disobey the regulations; quite the contrary. However, they do generally regard them with suspicion, and with good reason. After concluding that the governance regime in the Dutch flatfish fishing industry seems to be 'proficient though not completely robust', fisheries economist Luc van Hoof points out that 'new EU regulations affect this segment in a non-coherent way' and that 'the rather well adapted system to the specific Dutch situation may become lopsided' (van Hoof 2005). Nonetheless, under the co-management regime the flatfish fishermen sought their own ways of adjusting to the problems they encountered. They committed themselves to a tie-up scheme of four weeks in 2004 and of eight weeks in 2005, a self-regulatory scheme implemented through the co-management groups and monitored by the national authorities. In the latter year, this measure was undercut in two ways.

Firstly, the uptake of sole quotas proved to be difficult so that the fishermen needed more time and, secondly, owner-operators from some fishing ports initially did not honour the agreement, creating a free-rider problem and undercutting the willingness of others to continue their participation. For example, the Texel fishermen – who had collectively respected the measure – announced that they would not voluntarily stay in port in 2006.

The eroding willingness of restricting effort originated in a rapid deterioration of the flatfish fishing industry's economic situation. Following the implementation of the co-management regime, it had experienced seven plentiful years. Seven biblically lean years would follow. Initially, the decline did not give rise to great worries. Fishermen were used to the vicissitudes of conjunctures and fluctuating revenues. However, owners began encountering difficulties in breaking even. Net results became increasingly negative. Quota reductions, less-than-average plaice and particularly sole catches, occasionally low market prices and higher costs – particularly of fuel – were the chief reasons. The owner-operators used their time-honoured strategy of cutting back on incomes, postponing investments and eating into their capital. The solvency of owner-operators declined from forty-four per cent in 2001 to zero three years later, meaning that they were dependent on credit and loans from financial institutions. This undermined the financial elasticity and the economic resilience of many firms. As a board member of the fishermen's association, a retired skipper, said in a March 2005 presentation: 'The mood is grim, the reserve-capital has been exhausted and there is no meat on the bones.' But the worst was yet to come.

Fuel prices had been high throughout the early 2000s, but they began soaring in the course of 2005. In January, the price of red diesel was still €0.23 per litre, but it had almost doubled by September. From then on, it would remain high. With a weekly consumption of around 30,000 to 35,000 litres for the beam trawlers' 2,000 h.p. segment and approximately 7,000 to 10,000 litres for the Euro-cutters, this implied staggering fuel bills, in many cases subtracting forty per cent and over from the gross proceeds. When in August 2005 the government announced that thirty-eight million euros would be available for voluntary decommissioning of fishing vessels, the enthusiasm to apply was substantial, even though it prohibited any commercial use of decommissioned vessels. A few months later, no less than twenty-eight cutters were decommissioned. Three of them were Texel-registered beamers. One Texel owner-operator downsized his family firm and bought a Euro-cutter. The other two family firms owned and operated two boats each. Given the high operating costs and the uncertainties regarding future fishing opportunities, they deemed it wise to concentrate their quotas on one vessel. To add to the misery, the catchability of sole had been poor, and with the number of days-at-sea restricted, uptake was well below targets (it dwindled from ninety-five per cent in 2004, via seventy-nine per cent in 2005 to only sixty-five per cent in 2006, rising again to eighty-seven per cent in 2007). It is important to mention in this respect that around the millennium, in the mixed flatfish fisheries sole landings on average accounted for half of the gross revenues with full quota uptake (with plaice bringing in a quarter). The situation also had implications for the value of quotas, which had been decreasing considerably on both the sales and the rental market. For example, in May 2006 the lease of the landing rights for a kilo of sole was €0.80, a quarter of the price paid twelve months earlier. Over the past five years, the Dutch fishing fleet's cutter segment has faced negative net results, amounting to as much as eleven million euros in 2005. The owner-operators' equity capital has vanished completely. Fishing industry pleas for yet another round of decommissioning were getting louder and louder, and in early 2008 another twenty-three big beamers were decommissioned, although this time none hailed from Texel. The successive decommissioning rounds have cut overcapacity to a considerable extent, but many fishermen fear that a substantial part of the Dutch fishing fleet will nonetheless go bankrupt. Although banks are likely to accept a brief postponement of redeeming loans and paying interest, their patience will not extend beyond a year or so. Moreover, they will hesitate to furnish skipper-owners with new loans. The depression also affects the crewmembers because they are co-adventurers. Apart from that, they have been facing their own particular difficulties for quite some time.

The Deckhands' Bones of Contention

On the island of Texel, it became increasingly difficult to recruit crew in the course of the 1990s. As long as revenues were good and non-propertied fishermen benefited, the partnership contract and the share system of remuneration worked well. However, as we have seen in Chapter 4, following the oil crisis, the North Sea fleet's skipper-owners changed the share convention. They argued that the increasing costs of a vessel and equipment had altered the balance of risks and necessitated a greater boat share. This meant a shift away from the fifty–fifty arrangement that had been common for decades in the offshore fleet to a fifty-five per cent vessel share as of the mid-1970s. The figure allocated to the crew has thus become smaller. The unilateral decision went against the grain for the deckhands, who felt the old convention to be a moral duty. Although they opposed the change, they were powerless. Compared with onshore jobs, the remuneration was still good although work hours were long. Deckhands obtained excellent incomes. As there was no shortage of hands, boat owners tended to tell protesting crewmembers: 'If you don't like it, just leave'. They could pick other crewmembers from among a pool of stand-in deckhands who were willing to work for lower shares and still earn good money. These events affected the mutual relationships between skipper-owners and deckhands and the social distance between them increased. Although many crewmembers deemed a temporary change of the share division acceptable to enable owners to weather the depression, what aggravated them was the fact that the latter did not redress it when in the late 1970s things started to improve. On the contrary, renewed discussions pertaining to a revised partnership contract came about, with skipper-owners arguing that their capital input had increased following the introduction of quotas. Deckhands distrusted it and many initially refused to sign the contract. One-sidedly changing the share convention was indicative of asymmetric relationships. The egalitarian rhetoric notwithstanding, the power balance was clearly lopsided in the owner-operators' favour. Even trivial signs of hierarchical relationships sometimes gave rise to cantankerous remarks. During a fishing trip, a twenty-eight-year-old crewman-cum-cook told me of his 'old-style' skipper: 'He never clears away his plate and cutlery or banana peel and things like that. He expects you, as the cook, to clear the table, but other skippers do it themselves.' On the other hand, it was no exception that deckhands remained loyal to a firm for twenty or more years and appreciated relations aboard.

If mutual trust suffered from the bickering regarding the partnership contract and the share system, it received another blow when the boat owners changed the percentages once again in the 1980s. The acquisition of more powerful vessels and additional quota entitlements required considerable sums of money. Therefore, most owners demanded a fifty-eight per cent vessel share. Officially, the partnership contract gives joint responsibility to skipper-owners and deckhands when it comes to fishing, but in practice the former usually decide on when, where and how to fish. Deckhands have no influence in the running of the business such as decisions on investments and quota transactions. The skipper-owners argued that with the altered share division the net incomes of crewmembers would still be high since increased landing entitlements would imply greater revenues. Even though elsewhere in the Netherlands an even worse sixty–forty arrangement became common, the Texel deckhands were not amused. For them, it felt like déjà vu, leading to a further deterioration of their trust in and respect for the skipper-owners. A conflict was smouldering. The deckhands perceived it as a grave injustice that they were 'partners', yet had no say whatsoever when an owner changed the share convention. Many believed it unfair that owners and not vessels received quotas. This way, owner-operators became 'quota millionaires' while deckhands – who contributed their labour, knowledge and assiduity and thus were co-responsible for the results on which entitlement allocations were based – stood empty-handed. Although the share convention changed, they had to pay their part of the costs that were subtracted from the gross proceeds. Part of the crewmembers' dissatisfaction was also linked with the partnership contract's terms that specified that the rent of additional quotas and fines for rule beating would be subtracted from the revenues and hence partly from the crew share. Fines were tax deductible for owners, but the deckhands paid their relative share without having the tax-deduction opportunity. There were also instances of abuse by skipper-owners. Some of them first rented out a part of their entitlements – cashing the money without any redistribution to the crew – and then later had to lease additional quotas for which crewmembers had to pay part of the bill. One of the skipper-owners transferred part of his quotas to a small boat and then rented it back to the beamer (see below). He deducted the rent from the crew share, in fact making the deckhands pay for quotas he already owned. Two other skipper-owners sought to subtract the costs of quota entitlements they purchased from the crew share. It must be emphasized, however, that the vast majority of skipper-owners and the local fishermen's association's board unequivocally condemned such practices.

Another bone of contention was that the item 'costs', subtracted from the gross revenues, often tended to be rather diffuse and varied from cutter to cutter. Some skipper-owners were transparent and put bills on the table so that the crew could check them. 'You need each other, so if the deckhands complain about the net proceeds, you have to show them the bills. You should pay attention to their grievances,' explained a skipper-owner. Others, however, were less clear and subtracted standard sums of money. This gave rise to quay talk and gossip. The deckhands demanded more clarity and more uniformity. Progress through Unity chairman Bas van der Beek warned: 'One could almost forget that the partnership is a cooperative venture and start thinking in terms of a boss-servant relationship, with all consequences this entails.' Keeping the troops united and closing ranks was his aim. He cherished the cooperative ideology: all *should* work together. In order to subdue the deckhands' mounting discontent, in 1995 the fishermen's association's board began working on a new partnership contract that would be supported by owners and crew alike. The revised contract stipulated that owners should use up their entitlements before leasing additional landing allowances and could not subtract costs for the acquisition of additional quotas. In addition, the association provided for an arbitration by-law and an arbitration committee with representatives from both parties. The committee almost immediately had to address some moot points following complaints by a few deckhands. It referred some cases to the Federation for arbitration. The Progress through Unity board also commissioned an investigation into the issue of cost subtractions on the settlement. Accountants compared the ways in which owners handled the settlements, but found no irregularities.

A conflict between a deckhand ('Jan') and his skipper ('Dirk') would cause considerable turmoil in the occupational community. For a number of weeks, Dirk had been subtracting the rent for leased quotas from the gross revenues without deliberating with the crew. It amounted to a considerable sum of money. When it transpired that he had transferred part of his own quotas to a small boat and then 'rented' it back, a conflict was born. Jan set himself up as the spokesperson for the crew. He wrote a letter of protest, had the boat's entire crew sign it, and handed it over to the skipper. The latter responded angrily and threatened to sack the lot. Consequently, all deckhands apart from Jan dropped the complaint. According to

Jan, feelings of fear and guilt underlay their submissive behaviour. They even chose the skipper's side when Jan sent a letter to the editor of the national fishery weekly, *Visserijnieuws*, in which he expounded his view on what had been going on. Most fishermen considered washing one's dirty linen in public as treason, and Jan was more or less socially ostracized. Dirk sacked him. With his sense of justice breached, Jan continued fighting for justice to be done. He contacted a solicitor, wrote letters to the Progress through Unity board, phoned its chairman, Ben Daalder, and during the association's 1997 annual meeting, took the floor. The mood grew tense. He explained that he took pride in the fishing fleet, the growth of which the partnership contract had helped accomplish. However, the contract should be based on mutual trust and unity, but this was obviously lacking in the case at hand. Daalder replied that he agreed that the dubious practice of hiring back one's own quotas was wrong but that this practice had since stopped. Two skipper-owners exclaimed that 'it is all in the past now'. A board member, one of the deckhands' representatives, supported Jan and said that the crew should protest in unison if things like that happened again. After having crewed on another boat for a while, Jan eventually left the fishing industry, deeply disappointed and empty-handed. The Federation's appeals committee turned down his case. Although very few deckhands supported him in public, they often discussed his case on the quay and aboard the beamers. The crewmembers had a sneaking sympathy for Jan and their view of the skippers' attitude and mentality was changing.

With the local partnership contract revised and the association's consensus opinion of ending questionable cost subtractions, it would seem to be a case of all's well that ends well. However, the events had driven a wedge in the ranks of Texel's occupational community of fishermen. Owner-operators and deckhands began drifting more and more apart. The latter remained disgruntled for various reasons. When they asked to set a minimum crew percentage of forty-two per cent in the contract, the owner-operators refused, arguing that the acquisition of additional quotas meant higher exploitation costs. They also had to mind their competitive edge, since in other fishing ports the boat share was sixty per cent while on Texel it was fifty-eight per cent. They felt they had no financial slack and could therefore not contractually guarantee the fifty-eight–forty-two deal. Crewmembers further pointed out that contractual certainty was lacking. Skippers could sack them without much ado. It only took a four-week's term of notice to fire a deckhand, who then became unemployed without any kind of benefits. A particular case in point was when boat owners decided to sell or decommission their vessel. What bothered the deckhands was that an owner usually did not inform them about his plans until the notice term. The skipper knew that the deckhands would probably attempt to find a berth on another beamer long before this date if he told them any earlier. He therefore used the deadline of telling his crew in order to be able to continue operating the boat until the very last day. Although there were compensation measures for young crew of boats that were decommissioned, usually consisting of money for vocational training, it was yet another reason for growing mistrust and grumbling.

The state of affairs had a negative effect on the outlook of crewmembers on their occupation. Their pride in being a fisherman began withering away and they mentally dissociated from the occupation. Whereas in the past they identified with the boat they worked on and the skipper-owner they crewed *with*, they currently consider themselves to be working *for* a skipper-owner. They no longer perceive fishing as a way of life but simply regard it as backbreaking work to earn a living. The partnership contract stipulates that crew can also leave a boat after a four-week's notice and of late, many a skipper-owner has faced the problem of finding sufficient deckhands. The pool of stand-ins, which still numbered twenty to twenty-five men in 1992, declined to a handful. It has subsequently vanished. Whereas in the early 1990s most beamers were overstaffed: instead of the legally required crew of six they often had seven or eight men who all shared in the gross revenues, by the end of the decade this was no longer the case. Job satisfaction diminished when the deckhands saw their incomes declining and their labour hours increasing. To ensure their quota uptake, most skippers began sailing on Sunday evenings at around 10 p.m. instead of Monday mornings at 8 a.m., which ate into the crew's weekend leave. Upon their return on Friday, small repairs and preparations for the upcoming fishing week would last for the remainder of the day. Having had few opportunities to get sufficient sleep aboard, the crewmembers returned home exhausted and had hardly recuperated before sailing again. In 1998, several young Texel deckhands protested against sailing on Sunday evenings. They considered it 'socially unacceptable' and demanded 'a real weekend'. Some said they were beginning to watch the clock by Sunday afternoon. The issue remained under discussion during subsequent years, but the skipper-owners did not budge. They argued that the deckhands benefited financially through the share system and were compensated with extra time at home through the days-at-sea regime that restricted fishing time. However, some of that time went to maintaining the boat, an unpaid chore the deckhands were contractually obliged to perform. Deckhands began voting with their feet and fewer and fewer newcomers joined fishing crews.

Research into the 'crew problem' indicated that of fifty-seven Texel respondents more than half had difficulties with the time spent away from home, while over a quarter stated that (fluctuating) incomes were a problem (Productschap Vis 2001:13). The latter may come as a surprise. At the time of the investigation, many deckhands of offshore boats were earning annual gross incomes in excess of 100,000 guilders, which skilled labourers would consider an extremely good remuneration. However, as self-employed entrepreneurs the crew had to pay a considerable sum of money in taxes and social security and other types of insurance, leaving them with a rather modest net income relative to work hours and compared with what they had received in previous years. Other reasons for deteriorating job satisfaction and recruitment problems were the deteriorating mood aboard and the uncertainties inherent in the annual delimitation of national and thus individual quotas, with all the concomitant financial insecurities entailed, and the bad public image of fishing as an occupation. The fishermen resent being stereotyped as brigands of the sea. As an experienced deckhand in his fifties intimates:

> I used to take pride in being a fisherman. On the island, there was public esteem if you told people you were working on a North Sea cutter. You really were someone then. Now people consider you a pariah and blame you for over-fishing, for using six litres of fuel to catch a kilo of fish. I do not know how to say this. Perhaps it is because I am getting a bit older now and it changes your view, but in my experience, the image has become more and more negative. Today I can no longer say that I take pride in being a fisherman.

For some, the impossibility of upward mobility – that is, becoming a skipper-owner – was also a cause for diminished job satisfaction. Last but certainly not least, net incomes declined due to rising social security and additional insurance fees. Largely, this was a scale problem. With the fishing industry becoming smaller and the crewmembers older, the risks for the Social Fund for the Share Fishery increased. Older crewmembers are more susceptible to health troubles and more frequently fail to obtain medical approval to sail. With fewer fishermen contributing to the Social Fund for the Share Fishery and the insurants' mean age rising, insurance claims swiftly sapped the fund's resources. It was and is internally weak and vulnerable because of its small and homogeneous membership and accumulation of risks (de Swaan 1988:145-146). To enable the fund to survive despite the problems of scale, insurance fees had to rise considerably, further deducting from the crews' incomes.

The recruitment problems were not restricted to Texel. Dutch fishery schools faced declining numbers of enrolling students and several boat owners had difficulties finding replacements for deckhands who left their vessel. This happened despite the fact that the number of vessels and hence employment on the fishing fleet declined considerably in the 1990s. For those who left the fishing industry, the booming Dutch economy created many alternative job opportunities. Almost a quarter of the Dutch crewmembers turned their backs on fishing in 1998 and 1999 alone (Productschap Vis 2001). Many were younger than thirty-six. Some vessel owners began hiring Polish crewmembers to cope with the shortage of hands. Pessimism about the fishing industry's future was clearly on the rise. By the millennium, even owners and their wives allegedly encouraged their sons 'to learn a trade' instead of following in their fathers' footsteps. Occupational inheritance was no longer a matter of course. The difficulties pertaining to recruiting crew were evidence that the owners' power was limited. They depended on the labour provided by non-related share fishermen – at least more so than in the past, when nuclear families were much larger and crews composed largely of agnatic kin and in-laws were more common. The post-war demographic transition towards smaller nuclear families implied that in most cases this was no longer feasible, although on Texel it was rare to find a beamer that did not have any relatives at all aboard.[6] Still, such examples existed. Being a sole owner without any kinsfolk crewing, one young skipper-owner decommissioned his beamer in 2002 for want of a motivated crew. He also found it difficult to cope with the anxiety of running the business alone and his health suffered.

There was one potential solution: to solve the crew problem by attempting to interest women – daughters of owner-operators in particular – in a fishing career. However, both fishermen and fisherwomen simply dismissed it as being unfeasible. They said it would be physically 'too demanding' and shipboard life was 'a men's world' where a 'male culture' was dominant. Only two women have crewed on a Texel beamer for a substantial period of time: one in the 1980s and the other in the 2000s. They had their skipper certificates, but crewed as cooks-cum-deckhands. In general, cultural constructions and constrictions made for disapproval by both men and women of female crewmembers working on fishing vessels. The gendered division of labour is therefore still firmly in place. This does not mean that women's roles are insignificant: on the contrary. They contribute to the family firm in many respects, but going to sea is not one of them. Owner-operators with daughters but not sons even prefer to sell their boat, rather than having their female offspring skipper or crew. This is so even if a girl expresses an interest in assuming responsibility at the helm, as was the case in a Texel family firm. 'She's capable enough, but it's just too difficult to handle with all the men,' said the owner-operator in 1990. Ten years later, he sold his beamer. At least one other Texel skipper-owner sold his boat in the early 2000s for want of blood-related male successors. So, with the gender boundary deemed insurmountable, and in-laws being considered second-best successors and then only in exceptional cases, the dependency on non-related crewmembers continued to be a given.

Some Texel owner-operators regarded deckhands who left their boat as 'defectors', considering it a personal affront when one of 'their' crewmembers opted for another career or – worse yet – another fishing boat. They also resented the deckhands' criticism about the share division and the personal incomes of owners and quickly dismissed it as slander or 'quay talk'. Other owner-operators seemed indifferent. For example, a skipper who saw five experienced crewmembers in their thirties leave in the span of half a year just shrugged his shoulders when I asked him about whether he was worried about this phenomenon. 'I cannot force them to stay,' he replied in a rather down-to-earth and resigned tone, 'they are free to go.' Once a crew starts dissolving, it may create a 'flywheel effect'. Deckhands who have been crewing together for a number of years often enjoy the camaraderie of each other's company. They get to know each other thoroughly and spend more time with each other than with their spouses. If, for any reason, someone leaves the crew, this occasionally detracts from the pleasure the remaining crewmembers derive from their work. The fifth crewmember to leave in the example just given, a thirty-two-year-old engineer, related:

> We had a golden team. We could trust each other blindly, but when Jan left and then Peter left as well, I did not enjoy [being on the crew] so much anymore. That is when I began to think of leaving also. I really liked them a lot and we got along wonderfully.

The Texel deckhands continued to air their grievances in front of the skipper-owners, and especially among themselves, concerning the share division's alteration. Some skippers still allocated smaller shares to the crew than others. The issue of cost subtractions would also stay on the Progress through Unity agenda. A new investigation in 2001 disclosed that some skipper-owners subtracted considerably more than others did. For the big beamer category, the sums of money ranged from 1,900 guilders to 4,230 guilders per week (excluding fuel). Again, the matter was discussed in the local fishermen's association, but the owner-operators still argued that they did not have the financial slack to do something for the deckhands. They only agreed that crewmembers should receive a weekly overview of the gross proceeds and the costs. Although this increased transparency, it of course did not alleviate the pain. With revenues declining, the crewmembers' incomes fell, while social security, health and additional insurance fees continued to rise sharply. On top of these problems, tax inspectors – again – began to reject tax deductions of share fishermen who as self-employed entrepreneurs should be entitled to them. It took years of litigation to redress this change of affairs, but in the meantime the fishermen stood empty-handed. The crewmembers had grown used to good remunerations in the 1990s and had adjusted their spending patterns to them. Many had bought a luxurious house and other expensive items. Despite declining incomes, they had to pay off mortgages and meet the costs of living. This proved to be increasingly difficult after the millennium. With operating costs rocketing due to high fuel prices, the deckhands sometimes returned home after a week's hard work with ridiculously low remunerations. During the summer of 2005, when sole catches were well below average, scores of crewmembers earned a gross income of less than €150 per week. In many cases, spouses had to find a job to meet financial commitments. Later that year, results improved considerably, but in many nuclear families wives put pressure on their husbands to find other employment, if they had not themselves already decided to do just that. Twenty-five young deckhands switched careers in 2005 alone, forcing some skipper-owners to work with understaffed beamers: instead of the obligatory crew of six, they sailed with a crew of five. They risked a fine, but it also enabled dividing the crew share with five rather than six crewmembers, improving their incomes somewhat.

Deckhands of fifty-five and over could use a state arrangement that aimed at facilitating early retirement. Several older Texel crewmembers used this option. Many continued to crew, however, and, if need be, they ate into the savings they had intended for a rainy day. As one of them, a fifty-six-year old deckhand, told me:

> I have requested those papers to apply for the retirement allowance, but it just did not appeal to me to stay home. I would be bored stiff and I really like fishing. [...] At the moment, we are eating into our capital. We are living a rather frugal life and we cleared the mortgage, so we can make do with what I am earning without my wife having to find a job. I do not desire to have a yacht in the harbour. Mind you, we have had golden years. Take 1992. That was a top year. I earned 165,000 guilders then. There was one week when I brought home 17,000 guilders. *S-e-v-e-n-t-e-e-n-t-h-o-u-s-a-n-d guilders*! We were catching so much that we were on our feet nearly all the time. They had to pull my gumboots off when I got home. Those young lads, they do not even stand a chance of saving up for the future. That is why they leave. In Fishery School, they take extra courses so that they can switch jobs if they want or need to.

His wife added, passionately:

> For those young share fishermen, these are really hard times. I feel sorry for them. Their wives just *have to* take on a job. It is impossible to make a living without. They cannot save up for a pension and things like that, while they should. By the time you leave the cutter, you should have paid off the mortgage on your house. That is impossible now. So many costs subtract from your income. It is difficult, particularly if you cannot buy anything at all for your kids.

Whilst interviewing this couple, their son – pushing thirty – entered the living room. Having been a deckhand for seven years himself, he immediately joined the conversation. He said he was glad to have left the fisheries for a steady job with the oceanographic institute. Father and son argued that the owner-operators have an awkward way of dealing with the deckhands. The latter elucidated, 'When they decommission their vessel, you're kicked out just like that. That's impossible anywhere else in the Netherlands save for the fishing industry.' His parents simply nodded in agreement. In their view, the times have clearly changed.

Although there are no capitalist relations of production in the share-based fisheries – wage labour not being part of the deal – the deckhands currently do feel exploited. They are simultaneously workers and self-employed entrepreneurs, but under the extant conditions, they believe the weight of the partnership contract and the share convention rests on their shoulders. The share system has always been important for the resilience of family firms. Today, crewmembers feel that with fluctuating quotas and soaring expenses, they bear the brunt of entrepreneurial risks. They are of the opinion that as rights holders, the owner-operators sit in a comparatively comfortable position. Selling the firm with its quotas will yield them a considerable sum of money, none of which will accrue to the crewmembers. The social and mental distance between skipper-owners and non-propertied deckhands has increased tremendously. On board beamers or at the quay, some deckhands are quick to volunteer their profound disappointment in and dissatisfaction with the share system. With windfalls rarely coming their way these days, they indubitably experience the drawbacks of co-venturing. They are well aware that many an owner-operator is up to his neck in debt, but this offers no comfort. Nor does it lead to pity. The solidarity and the mutual empathy – once important aspects of the share system of remuneration – seem to have evanesced over the past three decades. This does not mean, of course, that affectionate bonds between individual owner-operators and crewmembers do not exist. For example, the deckhand just quoted has been with the firm of his skipper for nearly thirty years and he holds him in high regard. At a general level, however, the crewmembers' esteem for owners has been withering away.

From Entrepreneurial Skippers to Skippering Entrepreneurs

To sum up, following the severe compliance problems of the 1980s and the lack of legitimacy of the extant fisheries management regimes, the Dutch state created incentives and coercive measures in order to ensure that fishermen would cooperate in co-management groups and comply with rules and regulations. This commissioned cooperation brought back peace and stability in the fishing industry, which finally returned to calmer waters. It also led to changes in the occupational practice and culture of fishermen, particularly owner-operators. Entrepreneurial skippers have turned into skippering entrepreneurs. Modern-day owners of beam trawlers have to be businessmen and managers, who invest in fishing opportunities, draft fishing plans, and cooperate with colleagues in co-management groups and Producer Organizations. Skipper-owners now have to plan their fishing year thoroughly. Obviously, based on their personal experiences, they will try to use their tie-up weeks when they expect catches to be below average or the market to be at a low and their days-at-sea when catches and prices should be good. However, they can gamble wrongly, and there is always the risk that engine trouble or other such mishaps will prevent them from fishing. In addition, renting out part of one's quotas may yield nice revenues or, conversely, turn out to be unwise because fish prices are soaring or there is no demand for additional landing allowances. Investment decisions are also precarious. Many skipper-owners have invested large sums of capital in acquiring landing rights. If the next year, the EU reduces the quotas by, say, twenty per cent, the value of their investments is written off. Alternatively, entitlements may also increase if quotas are set higher – but since 1993, this has hardly ever happened (see Appendix D). In addition, quota values fluctuate. Operators may have garnered them at high costs while their current value is much lower. Therefore, the co-management system has led to more certainty when it comes to quota uptake and being able to fish out the year, but there are still many uncertainties. These new realities of the trade have affected some of the occupational culture's key values. For example, the enjoyable competition for recognition as a top skipper has made way for less exciting managing and harvesting of quotas. To be a catch king today, a skipper-owner needs to be a 'calculating quota-manager' (Davidse, McEwan and Vestergaard 1999:543) – although many fishermen would add that 'you still have to catch the fish'. To be a quota baron means complying with the state's regulatory regime and with the rules of the group, whose social control is tight. Unlike two decades ago, today Dutch sole and plaice fishermen generally seem to acquiesce in European Union and national quota rules and regulations, which is in large measure due to the co-management of individual transferable quotas.

Nevertheless, the Biesheuvel regime is certainly not a panacea. The economic and political advantages sought with the co-management regime have largely been fulfilled, but at considerable ecological and social costs. The public-private governance system is not about managing fish stocks, since this is done at the supranational level, but primarily designed to maintain the social peace. It solved the quota-busting problem through a mixture of input, output and technical measures and devolved management tasks. This brought an end to political turmoil and fishermen's overt resistance to the rules and their enforcement. Rights holders can be relatively sure of catching what they are legally entitled to, while fishing plans have led to more continuity in landings and higher prices. There are years when national quotas for sole and plaice are even under-utilized. However, there is still fishing industry pressure to set total allowable catches above maximum sustainable yield levels, while the rights-based beam-trawl fishery leads to a degree of ecologically undesirable discarding and high grading. Alternatively, landing allowances may be out of kilter with the fishermen's experience, in that catchability is much better than was to be expected on the basis of the quotas that were allocated. We also have seen that engine power limits have led to particular structural changes in the fishing fleet's composition, with strong concentrations of boats with either 300 h.p. or 2,000 h.p. engines. Fishing within the 12 nm zone with Eurocutters has increased considerably, leading to pressure on inshore sole and plaice stocks, while bigger beamers tend to concentrate in fishing areas just outside the limits of this zone, especially when facing high fuel prices. Such congestion leads to intensive trawling in particular locations.

The individual transferable quota system, the Biesheuvel regime and other national and European fisheries policies also have had a number of – partly unplanned, unanticipated and unintended – economic and social consequences (see also Venema 2001:149). There has been a concentration of quota entitlements in fewer hands and it has become virtually impossible for newcomers to join the fishing industry. There are even considerable problems with occupational succession within family firms. Young fishermen, especially, experience difficulties, while older slipper skippers use their entitlements as an old-age pension. Some have used the escape route of re-flagging and quota hopping, but they are up against the European integration project's contradictory tenets. With individual transferable quotas firmly in place, it will be difficult to alter the management system's nature. Many fishermen – the Texel owner-operators being a prime example – have developed a businesslike attitude and accumulated rights, precisely as neoclassical economists said would happen under an ITQ regime. They will defend these rights if confronted with alternative management options. Overall, the introduction of individual transferable quotas has been socially divisive in that the balance of power between skipper-owners and deckhands has become more skewed towards the former, as they are the rights holders. Non-propertied crewmembers face the new regime's consequences in that owner-operators have altered the division between the vessel share and the crew share, lowering the percentages given to deckhands. In addition, owner-operators subtract new costs from the gross revenues: the rent of quotas and fines. The shift in the balance of power between skipper-owners and deckhands has made recruiting crew increasingly difficult. Having little bargaining power, crewmembers can only vote with their feet. With the turning of the economic tide, many have done so.

It is uncertain whether the co-management system can withstand the major shocks that may assail the flatfish fishing industry: 'Resilience of management systems, including their flexibility and adaptability in the face of uncertain and changing social, economic and ecological conditions, is critical' (McCay 1995:18). In the 1990s, the conditions for the Dutch co-governance regime's economic and political success were extraordinarily favourable, in spite of occasional problems and setbacks. This could change if persistent storms undermine its legitimacy among the fishermen. These storms might come in the form of a stock collapse, additional reductions of quotas and days-at-sea, area or seasonal closures, mandatory decommissioning, or structurally high fuel prices. Susan Hanna hypothesises, 'even against a background of ongoing industry participation, participatory management processes increase in difficulty as resource scarcity increases' (1995:42). Indeed, a group-board administrator recently remarked that with the extant depression the group members are much more inclined to go their own individual ways and to neglect agreements. 'The tolerance has declined,' he says. Another weakness is unequal enforcement and inequity of inspections across member states. Despite the

European Union's promise pertaining to a level playing field, Dutch fishermen are of the opinion that enforcement is much tighter in the Netherlands than in other member states. They believe and often contend that Dutch regulators and inspectors want to be 'top of the class'. The weak point of co-governance – or any management regime for that matter – is that under duress, the temptation of cheating becomes more attractive. I do not have a shred of evidence, however, that this is currently the case. The fishermen are still highly satisfied with the co-management system, although they feel that their involvement in governance matters is extremely limited. They do perceive their relationship with the regulatory system as antagonistic, and they are experiencing difficulties in coping with the precariousness these institutions have created for the fishing industry's future. They believe themselves to be at the mercy of bureaucrats and regulators, over whom they have no control whatsoever. They view state and European Union authorities with extreme suspicion and consider them ignorant or ill informed, influenced by political considerations, and generally uninterested. The fisheries bureaucrats issue new rules and regulations continually, making for much uncertainty and anxiety in the fishing industry. These measures are often ill adapted to the realities of fishing, but strongly impact upon the livelihoods of fishermen. Total allowable catch and quota reductions may make their firms economically unviable.

The owner-operators deem fisheries policy haphazard and inconsistent, which makes medium- and long-term business planning rather precarious. They believe the biologists' stock assessments to be often flawed and underestimated, and yet the biologists have a disproportionate influence on the fixing of total allowable catches, whereas the fishermen – the experts par excellence – have an extremely limited say in the framing of policy. Only recently, some fisheries biologists have admitted that they focused too narrowly on biological processes and that this 'narrow focus has led to management strategies that ignore the dynamic response of fishermen to developments in the stock and to management regulations itself' (Rijnsdorp et al. 2000b:927). Skipper-owners feel policymakers should at least consult them, but the authorities tend to ignore their views and experiences. In general, the fishermen resent being haughtily treated as putative ignoramuses. They are convinced that the fishing industry would be better off if their voices were heard and their advice were heeded. They hate being patronized and stress the fact that they have sensible things to say. As proof of this, they often refer to the 1960s and 1970s horsepower race, when their opinion obviously did not count. The result was overcapacity, with all the accompanying difficulties that would subsequently assail the fish stocks and the fishermen.

Despite the weaknesses mentioned above, the individual transferable quota and co-management regime have meant an important improvement in comparison with the preceding governance system, in that compliance with quota regulations has improved tremendously. Fishermen generally acknowledge this benefit. Group management of quotas also gave them a say in governing the sea's resources, although their influence is extremely limited. Individual transferable quotas would seem to provide precisely the benefits that advocates of this management system usually point out: on the face of it, rights – or rather, the relative shares of national entitlements – were exclusive, permanent, secure and transferable (Brubaker 1997:161-162). The alleged additional benefit is that owners have a strong incentive to take better care of resources. Fishermen would therefore internalize considerations to operate in an ecologically sustainable way. With the Dutch near monopoly on the European flatfish production and market chain, this should apply even more. One would be inclined to attribute the exceptionally good results in the Dutch flatfish fishing industry of the 1990s to these professed benefits. However, there were many imperfections and things changed dramatically in the new millennium, when rights proved to be less permanent and less secure than everyone – including in particular the owner-operators – had hoped and expected them to be. The Dutch sole and plaice fishermen's strong position in Europe's flatfish fishing industry is simultaneously their weak point: being almost entirely dependent on flatfish species for which they have acquired quotas, they have little room to manoeuvre. It is telling that the Dutch flatfish fleet's owner-operators currently feel 'imprisoned' in the ITQ system (Task Force 2006:22). For all the politicians' rhetoric, the responsibilities devolved to fishermen are actually few and limited, so that participatory management is a grand designation for what essentially remains a command-and-control type of regulatory regime: Brussels rules. The technocratic approach's hubris entirely ignores *metis*: 'the indispensable role of practical knowledge, informal processes, and improvisation in the face of unpredictability' (J.C. Scott 1998:6), seriously impeding the fishermen's adaptive performance.

When I joined the Monday morning meetings of Texel shore captains again after more than fifteen years, it struck me how deeply pessimistic the skipper-owners are about the future. During my first stint of fieldwork, they had been prone to lamentation, but it was moaning with a militant fighting spirit in conditions of prosperity. Today, the mood is much more resigned. At times, there is a sense of despondency. The owner-operators feel they are up against the wall. They seem to think that their way of life is doomed, although they refrain from saying so in public. Primarily, the high fuel prices are a thorn in the side of the entrepreneurs and the deckhands. It has become increasingly difficult to break even. 'We are taking our money to the oil mafia,' is one of the frequently volunteered views. 'The greater part of our gross revenues disappears straight into the fuel tank.' But that is not all. The fishermen are still bitter about the cascade of petty regulations and the fault-finding inspectors and police and customs officials, who seem to take pleasure in trying to fine them for trivial technicalities. 'They are nitpicking and always find something that is an infraction,' many say. The fishermen opine that they have not been rewarded for their good behaviour over the past dozen odd years. They feel forsaken. Acquiring additional landing rights has proven to be an extremely risky investment, because quota cuts and fishing-time restrictions are the order of the day. Between 1998 and 2005, fishing time has been reduced by thirty per cent (Task Force 2006:14). There is unprecedented demoralization in the local fishing industry, and this goes for skipper-owners as well as deckhands. On a Saturday in March 2006, the Protestant church organized a meeting to discuss the deteriorating situation in the fisheries and in the fishing communities. More than a hundred people – predominantly owner-operators, deckhands and their spouses – attended the event. The intention was to address issues including the financial problems of family firms and fisher families, solidarity, pastoral care and so on. However, several Netherlands Institute of Sea Research biologists who were in the audience diverted attention away from social issues to ecological ones, forcing the fishermen onto the defensive, instead of giving them an opportunity to exchange views amongst themselves. The following Monday, several shore skippers gave voice to their disappointment about the meeting. Again, they came under fire, and with a World Wildlife Fund anti-fisheries campaign being broadcast at the same time, they felt unduly targeted once more. 'When will they leave us in peace?' a skipper-owner wondered. 'We have already given up so much, but it never seems to be enough.' The owner-operators continued contemplating ways of getting out of the crisis, but the atmosphere turned increasingly bitter. A few weeks after the March 2006 event, Progress through Unity leader Ben Daalder postponed the informal Monday morning meetings. A grumpy shore skipper's remark triggered his decision: 'If everything has to go to hell, we'll make sure it'll go to hell.' Daalder was fed up with the negativism, which he deemed detrimental to finding solutions for the extant problems. The meetings were only resumed after four months. Many skipper-owners fear that they will go bankrupt if fuel prices continue to stay high and the state does not come up with support measures. They attempt to cut down on fuel consumption through all sorts of technical inventions, by reducing steaming and fishing speed, by using lighter or alternative gear, by making efficient use of tidal currents and so on, but such measures only lead to relatively marginal advantages. Put simply, the beamers' engines are too powerful to make substantial gains with current fuel prices, and switching to smaller boats – which many expect will be a future development – is not a transition that can be materialized overnight. The problem is also that the sole fishery requires using rather heavy gear and horsepower. 'The bank still allows us to put fuel in the tank each week,' says a shore skipper in his early sixties, 'but it cannot go on this way much longer.' Some owner-operators have even taken additional mortgages on their property to cover for losses and stay in business. They realize that they are entrepreneurs who cannot just sit still to weather the depression. However, their room for manoeuvre is severely restricted. Although their dexterity and resilience in navigating regulations have been remarkable, they have gradually and unwittingly been funnelled into a trap from which there would seem to be no escape. They now feel indissolubly entangled in the nets of the national and particularly the European bureaucracies. They attempt to 'muddle through', as there are few alternatives. They are well aware that should they decide to sell their firm and its entitlements, this will yield only a tiny fraction of their initial investments. Nowadays, vessels sell for scrap prices and landing rights have saturated the market, diminishing their value significantly. Even more importantly, perhaps, for many owner-operators giving up as a fisherman would mean giving up their dignity and pride.

Conclusions: Seas of Trouble

I ended the previous chapter on a rather pessimistic note. Currently, the local fishing industry is in jeopardy, the owner-operators are exasperated; the deckhands have their own particular causes of disagreement, the fishing communities are in decline and more generally, the meaning of what it is to be a fisherman has changed profoundly. Just days after I wrote the chapter's concluding remarks, the Fishery Cooperative celebrated its seventy-fifth anniversary – a remarkable age for a voluntary association in an occupational world in which independence and individualism are important self-referents and factionalism and schisms are rife. Given the bad economic situation, the Co-op board had not initially planned for any festive celebration. They did eventually decide that this milestone should be marked symbolically and they organized a range of activities for Saturday 9 September 2006. The day attracted hundreds of fisher families, islanders and holidaymakers and was a huge success. There were official speeches, video presentations, photo galleries of boats and crews past and present, net-mending, fish-cooking and sole-smoking demonstrations, a Fishermen's Choir performance and various other events. The zenith came in the afternoon, when fishing-boat crews competed in a 'bathtub race' in the harbour. It involved the most curiously constructed craft, manned by sturdy deckhands in even more curious attire, including miniskirts, 'sexy' stockings, oversized bras, pink stoles and colourful wigs. This show of travesty met with great enthusiasm, roars of laughter and several rounds of applause from the crowd that had gathered at the quay to watch the event. The men in their makeshift boats doubtlessly had the most fun. In the evening, there was a plenteous barbecue for the families of owner-operators, deckhands, retired fishermen and others involved in the local fisheries arena. The following Monday, the mood at the fishermen's informal meeting was exceptionally cheerful. With smiley faces, the shore skippers rehashed the events, agreeing that it had been a perfect day that had contributed to bringing the occupational community closer together again and presenting a favourable image of the fishing industry. On a less functional level, the feast reasserted that there is a special meaning to 'being a fisherman'. It was a brief moment of relief in the midst of a formidable crisis.

I will not attempt to summarize here what has been going on in the local fishing industry over the past three centuries. I would simply fail to do justice to all the complexities, heterogeneities and dynamics that I have tried to unpack in the preceding chapters. Moreover, each chapter contains a final section that is more or less a conclusion in its own right, concerning a specific span of time. However, I would like to return briefly to a number of issues and quandaries for further scrutiny. These concern the dilemmas and paradoxes of independence, rivalry and competition on the one hand and collective action, social cohesion and cooperation on the other; the logic and dynamic of the kin- and share-based family firm versus the rational logic of capitalism; the adaptive performance of fishermen; and the unforeseen and counterproductive effects or 'pathologies' of regulatory regimes. Although unique in their specific configuration, the processes and events in the occupational community of Texel fishermen have a broader significance. Analyzing these processes and events can shed light upon fishing as an adaptive and evolving regime that is embedded in and affected by wider societal systems and dynamics. On an abstract plane, we can come across the resultant modes of response and patterns of behaviour in many regionally distinct settings. At the same time, we also need to do justice to the perceptions, experiences and practices of local fisherfolk, and carefully balance empirical precision and generalization.

Cooperating Competitors and Other Oxymorons

'It is easier to teach a cow how to pray than a fisherman how to listen,' Dutch Fishermen's Union leader Johan Nooitgedagt exclaimed in an interview, referring to the problem of cooperation among Dutch fishermen. 'It is frustrating. They fail to take up so many opportunities. [...] Fishermen remain first and foremost individualists.' As if to somewhat mitigate his harsh judgment, he added, 'Despite the individualism it is an occupational community close to my heart. They work hard and are honest. You can sympathize with them' (*Leeuwarder Courant*, 7 May 2005). For everyone who is even slightly familiar with what has been achieved in the Dutch fishing industry since 1993, Nooitgedagt's remarks must come as a surprise. Fishermen would seem to have learned how to cooperate in groups and how to manage their own affairs when it comes to quota transactions and uptake, compliance with the rules and so on. The fishermen entertain self-images of individualism that are, however, belied by their actual conduct, which proves to be deeply social.

More generally, there is an apparent paradox in many descriptions of occupational communities of fishermen. They are often portrayed as rugged independent and secretive individualists who compete ferociously with each other but whose shipboard relations are characterized by an ethos of egalitarianism and close cooperation and who live in densely knit communities of place (Löfgren 1977:234; see also Pollnac 1991:265). Moreover, in adjusting to ecological, economic, social and political perturbations, fishermen would seem to face perennial problems that structurally pit the individual entrepreneur against the collective of marine resource users, and the self-serving maximizer versus the public good. At face value, this seems to be true of the adaptive strategies of Texel fishermen as much as those of their colleagues elsewhere. There appears to be an 'inescapable tension ... between the desire of parties to maximize their own advantages and/or seek cooperative solutions to mutual problems' (Jentoft, McCay and Wilson 1998:431; also see Löfgren 1977:235). Still, there is no need for binary reasoning in the sense that fishermen are either competitive or cooperative. Competition and cooperation are not incompatible and mutually exclusive activities (Hanna and Jentoft 1996:47-48). Rather, there is no competition without cooperation and no cooperation without competition: they presuppose each other and they are interdependent, albeit different, ways of handling problems that arise with any change in the encompassing social system or mode of production. Fishermen must continually solve dilemmas that are, *ipso facto*, of a social nature: '[t]he critical point about "the fisherman's problem" ... is that it is by its very nature a *social* problem and thus beyond the control of individuals qua individuals' (McEvoy 1986:xii).

This is so on a number of levels. Firstly, fishermen rarely operate a boat alone. Even if they skipper a vessel single-handedly, they usually compete and cooperate with other owner-operators for resources and markets, making fishing a deeply social endeavour. Secondly, finding and capturing fish is a highly competitive game that is often characterized by a climate of secrecy, avoidance, opportunism, distortion of information and deceit.[1] Again, such stratagems of guarding one's own information, while attempting to discover that of others, make sense only if fishing is a socially meaningful venture. Moreover, the management of information is hedged in by economic considerations, social duties and moral restrictions. Wilson hypothesizes that 'fishermen can be expected to resort to small numbers cooperative arrangements in order to obtain the benefits of coordination without the high costs of opportunism. The hazards and costs of cooperation in the face of potentially opportunistic behavior tend to be mitigated by repetitious, reciprocal exchanges' (1990:21). Refraining from the exchange of information may mean being barred from networks that share knowledge and eventually being socially ostracized from the occupational community of fishermen: 'To guard a secret is to withdraw from social exchange. The guardian of secrets and treasures becomes asocial [...] Long-term secrecy is neither socially or economically advantageous [and] the unwillingness to exchange information results in loss of social esteem which has also economic consequences' (T. Vestergaard 1992:172). In this connection, McGoodwin points out that there is a 'skipper's paradox': skippers must maximize their gains of valuable information and minimize their loss of it. At the same time, they must be 'socially participant cooperators and competitors' (1990:137). In other words, they have to balance the dilemma of being competitive and secretive on the one hand, and being – or appearing to be – helpful and cooperative to peers on the other hand. As Pálsson maintains, 'One way to solve the tricky problem of co-operating under conditions of competition is to participate in an informal club of skippers, a network with relatively stable membership, thereby exchanging information on a regular and reciprocal basis' (1994:914; see also Gatewood 1984; Wilson 1990). In this respect, Barth (1966) refers to 'clusters' or reference groups, in which information regarding the proper ways of fishing, the location of species and markets, and technological and economic innovations is exchanged in the hope that there will be reciprocation in the future. Should reciprocity fail to come, then cooperation will be short-lived. Crosscutting social ties at the community level also to some extent mitigates inter-crew rivalry (Byron 1986:102; LiPuma 1992:57). Competitive interaction is thus contained within the rules of the encompassing community or society and may therefore not cause disruption and disorder (Hanna and Jentoft 1996:47-48).

We have seen that Texel fishermen past and present have also attempted to strike a balance between competition and cooperation. Being each other's competitors, there seems to be a 'subjective independence', but this has been coupled with an 'objective interdependence' as they have many interests in common (Bowman 1982:575). They have usually dealt with the dilemma through the formation of informal clubs, which have often been long-lived and based on the reciprocal barter of information. Empirical evidence suggests that there has been considerable continuity in this mode of solving the 'skipper's paradox', although there has been some variation across particular types of fishery and with the emergence of new technologies such as navigation and communication devices and scramblers. Even the commended cooperation under the co-management regime has not brought about a major change in this regard. Importantly, the occupational community of Texel fishermen is rather small and, moreover, to a considerable extent crosscut by relations of kinship and affinity. Clubs based on such multiple local and kin ties have certain advantages, as the accounts in reciprocal barter need not always be in perfect balance. Nor do skills have to be strictly equal to allow for information exchange. At the same time, though, club memberships sometimes overlap and their boundaries are consequently permeable, making for eventual seepage of knowledge. Since the most valuable information is highly ephemeral – for the simple fact that fish are mobile – the selective dispersal of idiosyncratic knowledge is able to maintain the club of cooperating fishermen through the system of proximate reciprocity unless free riding and opportunism undercut its *raison d'être*.

Although not entirely unproblematic, there has also been sustained cooperation in larger and formalized social formations. From the history of Texel's fishing industry, it is evident that prior to the establishment of official fishermen's associations, Texel fishermen cooperated informally in various ways. In addition to supporting each other in emergencies, they banded together when they collectively faced a problem. They would determine a strategy of action, which usually consisted of addressing the appropriate authorities. Such interest coalitions had a single short-term purpose and did not formalize, and they dissolved once the problem had been dealt with. Local attempts at formal organization emerged from the second half of the 19th century onwards. Repeatedly, however, voluntary associations were quickly established but were also swiftly liquidated, while institutionalized cooperation between Oudeschild and Oosterend fishermen did not come about until the 1970s. In the Texel fishermen's rhetoric and folk myth, problems pertaining to sustained collective action were and are often 'explained' by pointing to their independence and individualism, traits that seem ingrained in their self-image. As I have shown in Chapter 3, some social scientists also use such tautological arguments of 'a psychology of autonomy' that allegedly inhibits cooperation, but rather than having psychological reasons, problems of cooperation emanate from socio-economic and cultural diversities. What often impedes formalizing cooperation into voluntary associations and other organizations is the fact that locally, regionally or nationally, fishermen are a heterogeneous grouping expressive of contrasting and sometimes conflicting ideologies, interests and behaviours (see also Jentoft and Davis 1993). To understand (the lack of) cooperation, especially at the national level, the wider socio-economic and political contexts and dynamics must therefore be taken into consideration. It is not sufficient to reduce organizational problems to the structural characteristics of an economic system or to an alleged collective behavioural disposition leading to social atomism. Shared interests have often brought fishermen together, although there appear to be fundamental ambiguities in individual actors' choices based on costs-benefits considerations, while opportunism and free riding may undercut sustained solidarity. In this respect, small social formations stand the best chance of achieving long-term cooperation. Face-to-face interactions and multi-stranded relations create social pressure to participate in voluntary associations and social-control mechanisms to maintain cooperation.

The Texel Fishery Cooperative, Progress through Unity and several other voluntary associations evidence the willingness of cooperating at local level. In response to growing external interference that affected the fishing industry, Texel owner-operators have organized more tightly and the number of local organizations has even multiplied in an attempt to regain some control over their occupational world. In addition, there is the commissioned cooperation in a group following the introduction of the early 1990s co-management regime. A well-seasoned Texel administrator of local and national voluntary organizations remarks that at local level, things work out fine. This is, however, not the case at national level:

> It is very hard to strike agreements. You make a deal, go home and then they seem to have forgotten all about it. It is part of the culture. In addition to being board members, they are also entrepreneurs. They strike a compromise deal in the national organizations, but their grassroots simply will not buy it. The entire administration [of voluntary associations] in the fisheries is unprofessional. You ought to have autonomous administrators. With the kind of financial capital you are dealing with, it should definitely

> be more professional. As it stands, they only mind the interests of their own individual firm.

This is the crux of many problems. The larger and the more heterogeneous the social configuration, the harder it is to bring diverging interests together under a common denominator. Consequently, the small co-management groups in the Dutch fishing industry tend to work better than national interest associations. The fact that there are two such national organizations of course only exacerbates the situation, and seriously undermines the fishermen's political power. Although talks about reuniting began soon after the Federation's fission from the Dutch Fishermen's Union, economic and ideological differences continue to thwart any attempt at a merger. With the current crisis deepening, many an owner-operator would like the fishing industry's leadership to be more militant:

> Our leaders ought to put up a much fiercer fight. There is no unity among the fishermen. They blame each other and they are individualists. As if things weren't bad enough already. We have to work together, we need each other, particularly so in the current crisis. We have to tell each other the truth, and we should fight. It seems, however, that everyone is acquiescing.

Fishing industry leaders are in a difficult position. They have to weigh the interests of the various fishing communities and fleet segments. Ben Daalder intimates:

> We have to knit a sweater together, but everyone says 'In *my* opinion.' Mind you, tempers are fraught. Therefore, I have to perform a balancing act and phrase things slightly differently for each audience while still communicating the same message.

In a progressively complex world, with shifting power and dependency relationships and growing state intervention, organization and political participation would offer fishermen a counterbalance to developments that threaten to push them into a position of increasing powerlessness and marginalization. Yet it proves to be hard to heal still-oozing sores and to overcome old differences of opinion. However, under certain circumstances enduring institutionalized cooperation of fishermen is likely. In general, this is the case if problems of scale create strong incentives to work together; an organization receives the (tacit) support of most fishermen in an area and the membership is relatively small and homogeneous in a socio-cultural sense (although in the case of mutual-aid funds this harbours the danger of a concentration of risks). Other prerequisites for successful cooperation are: mutual trust, confidence, social solidarity, loyalty, shared economic interests and goals (for example, product and producer homogeneity), limited competition from outsiders, difficult access to the industry and few product alternatives. Corporate results or operating and managerial successes have to be sufficient to satisfy the expectations of the organization's members (conversely, dissatisfaction will lead to reduced loyalty, intolerance and increasing detachment) and it must be possible to sanction uncooperative behaviour in order to avoid free-rider problems. If all or most of these terms apply, independent fishermen and other petty capitalists will often be cooperating competitors. Arguably, these conditions are easier to accomplish at local level in small and homogenous groupings than at higher levels of integration with large heterogeneous social formations (Olson 1971[1965]:61-62).

The Logic and Dynamic of the Share-Based Family Firm

The myriad management measures have impinged upon the family firm's importance. The involvement of kin and family in fishing has been a common pattern for many decades. Some prominent Texel family firms can even trace their roots back to the 18th century. However, the current involvement of kinsfolk in fishing crews should not be regarded as a relic of an ancient mode of production. Newcomers to the 1960s and 1970s local fishing arena were generally more successful if they worked with agnatic relatives, suggesting that the family firm is not an archaic form of social organization, but a well-adapted institution in the days of high capitalism. Rules of inheritance insufficiently explain the preference for relatives: 'Among kinsmen, membership of the crew is a complex relationship of interwoven strands of economic interest and social obligation within and across the generations' (Byron 1994:287). Pooling economic, social, cultural and cognitive resources provides a common fund to cope with risks and to accommodate to shifting conditions. The family firm is a relatively fluid and flexible unit, which is highly adaptive under circumstances of uncertainty (Durrenberger and Pálsson 1985:114ff.). Relatives are prepared to postpone (returns on) investments, work longer hours, defer gratification, cut remuneration, adjust household budgets and eat into their capital in times of duress or to speed up the redemption of debts and intensify investments in times of prosperity. They can do so because the family firm is at the same time a unit of (re)production, consumption and (re)distribution, providing the family firm with an adaptability not usually found in company-owned firms that operate under the capitalist mode of production. The 'family firm logic' also enables accumulating capital that owners reinvest in the family firm to allow for expansion. Furthermore, a pervasive labour ethos where work permeates the entire existence of fishing households greatly enhances adaptive performance.

Through their contributions to running the firm and their specific patterns of expenditures, skipper-owners' wives have usually played an important role in materializing the family firm's ideals and objectives. This is still the case today. Many skipper-owners' wives take part in decisions concerning how a firm is managed financially: such as, for example, whether or not to invest in additional landing rights. They also take on various tasks and chores linked with the firm, including bookkeeping and settling accounts, ordering spare parts and provisions, doing the ship's laundry or cleaning the vessel's galley and dwelling quarters. They feel committed to and emotionally involved in the firm and appreciate their contribution. For some small firms, the unpaid work is important in keeping them financially afloat. The wives' organizational, economic and emotional contribution to the fishing household and their accommodating role can be crucial for a firm's flexibility, versatility and resilience (Hoefnagel and Smits 2000). Husbands still appreciate their wife's contribution. 'She's my main sounding-board, we even discuss details about the rigging of the nets,' relates a Texel owner-operator. Some women I interviewed denigrated their own roles, in saying that earlier generations of fishermen's wives had to be even more independent as there were no means of ship-to-shore communication. Telecommunication (mobile phone, e-mail and particularly fax) has facilitated contact with a vessel's skipper and crew and, consequently, has alleviated fears and worries at home. 'We are always in touch at least once a week,' says a spouse of an owner-operator. 'Still, the week can be long when the kids are ill or when there is some other crisis at home.' The very same means of communication are probably also important in putting additional workloads on women's shoulders. As a skipper's wife told: 'I always get a fax on Thursday with information on whether I need to call mechanics, the auction, the GIS and so on.'

The family firm is a system of self-motivated labour and a form of self exploitation (Thompson et al. 1983:156; Jorion 1983:10) characterized by an ethic of deferred gratification. For close agnatic kin, the share system usually works well, although it is not necessarily devoid of conflict. Agnatic kinsmen who are co-owners often accept receiving a fixed sum of money out of the revenues while keeping the rest of their share in the firm. Therefore, having sons makes for an important socio-economic asset and the preferred social organization aboard ship usually hinges on a core of father-sons or brothers. Ideal-typically, fissions are common in the stage when brothers have adult sons (patrilateral cross cousins) who are generally reluctant to work for their uncles. This is inextricably connected with the fact that part of their shares remain in the firm. Deferred gratification is based on mutual trust in achieving a future goal, including, for example, reinvestment, expansion and, eventually, succession. This gives family firms a resilience that would be unthinkable for profit-oriented capitalist businesses. The ethic of cooperation for the common purpose of maintaining and continuing the family firm lends it a rationale that deviates from the goals of business units working under capitalist relations of production, where wages and profits are the main aim of labour and capital, respectively. The predominant goal of (prospective) co-owners of a family firm, on the other hand, is to keep the firm afloat even in the face of formidable and enduring adversities. This is so because the firm is much more than a vehicle to earn an income. It is at the same time a source of pride and identity, intrinsically a *raison d'être*. It is for this reason that owner-operators often tenaciously hang on to continuing the firm so that they can hand it on to the next generation.

The partnership contract and the share system of remuneration tie non-propertied fishermen to the family firm, and they benefit directly from the proceeds of the owner-operators' capital assets. Skipper-owners, in turn, have highly motivated and dedicated crewmen in whose self-interest it is to work relentlessly and put in long hours, to maximize the gross proceeds and minimize the costs, and to keep the boat and the equipment in good repair. An additional advantage for owners is that they do not have to pay fixed wages and can distribute risks to the crew as a whole, which provides the firm with considerable flexibility in times of meagre revenues. However, under prolonged duress, non-propertied share fishermen tend to exit the fishing industry for more secure jobs. This is the share system's fundamental weakness: it only loosely ties co-venturing labour to a firm. Ever since the share system turned into common practice on Texel, skipper-owners have faced problems in keeping or recruiting crewmembers in times of depression, regardless of the share division's exact details. For the former, fishing is an existential matter that goes beyond a simple account-book reckoning of proceeds and profits. Even though share fishermen may be considered independent petty entrepreneurs before the law, their 'independence' is in fact limited to their supply of labour. Through the share system, they are directly linked with and vulnerable to ecological shifts, economic cycles and political interventions. There may be windfalls, but also hardship. During crises, share fishermen feel they carry the brunt of the fishing venture's risks. This is profoundly evident at present. There is a reluctance to join fishing crews and even owners allegedly encourage their sons 'to learn a trade'.

In general, however, owner-operators still highly appreciate it if their male offspring continue a firm. When an owner's son is old enough to join the crew, one of the deckhands usually has to leave the vessel to make way for the newcomer. Similarly, an owner's son will eventually skipper. This pattern of preferring to work with one's kin is a time-honoured practice. Owners prefer working with 'one's own folk' because they believe this automatically means a maximum job effort and commitment to the firm. In addition, the family boat means more to them than a material vehicle to earning an income; it is also a patrimonial source of pride and social and individual identity. Until the early 1980s, the ideal of owners was to set up each son with his own vessel. Due to the recent developments in the industry, this ideal of succession *and* expansion is no longer feasible. Supranational and national policies aim at reducing the number of vessels and aggregate engine power in an attempt to tackle overcapacity. In addition, due to the high costs involved it is currently almost impossible to take over a firm, let alone begin one. The fishing industry has faced a process of unprecedented capital intensification. Today, a vessel including landing rights costs many millions of euros. The introduction of transferable fishing rights has created enormous barriers for potential newcomers. Even continuing a family firm is extremely difficult due to the high prices and succession duties that have to be paid (Dubbink, van der Schans and van Vliet 1994:33). Moreover, there are additional uncertainties concerning the allocation of landing rights in terms of the real quantities of fish quota holders are entitled to land. Along with the congeries of restrictions that fishermen have to live up to, over the past decade this has affected the enthusiasm to enrol in the fishing industry even among members of the local fishing elite. Whether occupational inheritance will be dominant in future is questionable. The continuity of many family firms is at stake. In part, it is also connected with demographic developments that tend to undermine the social asset of having several sons from among whom skippers can recruit crew and successors. However, many owner-operators obstinately adhere to their occupation even when facing declining stocks and catches, and substandard incomes. This has a lot to do with the fact that fishing is not just a job but an all-pervasive métier and a matter of identity.

In the Final Analysis...

Fishermen generally acknowledge that without any regulation whatsoever, tragedy is imminent, though they find it difficult to grapple with the uncertainties external authorities have created for the fishing industry's future. They have always adapted to the vicissitudes of the ecosystem and markets, but currently many are up against erratic powers beyond their adaptive capabilities. Usually, fishermen have short-time horizons. Due to high levels of uncertainty, they prefer short-term planning. The incentives to adopt a new mode of production are inextricably intertwined with opportunities for change and perceived costs and benefits. At face value, fishermen seem to be conservative. They often stubbornly resist change in resource management regimes, and over time they have occasionally given vent to their discontent through strikes, protests, blockades, non-compliance and confrontations with (foreign) competitors and external authorities. From the fishermen's point of view, they usually have good reasons to behave like this. This book has shown several contradictions of policy and management regimes, the fishermen's ambivalent coping responses, and the unplanned for and perverse outcomes of the process. Between intended policy goals and desired results there seems to be a yawning gap. New management regimes often impede time-honoured fishing strategies, adaptive performance, flexibility and switching behaviour. Due to the cultural emphasis on independence and individualism, fishermen are often suspicious of and resent interference from fisheries policymakers and regulators, especially if measures are believed to be flawed or unjust: 'The motivations of actors in the fisheries and the moral legitimacy of management measures are closely linked with meaning and values' (T. Vestergaard 1996:87). At best, they are ambivalent about external interventions.

However, it would be erroneous to assume that fishermen are merely powerless victims of state intervention. They try to make the best of a world full of structures and strictures and they navigate seas that are treacherous in a real and metaphorical sense. Despite the fact that fishermen can hardly be called autonomous actors – being embedded as they are in wider social configurations – there is agency. They plan, scheme, plot, and adjust their lives to the new realities of encompassing management regimes, seeking to find the loopholes, contravening the rules if they deem so fit or complying and cooperating if they perceive benefits in such behaviour. Not all fishermen merely react to 'the rules of the game'; there are also active innovators who attempt to stay at least one step ahead of state-induced constraints to remain in control. In both cases, fishermen confront national and supranational governments with new situations to which the authorities subsequently have to respond. This two-way dynamic has seriously affected the nature of the policymaking process. A case in point is the state's policy concerning the fishing fleet's development and modernization. Ironically, it first used subsidies to fund overcapacity and then used public resources to reduce excess capacity, putting a double tax burden on its citizens. Understandably, such dialectic often holds little water with fishermen who first received incentives to invest and next were severely restricted in their operations. Adding insult to injury, they were blamed for over-fishing, while in the early 1970s they begged the state to cap engine power. The dialectic is also evident at a higher level of integration.

In the European Union, 'both governments and non-governmental actors no longer have a monopoly over the policy agenda. They define public policies through permanent interactions, negotiations and compromises' (Lequesne 2000:791). Although fishermen resent the myriad of rules and regulations that Brussels imposes on them, it is ironically to a considerable extent their own call for exemptions, clarifications and specifications that has led to a reactive multiplication of bureaucratic procedures and rules. This is because changes in regulatory regimes and property rights often have profound distributive effects and may therefore meet with vehement opposition. Such shifts do not emerge in an institutional vacuum, but involve political bargaining and manoeuvring by individuals and groups with varying resources of power who seek to influence the rules of the game and attempt to control, negotiate or contest the system's definition, legitimization and enforcement (Jentoft, McCay and Wilson 1998:431; see also Libecap 1989). The actors may do so through strategies of deliberation and cooperation or, alternatively, evasion, violation, litigation and/or other forms of rule beating and resistance. Processes of interest group mediation often interfere with policy objectives and the outcomes of policymaking are susceptible to manipulation by powerful stakes (Symes 1999:142). Mediation of conflicting interests and the resulting compromises might lead to suboptimal 'solutions', perpetuating or even deepening fisheries crises. The 'winners' support extant management regimes and attempt to ensure that they continue to sustain their benefits, rendering internal reform difficult (Jentoft 2006). There may also be the phenomenon of 'agency capture', where powerful individual actors and groups seem to be determining the definition of the rules, as regulators become inclined to completely identify with their interests (Singleton 2000:7). Due to these political processes, the rather straightforward means-to-an-end approach creates complexities of its own. Rather than being implemented, European Union policies are '*translated* by a range of national and local actors who remain constrained at the state level by specific arrangements between societies, markets and governments' (Lequesne 2000:785, emphasis in original). Any blueprint management regime therefore inexorably creates contradictions and unanticipated outcomes.

Successive management regimes eventually leading to the late 20th-century rights-based co-governance system in the Dutch flatfish fishing industry have increasingly stymied the fishermen's key adaptive feature: their polyvalent short-term switching behaviour. In the face of declining resource abundance or fluctuating markets, fishermen have for centuries been able to cope with these ecological and economic vicissitudes by targeting different species, diversifying or specializing, pluralizing activities, or temporarily abandoning fishing, with permanent exit usually providing a strategy of last resort. However, with the allocation of individual transferable quotas, their freedom to do so has been stifled to a considerable extent, which has forced them to develop a long-term perspective on their business. The upshot is that they find themselves in a situation in which specialization and intensification have become of paramount importance to their economic survival. According to the logic of the 'economics of flexibility', 'intensification requires greater investments and can lock people and their organizations into particular, "deeper" modes of response, becoming nearly irreversible' (McCay 2002:377). Investment in quota entitlements certainly leads to a commitment to exploit specific resources that is hardly reversible. At the same time, it has made quota holders profoundly susceptible to top-down interventions. Since the national offshore fishing industry – including the Dutch-owned re-flagged fleet – has come to depend on sole and plaice to such a vast extent that it virtually 'owns' the European total allowable catch for these species, it is extremely vulnerable to the fickle horse-trading of national and supranational interests in Brussels. 'We have no friends in Europe,' sighed Dutch fishing industry leader Ben Daalder, following more bad news for the Dutch flatfish fishermen in the summer of 2006. The Dutch fisheries negotiators in the European headquarters do not seem to have sufficient change for barter in their pockets to successfully participate in the tit-for-tat negotiations that spring from the contradictions between national fisheries interests and international and supranational rules aimed at regulating fishing effort in connection with biological and environmental concerns. Such bargains still characterize the Council of Ministers' decision-making. The 'weak "transnationality"' of European fisheries policy explains why many fishermen still regard the Common Fisheries Policy 'as an interstate bargain and do not consider that their occupation takes place in the context of a European public space' (Lequesne 2000:783).

The bureaucratic process that aimed at producing solutions for problems of overcapacity and overexploitation has instead become part of the problem. Formalized private property regimes may sever the resource from its socio-cultural context, 'reducing the social capital and ecological flexibility needed for effective management' (Jentoft, McCay and Wilson 1998:432). Imposing a rights-based fishery in the Netherlands has indeed resulted in rigidity and a loss of resilience. More generally, this is an outcome inherent in the command-and-control approach to natural resource management, which:

> implicitly assumes that the problem is well-bounded, clearly defined, relatively simple, and generally linear with respect to cause and effect. But when these same methods of control are applied to a complex, nonlinear, and poorly understood natural world, and when the same predictable outcomes are expected but rarely obtained, severe ecological, social, and economic repercussions result (Holling and Meffe 1996:329).

The unforeseen and undesirable consequences that arise from such top-down interventions seeking variance reduction and standardization are 'less resilient and more vulnerable ecosystems, more myopic and rigid institutions, and more dependent and selfish economic interests all attempting to maintain short-term success' (ibid.:331). This is what Holling and Meffe term the 'pathology of natural resource management', a pathology that deepens if states respond to failures of command and control by yet tighter regulations and prohibitions. Clearly, the way in which the Dutch flatfish fishing industry was managed during the 1970s until the early 1990s was ailing from this pathology, and despite the state's devolvement of some responsibilities to owner-operators under the extant co-management regime, the European command-and-control management regime is still firmly in place. So is the pathology. For all the supra-state's rhetoric of the 'good-governance' principles of openness, participation, accountability, effectiveness and coherence, in fact its Common Fisheries Policy remains top-down with in camera decision-making, token fishermen involvement, obscure responsibilities, grave inefficacies, and many incoherent and ad hoc adjustments. It is clearly pervasive, but does not take into account the fishermen's motives, interests, opinions and values.

What is even worse from the fishermen's point of view is that entitlements are utterly insecure. This has been true from the onset of the quota system. Initially, the state withdrew landing privileges once the national share of the total allowable catches was exhausted. This implied that, de facto, individual transferable quotas did not amount to a secure property right, since owner-operators could not be certain of taking up the amounts of fish to which they were entitled. In fact, under these conditions ITQs exacerbated rather than countered the 'race for fish', particularly because enforcement was haphazard. Fishermen who held sufficient quota rights called for tighter supervision, but as long as it paid to circumvent the law they opted for illegal fishing and landing. Their definition of the situation was that since they could not trust the state to enforce its rules, they simply had no alternative but to behave as maximizers. The underlying social dynamic was that in order to secure as much as possible of its rights, each firm had to land as great a portion of the national TAC as possible in the shortest time possible, regardless of its individual limits. With tightened enforcement, and particularly following the introduction of co-management groups, this type of behaviour changed. However, it took considerable effort to bring the fishermen into the fold of fisheries politics. Legitimacy and transparency have been the basis of the relatively large measure of success of recent governance regimes. The state also encouraged owner-operators to invest in acquiring additional landing rights. Most Texel owner-operators have done so with alacrity, behaving in precisely the way neoclassical economists predicted the stronger and more efficient fishermen would do because accruing landing entitlements was in their 'enlightened self-interest'. The Texel skipper-owners now have a considerable portion of the national share of total allowable catches for sole and plaice at their disposal. They generally equate these entitlements to land a specified amount of fish as ownership of that amount of fish still swimming in the sea. Their catching efforts obviously tend to focus on the species they have 'bought'. If they were to refrain from doing so, their expenditures to purchase quotas would be wasted and thus be highly inefficient costs – a loss. *Rights-to-land* therefore tend to turn into *duties-to-catch*, with potentially detrimental ecological effects. In addition to profoundly thwarting switching behaviour and thus flexibility and resilience, the quota regime means that absolute quantities of fish that rights holders are entitled to land fluctuate from year to year, sometimes quite sharply. This makes purchasing additional landing rights a precarious investment that also lends itself to a considerable degree of speculation: 'a percentage-based right ... offers less security and marketability to the holders' (Rose 2002:242). Moreover, days-at-sea regulations contravene opportunities for quota uptake, which interferes with the market of transfers because if there is under-use, quota offers in the market outweigh demand and prices drop. Still, a major fear of the owner-operators is that the stroke of a Brussels's pen might replace their landing rights with a days-at-sea regime.

Many problems emanate from the EU's penchant for regulatory uniformity across member states. These rules and regulations do not always make sense for specific fisheries and they may create 'perverse incentives' and produce 'deleterious effects' (Acheson 2006:125-126). The point is that generic and homogenizing 'state simplifications' (J.C. Scott 1998) aimed at technocratic engineering of natural and social environments usually fail because they are at loggerheads with practical and situated knowledge and contextual skills that can be acquired only through local practice and experience. Composite, dynamic and discrete situations are manipulated into simplified, static, aggregated and standardized data for purposes of management and control. However, such grand planning designs usually fail because they ignore what James Scott dubs *metis*, 'a wide array of practical skills and acquired intelligence in responding to a constantly changing natural human environment' (ibid.:313). Against this background, it may be a reasonable strategy for management goals to devote closer attention to fishermen's knowledge, 'allowing for extreme fluctuations in the ecosystem, relaxing at the same time the modernist assumption of predictability associated with the ecological project of sustainability' (Pálsson 1996:75; also see Hornborg 1996:54). These knowledge systems need to be re-contextualized in policy and management frameworks, and ideally, resource users should be closely involved in the process because they have special stakes in the outcome. To be optimally adaptive, these strategies should provide for flexibility, since contingent and capricious remote forces make for uncertainties and risks that affect local resources and their users. We should bear in mind, however, that all governance structures 'involve tradeoffs between stability and flexibility, authority and representation, social and individual' (Hanna 1999:280), while their outcomes are hard to predict and control. Therefore, actors and communities will continue to depend on social resilience: their ability to cope with perturbations because of social, political and environmental change (Adger 2000:347). Currently, however, the Texel fisherfolk's greatest economic concern is how to survive the present depression caused mainly by unprecedented fuel prices. With no signs whatsoever of the crisis abating, the owner-operators just attempt to do what many of their predecessors have done in the past: brave troubled waters for love of their way of life.

Appendix A

Main Species Pursued by Texel Fishermen (1700-2008)

Sole (tong), *Solea solea*

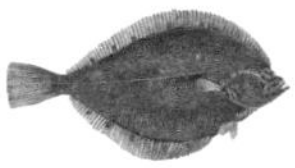

Plaice (schol), *Pleuronectus platessa*

Brill (griet), *Scophthalmus rhombus*

Turbot (tarbot), *Scophthalmus maximus*

Flounder (bot), *Platichthus flesus*

Dab (schar), *Limanda limanda*

Halibut (heilbot), *Hippoglossus hippoglossus*

Ray (rog), *Raja clavata*

Garfish (geep), *Belone belone*

Eel (paling), *Anguilla anguilla*

Cod (kabeljauw), *Gadus morrhua*

Whiting (wijting), *Merlangius merlangus*

Herring (haring), *Clupea harengus*

Anchovy (ansjovis), *Engraulis encrasicholus*

Brown shrimp (garnaal), *Crangon crangon*

Nephrops (Noorse kreeft), *Nephrops norvegicus*

Whelk (wulk), *Buccinum undatum*

Oyster (oester), *Ostrea edulis*

Mussel (mossel), *Mytilus edulis*

Cockle (kokkel), *Cerastoderma edule*

Periwinkle (alikruik), *Littorina littorea*

Starfish (zeester), *Asterias rubens*

Eelgrass (zeegras), *Zostera marina*

Appendix B

Development of the Texel Fishing Fleet, 1970-2006

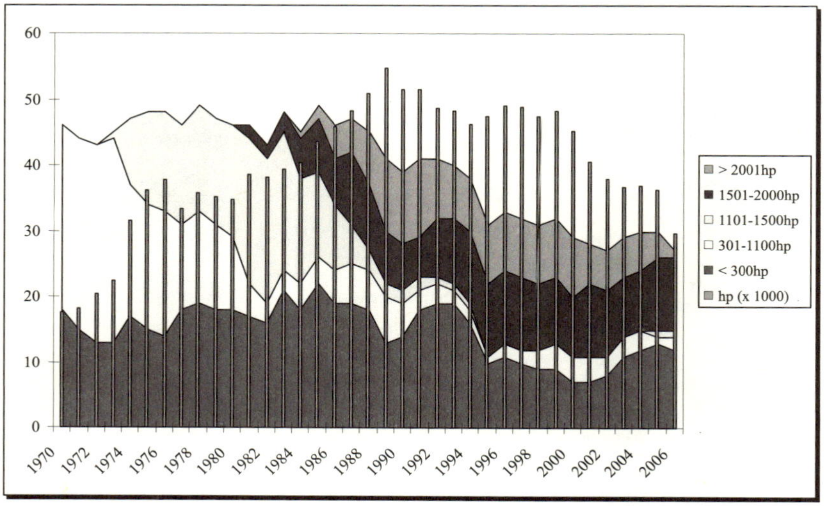

Source: Adapted from the Fishery Register.
Note: Development of the Texel fishing fleet in horsepower categories and aggregate engine power (x 1,000 h.p.). The figure pertains to all vessels on the Fishery Register, including shrimp-cum-tourist boats not used for commercial fishing.

Appendix C

Development of the Dutch Cutter Fleet, 1973-2007

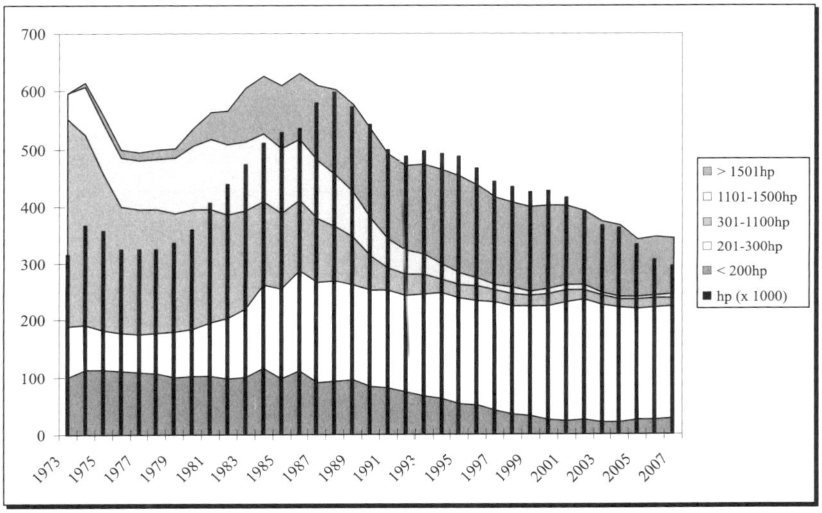

Source: Adapted from LEI's annual fishery statistics *Visserij in Cijfers.*

Note: Development of the cutter segment of the Dutch fishing fleet in horsepower categories and aggregate engine power (x 1,000 h.p.). The figure is slightly distorted because several firms operated second-hand side and stern trawlers in the 1970s. Statistically, these vessels were included in the distant-water-fisheries segment.

Appendix D

Dutch Sole and Plaice Quotas (x 1,000 metric tonnes), 1975-2008

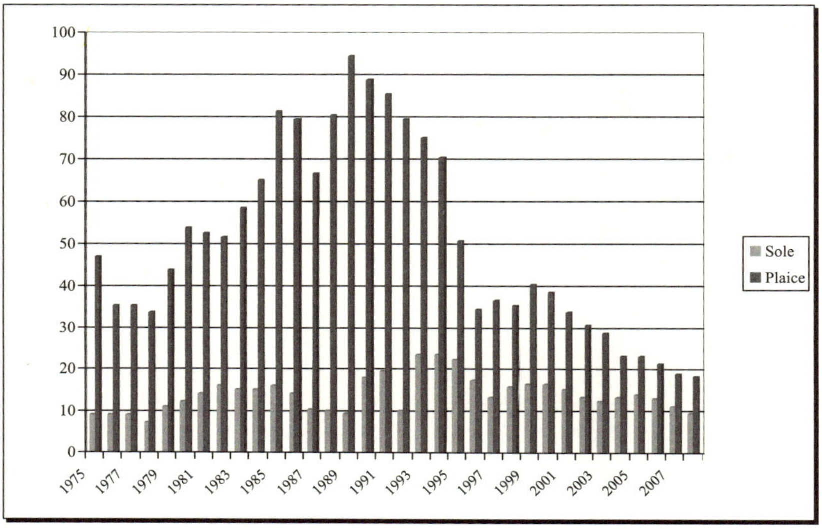

Source: LEI annual reports *Visserij in cijfers.*
Note: excluding additional quotas obtained through bilateral quota swaps.

Notes

Note to Introduction

1. I will refrain from using the gender-neutral word 'fisher', which is currently popular among social scientists. I deem it a more obfuscating than clarifying term, and prefer to use fisherman or fishermen when referring to male producers in the fishing industry. It was and is highly exceptional to find women among the crewmembers of Dutch fishing vessels.

Notes to Chapter 2

1. Except when quoting, I have refrained from referring to archival sources as they are all in Dutch. For fuller reports that include comprehensive references and sources, see van Ginkel (1993 and 1995a – in Dutch). For extensive articles in English on Texel oyster fishing, eelgrass harvesting, and fishermen cooperation, see van Ginkel (1996a, 1996b, 1997).
2. Municipal annual report 1814 (Texel archives, no. K-410).
3. Letter to the governor, dated 24 June 1841 (Texel archives, no. K 853).
4. 'Beschrijving van de oestervisscherij zoo als die op het Eiland Texel wordt uitgeoefend' ['Description of the oyster fishery as it is practised on the island of Texel'] (Texel archives, no. K 853).
5. Petition of Oosterend oyster fishermen to the King, dated 12 March 1839 (National archives, The Hague, RWS 2.04.07.02, no. 409-412).
6. At the same time, the number of fishing vessels hailing from other locales that exploited oysters in the northern Zuider Sea declined. By about 1850, only thirty-three vessels pursued oysters as opposed to eighty-five a century earlier.
7. Municipal annual report 1864 (Texel archives, no. K-412).
8. Appendix to a letter, dated 23 January 1859, from the Board of Sea Fisheries to the mayor and councillors of Texel concerning the cultivation of oysters (Texel archives, no. K 853).
9. Letter of Texel's mayor to the King's Commissioner in North Holland, dated 6 September 1852 (North Holland state archives, PB 1851 1943 werkdossiers, no. 46).
10. Request by the mayors of Wieringen and Texel to the King's Commissioner of North Holland, dated 18 and 19 August 1867 (North Holland state archives, PB 1851 1943, werkdossiers, no. 46).
11. Eelgrass workers' petition to the council of Texel, September 1882 (Texel archives, no. K 1120).

12. In the entire North Atlantic region, fish merchants and processors have dominated and sometimes exploited fishermen (Smith 1977:4; Andersen 1979:21; Acheson 1981:282). The latter often responded by establishing cooperatives to further their position in trade networks, gain leverage vis-à-vis fish merchants and processors, and circumvent as many middlemen as possible (Orbach 1980; Pollnac 1991:284ff; Jentoft 1986:199-200; Prattis 1987; Davis and Jentoft 1989:195; Magnússon 1990:73; Durrenberger 1992a:77ff.).
13. Letter to the mayor of Texel, dated 8 August 1922 (Texel archives, no. K-864).
14. Letter of K. Vink to the local administration, 29 January 1921 (Texel archives, no. K-861).
15. See, for example, McCay and Acheson 1987a, 1987b; McEvoy 1988; Acheson 1989; Berkes et al. 1989; Pinkerton 1989; van Ginkel 1989; McGoodwin 1990; Ostrom 1990, 1999; Pálsson 1991; Bromley 1992; McKean 1992; Feeny, Hanna and McEvoy 1996; Agrawal 2003.
16. See, for instance, Gordon 1954; Agnello and Donnelley 1976; Anderson 1976; Santopietro and Shabman 1992; Mace 1993; Ajuzie and Altobello 1997; de Alessi 1997; Jones and Walker 1997; Grafton, Squires and Fox 2000. I will return to the neoclassical viewpoint when dealing with individual transferable quotas in Chapters 5 and 6.

Notes to Chapter 3

1. Measures were also taken with regard to mussel fishing. Texel fishermen had not been fishing mussels from 1 June 1934 to 1 June 1935. The Mussel Crisis Decree stipulated that, in order to get a mussel fishing licence, this should be the case. For this reason, the Texelians did not get licences to fish mussels, despite their requests in 1936 (see below).
2. Petition of Texel's mayor and aldermen to the House of Representatives, 12 February 1938 (Texel archives, no. P-747).
3. On sailing vessels with auxiliary engines, which usually had a crew of two, the division was forty per cent for the vessel share and sixty per cent for the crew share. The skipper and the deckhand would each get thirty per cent. On sailing vessels without an engine, the share division was a third for the vessel, a third for the skipper and a third for the deckhand.
4. 'Head teacher Daalder' was the local secular school's principal, and was pivotal in various voluntary associations, including fishermen's organizations. He was the father of the author whom I quote here, Dirk Daalder (born in 1887), who was born and bred in Oosterend.
5. Second World War explosive devices would remain a fishermen's hazard for decades. A Texel cutter was damaged beyond repair in 1971 when a bomb that was caught and returned to the sea exploded underneath the hull. No one was injured. As late as 2005, three fishermen working aboard an Ouddorp beam trawler died when they netted a bomb that exploded on deck. It is not unusual for explosive devices to end up in the nets of fishermen, as I witnessed when joining the crew of a beam trawler in 1990. One of my interlocutors told me that fishermen had unwillingly been 'the best minesweepers of the post-war era'.
6. It is a tenacious stereotype. Later, several authors referred to the fishermen's 'deep-seated individualism and local chauvinism' (Schaper 1962:148), their

'weak organization' and 'individualistic attitude' (Bossaers 1987:63) or their 'fragmentation, and not organization' (Kerkhoven 1986:55). Kranenburg (1980:12) suggests that all Dutch fishermen behaved and thought individualistically.

7. In 1967, the auction finally closed down. Only two petty fish merchants were still active then. In the previous six years, the annual turnover had never surpassed 50,000 guilders.
8. See, for example, M.E. Smith 1977:5; Poggie 1980:21; Knipe 1984:98; Pollnac 1991:284; McGoodwin 1990:127; Valdés-Pizzini 1990:165; Pollnac and Poggie 1991:44; Thomas, Johnson and Riordan 1995:150-151.

Notes to Chapter 4

1. The EEC structural policy demanded an end to subsidies. Dutch fishermen believed that the Dutch state complied with the rules, but other member states did not, while Germany also provided low-interest loans. The conviction that European rules did not apply equally to fishermen from all member states would become deeply seated, in the Netherlands as well as elsewhere.
2. Their favourite auction was still IJmuiden, where they grossed 13.3 million guilders in 1972. This accounted for seventeen per cent of IJmuiden's turnover. In nearby Den Helder, they grossed 6.5 million guilders, accounting for nineteen per cent of the turnover (Anonymous 1973:14-15).
3. On the important role of women in fishing economies, see P. Thompson 1985; Nadel-Klein and Davis 1988; Cole 1991; Davis and Nadel-Klein 1992; Thiessen, Davis and Jentoft 1992; Jentoft 1993; Meltzoff 1995; Munk-Madsen 2000; Binkley 2002; Nadel-Klein 2003; Neis et al. 2005.
4. The partnership contract or *maatschapscontract* was officially recognized in 1923 (de Bruyn 2001). However, the share system as such is considerably older. In this section, I will use the present tense as much of what I write still applies. I will use the past tense only when referring to specific historic events and situations.

Notes to Chapter 5

1. See, for example, Grafton 1996; Jones and Walker 1997; Shotton 2000; Pascoe, Andersen and de Wilde 2001. For a critical economist's view, see Copes 1986, 1997.
2. In 1977, due to widespread dissatisfaction, the allocation rule was altered, giving equal weight to the basis of the historical catch record and a vessel's engine power. The implicit understanding was that these individual quotas – relative shares of the national part of the total allowable catch – were assigned to firms in perpetuity.
3. On 'luck' in fishing, see for example Tunstall 1962; Orbach 1977; Zulaika 1981; Lummis 1983; Byron 1986, 1988; McGoodwin 1990.
4. See Andersen 1973; Löfgren 1977; Acheson 1981; Knipe 1984; Cohen 1987; McGoodwin 1990; Pollnac 1991; Peace 1991; LiPuma 1992; T. Vestergaard 1992; Pálsson 1994; Hanna and Jentoft 1996.
5. The names are pseudonyms.

6. While some anthropologists attribute success to the leadership role and personal skills of skippers (Byron 1980; Thorlindsson 1988), others point to a combination of engine power, equipment, skipper performance and skipper-crew interactions (Barth 1966; Baks and Postel-Coster 1977). The issue has kindled debate about the so-called 'skipper effect', which has been critically assessed by Pálsson and Durrenberger for the Icelandic fishing industry. They consider it a folk myth that serves social functions (see, for example, Pálsson and Durrenberger 1990; Durrenberger and Pálsson 1986). In their view, success mainly depends on fishing effort, that is, boat size and number of trips.
7. See, for example, Pálsson and Helgason 1995; Symes and Crean 1995; McCay 1995, 2000; McCay et al. 1995; Wiber 2000; Shotton 2000.

Notes to Chapter 6

1. As I pointed out in Chapter 4, auctioning was mandatory between 1959 and 1975, but had to be abolished to meet Community rules. In 1982, these rules changed. There were calls to reintroduce mandatory auctioning, but initially the House of Representatives rejected a proposal to do so, with the argument that the system should be effective in the European Community as a whole. In 2001, the EU again changed its stance concerning mandatory auctioning (see below).
2. See, for example, Dubbink, van der Schans and van Vliet 1994; Directie Visserij 1995; Hoefnagel and Smit 1997b:175; Symes 1997b:111-113; Berg 1999b:159; Langstraat 1999; van der Burg 2000:48; van Geffen et al. 2002; Symes, Steins and Alegret 2003:124-126; Bavinck and Hoefnagel 2004:43; Hoefnagel, Visser and de Vos 2004:45-46.
3. On Dutch quota hoppers, see also Davidse 1998; Hoefnagel 1998; Lequesne 2000; Valatin 2000 and Hatcher et al. 2002. Between 1996 and 2003, thirty-three to forty Dutch-owned vessels were on the UK register, approximately twenty-four to thirty-three flew German flags and between twenty-two and twenty-nine were registered in Belgium. A handful of Dutch fishermen re-flagged their boats to Denmark, Norway, France and Ireland. If we were to include the landing rights of the re-flagged Dutch-owned fishing vessels, Dutch owner-operators currently have more than ninety per cent of the European sole TAC and sixty per cent of the plaice TAC at their disposal (Task Force 2006:14).
4. UK and Irish fishermen also endorse this 'powerful and deeply embedded' belief (E. Bennett 1999:10; see also Connolly 1997; Rossiter and Stead 2003:282; Daw and Gray 2005:190). They feel picked on by the authorities and are convinced that they cannot win in a system they consider unfair and biased because of unequal enforcement. More generally, there is a climate of mutual suspicion concerning proper enforcement of fisheries regulations across member states.
5. The one-net rule, introduced in 2002 to enable linking days-at-sea allocations with gear type used, stipulates that only one set of gear is allowed aboard. It obstructs flexibility at sea, since boats must return to their homeport to fetch another net should a skipper decide to switch to another kind of fishery.
6. In early 2005, twelve out of nineteen Texel North Sea cutters had at least two relatives who crewed. Six of these boats had three crewing relatives and in-laws and one vessel had four crewing relatives and in-laws. Kin involvement is considerably greater if one considers shore-based co-owners.

Note to Conclusions

1. See, among others, Löfgren 1972:87ff.; Andersen 1973:159; Byron 1975, 1986:101ff.; Acheson 1981; Moustgaard 1984:345-347; T. Vestergaard 1990; Pálsson 1991:122ff.; Peace 1996:88.

References

Acheson, J.M. (1981) Anthropology of Fishing. *Annual Review of Anthropology* 10:275 316.

—— (1988) Patterns of Gear Changes in the Maine Fishing Industry. *Maritime Anthropological Studies* 1(1):49-65.

—— (1989) Management of Common Property Resources. In: S. Plattner (ed.), *Economic Anthropology*. Stanford: Stanford University Press, pp. 351-378.

—— (2006) Institutional Failure in Resource Management. *Annual Review of Anthropology* 35:117-134.

Adger, N.W. (2000) Social and Ecological Resilience: Are They Related? *Progress in Human Geography* 24(3):347-364.

Agnello, R.J. and L.P. Donneley (1976) Externalities and Property Rights in the Fisheries. *Land Economics* 52(4):518-529.

Agrawal, A. (2003) Sustainable Governance of Common-Pool Resources. *Annual Review of Anthropology* 32:243-262.

Ajuzie, E.I.S. and M.A. Altobello (1997) Property Rights and Pollution. *Review of Agricultural Economics* 19(2):242-251.

Andersen, R. (1973) Those Fisherman Lies. *Ethnos* 38(1-4):153-164.

—— (1979) Introduction. In: R. Andersen (ed.), *North Atlantic Maritime Cultures*. The Hague: Mouton, pp. 1 28.

Andersen, R. and C. Wadel (1972) Comparative Problems in Fishing Adaptations. In: R. Andersen and C. Wadel (eds.), *North Atlantic Fishermen*. St. John's: ISER, pp. 141-165.

Anderson, L.G. (1976) The Economics of Marine Resource Management. In: D.M. Johnston (ed.), *Marine Policy and the Coastal Community*. London: Croom Helm, pp. 65-84.

Anonymous (1852) De Texelsche en Zeeuwsche oester-visscherij. *Tijdschrift voor Staathuishoudkunde en Statistiek* 7:354-369.

—— (1854) *Verslag over de Zeevisscherijen*. No. 57. 's Gravenhage: Van Weelden en Mingelen.

—— (1870) Verslag der Staatscommissie in zake de wiermaaierij. In: *Verslag aan den Koning over de Openbare Werken in het jaar 1869*. 's Gravenhage: Van Weelden en Mingelen. Bijlage V, pp. 199 231.

—— (1956) *De toekomst van de Nederlandse zeevisserij*. Amsterdam: Wiarda Beckman Stichting.

—— (1973) *Visserijnota Noord-Holland*. Haarlem: Economisch-Technologische Dienst van Noord-Holland.

—— (1993) *Vissen naar evenwicht*. Den Haag: Ministerie van Landbouw, Natuurbeheer en Visserij.

Baks, C. and E. Postel-Coster (1977) Fishing Communities on the Scottish East Coast. In: M.E. Smith (ed.), *Those Who Live from the Sea*. St. Paul: West, pp. 23-40.

Banks, R. et al. (2001) *The Impact of Technological Progress on Fishing Effort. Final Report*. The Hague: LEI.

Barth, F. (1966) Models of Social Organization. *Royal Anthropological Institute Occasional Paper* 23.

Bavinck, M. and E. Hoefnagel (2004) 'If You Can't Beat Them, Join Them'. In: T. Dietz, P. Hoekstra, & F. Thissen (eds.), *The Netherlands and the North Sea*. Utrecht: KNAG, pp. 36-44.

Bechhofer, F. and B. Elliott (1981) Petty Property. In: F. Bechhofer and B. Elliott (eds.), *The Petite Bourgeoisie*. London: Macmillan, pp. 182 201.

Bennett, E. (1999) What Do Fishermen Really Think? *Samudra* 23:9-13.

Bennett, J.W. (1976) *The Ecological Transition*. New York: Pergamon Press.

—— (1990) Ecosystems, Environmentalism, Resource Conservation, and Anthropological Research. In: E.F. Moran (ed.), *The Ecosystem Approach in Anthropology*. Ann Arbor: The University of Michigan Press, pp. 435-457.

Berg, A. (1999a) *Implementing and Enforcing European Fisheries Law*. The Hague: Kluwer.

—— (1999b) Shifting Boundaries between Public and Private Law Enforcement. In: J.A.E. Vervaele (ed.), *Compliance and Enforcement of European Community Law*. The Hague: Kluwer, pp. 151-170.

Berghahn, R. and M. Ruth (2005) The Disappearance of Oysters from the Wadden Sea. *Aquatic Conservation* 15(1):91-104.

Bergman, M.J.N. et al. (1991) *Protected Areas in the North Sea*. Texel: Netherlands Institute for Sea Research.

Berkes, F. et al. (1989) The Benefits of the Commons. *Nature* 340:91 93.

Binkley, M. (2002) *Set Adrift: Fishing Families*. Toronto: University of Toronto Press.

Borghouts, T. (1991) *Visserij in de jaren '80*. Den Haag: Uitgeverij Borghouts.

Bossaers, K.W.J.M. (1987) *Zuiderzeevissers*. Zutphen: De Walburg Pers.

Bromley, D.W. (ed.) (1992) *Making the Commons Work*. San Francisco: ICS Press.

Brubaker, E. (1997) Beyond Quotas. In: L. Jones and M. Walker (eds.), *'Fish or Cut Bait!'*. Vancouver: The Fraser Institute, pp. 151-178.

Buskes, J.J. (1946) *Het eiland, de stad en het koninkrijk*. 's Gravenhage: D.A. Daamen's Uitgeversmaatschappij.

Butler, C.F. (2005) More Than Fish. Vancouver. Unpublished Ph.D. dissertation, University of British Columbia.

Byron, R.F. (1975) Economic Functions of Kinship Values in Family Businesses. *Sociology and Social Research* 60(2):147 160.

—— (1980) Skippers and Strategies. *Human Organization* 39(3):227-232.

—— (1986) *Sea Change. A Shetland Society, 1970 79*. St. John's: ISER.

—— (1988) Luck and Leadership. *Maritime Anthropological Studies* 1(1):3-14.

—— (1994) The Maritime Household in Northern Europe. *Comparative Studies in Society and History* 36(2):271-292.

Caddy, J.F. and J.C. Seijo (2005) This Is More Difficult Than We Thought! *Philosophical Transactions* 360:59-75.

Clark, D. (1982) *Between Pulpit and Pew*. Cambridge: Cambridge University Press.

Cohen, A.P. (1985) *The Symbolic Construction of Community*. Chichester: Ellis Horwood.

—— (1987) *Whalsay*. Manchester: Manchester University Press.

Cole, S. (1991) *Women of the Praia*. Princeton: Princeton University Press.

Commissie Sanering Zeevisserij (1952) Rapport van de Commissie Sanering Zeevisserij. 's-Gravenhage. *Verslagen en Mededelingen van de Directie der Visserijen* No. 42.
Connolly, B. (1997) *Traditional Fishery Knowledge and Practice for Sustainable Marine Resources Management in Northwestern Europe*. Leiden: Institute of Cultural and Social Studies, Universiteit Leiden.
Cooper, M-C. (1999) The Common Fisheries Policy of the European Union – A Lesson in How Not to Make Policy. *Politics* 19(2):61-70.
Copes, P. (1986) A Critical Review of the Individual Quota as Device in Fisheries Management. *Land Economics* 62(3) 278-291.
—— (1997) Social Impacts of Fisheries Management Regimes Based on Individual Quotas. In: G. Pálsson and G. Pétursdóttir (eds.), *Social Implications of Quota Systems in Fisheries*. Copenhagen: Nordic Council of Ministers, pp. 61-90.
—— (2000) Adverse Impacts of Individual Quota Systems on Conservation and Fish Harvest Productivity. Vancouver: Simon Fraser University. *Institute of Fisheries Analysis Discussion Paper* 00-2.
Corten, A. (1996) The Widening Gap Between Fisheries Biology and Fisheries Management in the European Union. *Fisheries Research* 27(1):1-15.
—— (2001) Herring and Climate. Unpublished Ph.D. Dissertation, Rijksuniversiteit Groningen.
Daalder, D.L. (1978) *Fijn en grof*. Schiedam: Interbook International B.V.
Daan, N. (1997) TAC Management in North Sea Flatfish Fisheries. *Journal of Sea Research* 37(3-4):321-341.
Davidse, W. (1997a) ITQs in the Netherlands. In: G. Pétursdóttir (ed.), *Property Rights in the Fishing Industry*. Reykjavik: Fisheries Research Institute, University of Iceland, pp. 3-9.
—— (ed.) (1997b) *Property Rights in Fishing*. The Hague: Agricultural Economics Research Institute.
—— (1998) Property Rights and Regulatory Systems: The Strategic Response of Dutch Fishermen. In: D. Symes (ed.), *Property Rights and Regulatory Systems in Fisheries*. Oxford: Blackwell Science, pp. 57-66.
—— (1999) Fisheries Management by Property Rights. In: P. Salz (ed.), *Proceedings of the 10th Annual Conference of the European Association of Fisheries Economists*. The Hague: LEI, pp. 321-328.
—— (2000) The Effects of Transferable Property Rights on the Fleet Capacity and Ownership of Harvesting Rights in the Dutch Demersal North Sea Fisheries. In: R. Shotton (ed.), *Use of Property Rights in Fisheries Management*. Rome: FAO, pp. 15-25.
Davidse, W. and J.W. de Wilde (2001) *The Dutch Beam Trawl Fishery for Flatfish*. The Hague: LEI.
Davidse, W., L.V. McEwan and N. Vestergaard (1999) Property Rights in Fishing. *Marine Policy* 23(6):537-547.
Davis, A. (1996) Barbed Wire and Bandwagons. *Reviews in Fish Biology and Fisheries* 6(1):97-107.
Davis, A. and S. Jentoft (1989) Ambivalent Co-operators. *Maritime Anthropological Studies* 2(2):194-211.
Davis, D.L. and J. Nadel-Klein (1988) Terra Cognita? In: J. Nadel-Klein and D.L. Davis (eds.), *To Work and to Weep*. St. John's: ISER, pp. 18-50.
—— (1992) Gender, Culture, and the Sea. *Society and Natural Resources* 5(2)135-147.

Daw, T. and T. Gray (2005) Fisheries Science and Sustainability in International Policy. *Marine Policy* 29(3):189-197.
de Alessi, M. (1997) Technologies for Sequestering and Monitoring Ocean Property. In: L. Jones and M. Walker (eds.), *'Fish or Cut Bait!'*. Vancouver: The Fraser Institute, pp. 125-149.
de Boer, E.J. and C. van der Meulen (1976) *Een schip vis*. Bussum: De Boer Maritiem.
de Bruyn, M.A. (2001) *Arbeid kennis vlijt*. Rijswijk: SFM.
de Groot, S.J. (1984) The Impact of Bottom Trawling on Benthic Fauna of the North Sea. *Ocean Management* 9(3-4):177-190.
—— (1988) *Een eeuw visserijonderzoek in Nederland 1888-1988*. IJmuiden: RIVO.
de Groot, S.J. and H.J. Lindeboom (1994) *Environmental Impact of Bottom Gears on Benthic Fauna in Relation to Natural Resources Management and Protection of the North Sea*. Texel: Netherlands Institute for Sea Research.
de Jonge, L. (2005) *De laatste haringvissers van Nederland*. Amsterdam: Uitgeverij Balans.
de Swaan, A. (1988) *In Care of the State*. New York: Oxford University Press.
de Wolf, C.J. (2005) *Eb en vloed*. Oudeschild: Maritiem en Jutters Museum.
Dijt, M.D. (1961) Texelse oesters. *Visserij Nieuws* 14(7):99 102.
Directie Visserij (1995) *Samenwerking in Beheer*. Den Haag: Directie Visserij, Ministerie van Landbouw, Natuurbeheer en Visserij.
Dubbelaar, W. (1980) Sociale verzekering in de visserij. In: *Het Visserijschap*. Haarlem: De Boer Maritiem, pp. 116-123.
Dubbink, W., J.W. van der Schans and M. van Vliet (1994) *'Vissen is veel meer een berekening geworden, maar nog niet al mijn collega's zijn zich hiervan bewust'*. Rotterdam: Faculteit Bedrijfskunde, Erasmus Universiteit Rotterdam.
Dubbink, W. and M. van Vliet (1996) Market Regulation Versus Co-management? *Marine Policy* 20(6):499-516.
—— (1997) The Netherlands: From ITQ to Co-Management. In: *Towards Sustainable Fisheries: Issue Papers*. Paris: OECD, pp. 177-202.
Durkheim, E. (1982) *The Rules of Sociological Method*. London: Macmillan Press.
Durrenberger, E.P. (1988) Shrimpers and Turtles on the Gulf Coast. *Maritime Anthropological Studies* 1(2):196 214.
—— (1992a) *It's All Politics*. Urbana: University of Illinois Press.
—— (1992b) Psychology, Unions, and the Law. *Human Organization* 51(2):151 154.
Durrenberger, E.P. and G. Pálsson (1985) Peasants, Entrepreneurs and Companies. *Ethnos* 50(1/2):103 122.
—— (1986) Finding Fish. *American Ethnologist* 13(2):213-229.
Edwards, V.M. and N.A. Steins (1998) Developing an Analytical Framework for Multiple-Use Commons. *Journal of Theoretical Politics* 10(3):347-383.
—— (1999) A Framework for Analysing Contextual Factors in Common Pool Resource Research. *Journal of Environmental Policy and Planning* 1(3):205-221.
Elias, E. et al. (2003) Tidal Inlet Dynamics in Response to Human Intervention. *Coastal Engineering Journal* 45(4):629-658.
Elias, N. (1974) Towards a Theory of Communities. In: C. Bell and H. Newby (eds.), *The Sociology of Communities*. London: Frank Cass & Co, pp. ix-xli.
Elliott, E. and L.D. Kiel (1997) Nonlinear Dynamics, Complexity, and Public Policy. In: R.A. Eve et al. (eds.), *Chaos, Complexity, and Sociology*. Thousand Oaks: Sage, pp. 64-78.

European Commission (2001a) *Green Paper on the Future of the Common Fisheries Policy*. Brussels: Commission of the European Communities.
—— (2001b) *European Governance. A White Paper*. Brussels: Commission of the European Communities.
—— (2002) *Communication from the Commission on the Reform of the Common Fisheries Policy ('Roadmap')*. Brussels: Commission of the European Communities.
—— (2005) *Communication from the Commission to the Council and the European Parliament. 2006-08 Action Plan for Simplifying and Improving the Common Fisheries Policy*. Brussels: Commission of the European Communities.
—— (2006) *Proposal for a Council Regulation Establishing a Management Plan for Fisheries Exploiting Stocks of Plaice and Sole in the North Sea*. Brussels: Commission of the European Communities.
Feeny, D., S. Hanna and A.F. McEvoy (1999) Questioning the Assumptions of the 'Tragedy of the Commons' Model of Fisheries. *Land Economics* 72(2):187-205.
Finlayson, C. (1991) Notes on Chaos in Fisheries Management by Estellie Smith. *Maritime Anthropological Studies* 4(1):91-97.
Flyvbjerg, B. (2003) *Making Social Science Matter*. Cambridge: Cambridge University Press.
Garcia, S.M. and C.H. Newton (1994) Responsible Fisheries. *Marine Pollution Bulletin* 29(6-12):528-536.
Gatewood, J.B. (1984) Cooperation, Competition, and Synergy. *American Ethnologist* 11(2):350-370.
Giesen, W.B.J.T., M.M. van Katwijk and C. den Hartog (1990) Temperature, Salinity, Insolation and Wasting Disease of Eelgrass (*Zostera marina* L.) in the Dutch Wadden Sea in the 1930's. *Netherlands Journal of Sea Research* 25(3):395-404.
Godelier, M. (1986) *The Mental and the Material*. London: Verso [orig. French ed. 1984].
Gordon, H.S. (1954) The Economic Theory of a Common Property Resource: The Fishery. *Journal of Political Economy* 62(2):124 142.
Grafton, R.Q. (1996) Individual Transferable Quotas. *Reviews in Fish Biology and Fisheries* 6(1):5-20.
Grafton, R.Q., D. Squires and K.J. Fox (2000) Private Property and Economic Efficiency. *Journal of Law and Economics* 43(2):679–713.
Gray, T. and J. Hatchard (2003) The 2002 Reform of the Common Fishery Policy's System of Governance. *Marine Policy* 27(6):545-554.
Grift, R. et al. (2004) *Assessment of the Ecological Effects of the Plaice Box*. Brussels.
Hamilton, L.C. et al. (1998) Management, Adaptation and Large-Scale Environmental Change. In: D. Symes (ed.), *Property Rights and Regulatory Systems in Fisheries*. Oxford: Blackwell Science, pp. 17-33.
Hanna, S. (1995) User Participation and Fishery Management Performance within the Pacific Fishery Management Council. *Ocean and Coastal Management* 28(1-3):23-44.
—— (1999) Strengthening Governance of Ocean Fishery Resources. *Ecological Economics* 31(2):275-286.
Hanna, S. and S. Jentoft (1996) Human Use of the Natural Environment. In: S.S. Hanna et al. (eds.), *Rights to Nature*. Washington, DC: Island Press, pp. 35-55.
Hardin, G. (1968) The Tragedy of the Commons. *Science* 162:1243 1248.
—— (1998) Extensions of the 'Tragedy of the Commons'. *Science* 280(5364):682-683.

Hatcher, A. et al. (2002) 'Quota-Hopping' and the Foreign Ownership of UK Fishing Vessels. *Marine Policy* 26(1):1-11.
Healey, M.C. and T. Hennessey (1998) The Paradox of Fairness. *Marine Policy* 22 (2):109-118.
Hildebrandt, A.G.U. (1947-1948) De vissersbevolking vraagt om erkenning. *Het Gemeenebest* 8:439-442.
—— (1952) De Nederlandse visserij. In: *De Nederlandse volkshuishouding tussen twee wereldoorlogen*. Utrecht: Het Spectrum, pp. X.1-X.109.
Hildebrandt, A.G.U. and M.A.J. Visser (1947) *Onderzoek ten behoeve van de sanering van de Nederlandse garnalenvisserij*. Den Haag: LEI.
Hoefnagel, E. (1996) Trade in Fishing Rights in the Netherlands. In: K. Crean and D. Symes (eds.), *Fisheries Management in Crisis*. Oxford: Blackwell Science, pp. 61-70.
—— (1998) Legal but Controversial. In: D. Symes (ed.), *Property Rights and Regulatory Systems in Fisheries*. Oxford: Blackwell Science, pp. 80-91.
Hoefnagel, E. and W. Smit (1997a) Co-Management and Property Rights in the Netherlands. In: G. Pétursdóttir (ed.), *Property Rights in the Fishing Industry*. Reykjavik: Fisheries Research Institute, University of Iceland, pp. 11-24.
—— (1997b) The Netherlands: Experiences in Dutch Co-Management of Marine Fish Resources. In: *Towards Sustainable Fisheries: Issue Papers*. Paris: OECD, pp. 156-176.
Hoefnagel, E. and M.J. Smits (1999) Modern Dutch Fishermen's Wives. *Europaea* 5 (2):147-162.
—— (2000) *De Nederlandse vissersvrouw*. Den Haag: LEI.
Hoefnagel, E., L. Visser and B. de Vos (2004) *Drijfveren van vissers en duurzaam visserijbeheer*. Den Haag: LEI.
Hoek, P.P.C. (1878) Oestercultuur in den vreemde en bij ons. *Eigen Haard* 41:389 393.
—— (1902) De organisatie van de beroepsvissers. *Mededeelingen over visscherij* 9:113 122.
Hoekstra, A. (1939) Bijdrage tot de sociographische kennis van Wieringen. Unpublished manuscript.
Holden, M. (1984) *The Common Fisheries Policy*. Oxford: Fishing News Books.
Holling, C. and G. Meffe (1996) Command and Control and the Pathology of Natural Resource Management. *Conservation Biology* 10(2):328-337.
Holm, P. (1996) Fisheries Management and the Domestication of Nature. *Sociologia Ruralis* 36(2):177-188.
Hornborg, A. (1996) Ecology as Semiotics. In: P. Descola and G. Pálsson (eds.), *Nature and Society*. London: Routledge, pp. 45-62.
Ingold, T. (1992) Culture and the Perception of the Environment. In: E. Croll and D. Parkin (eds.), *Bush Base: Forest Farm*. London: Routledge, pp. 39-56.
Jentoft, S. (1986) Fisheries Co-operatives. *Canadian Journal of Development Studies* 7(1):197-209.
—— (1989) Fisheries Co-Management. *Marine Policy* 13(2):137-154.
—— (1993) *Dangling Lines*. St. John's: ISER.
—— (2006) Beyond Fisheries Management. *Marine Policy* 30(6):671-680.
Jentoft, S. and B.J. McCay (1995) User Participation in Fisheries Management. *Marine Policy* 19(3):227-246.
Jentoft, S., B.J. McCay and D.C. Wilson (1998) Social Theory and Fisheries Co-Management. *Marine Policy* 22(4-5):423-436.

Jones, L. and M. Walker (eds.) (1997) *'Fish or Cut Bait!'*. Vancouver: The Fraser Institute.
Jorion, P. (1983) *Les pêcheurs d'Houat*. Paris: Hermann.
Keesing, R.M. (1981) *Cultural Anthropology*. New York: Holt, Rinehart and Winston.
Kennelly, S.J. and M.K. Broadhurst (2002) By-Catch Begone. *Fish and Fisheries* 3 (4):340-355.
Kerkhoven, J. (1994) *Het net en de wet*. Amsterdam: P.J. Meertens Instituut.
Klug, H. (2002) Straining the Law. *Society and Natural Resources* 15(8):693-707.
Knipe, E. (1984) *Gamrie*. Lanham: University Press of America.
Knutson, P. (1991) Measuring Ourselves. *Maritime Anthropological Studies* 4(1):73-90.
Kooiman, J., M. van Vliet and S. Jentoft (eds) (1999) *Creative Governance*. Aldershot: Ashgate.
Kranenburg, H.A.H. Boelmans (1968) De maatschap in de visserij. *Sociaal Maandblad Arbeid* 23:550-557.
—— (1977) *Achter de Branding*. Bussum: De Boer Maritiem.
—— (1978) Visserij. In: G. Asaert et al. (eds.) *Maritieme geschiedenis der Nederlanden. Deel 4*. Bussum: De Boer Maritiem, pp. 272-301.
—— (1980) Organisatie van de visserij tot 1955. In: *Het Visserijschap*. Haarlem: De Boer Maritiem, pp. 9-16.
Langstraat, D. (1998) The Economic Contribution in Policy Process. In: P. Salz (ed.), *Proceedings of the 10th EAFE Conference*. The Hague: LEI, pp. 11-15.
—— (1999) The Dutch Co-Management System for Sea Fisheries. In: D. Symes (ed.), *Alternative Management Systems for Fisheries*. Oxford: Blackwell Science, pp. 73-78.
le Long, I. (1727) *De Koophandel van Amsterdam*. Amsterdam: Andries van Damme en Johannes Ratelband.
Lequesne, C. (2000) Quota Hopping. *Journal of Common Market Studies* 38(5):779-793.
—— (2004) *The Politics of Fisheries in the European Union*. Manchester: Manchester University Press.
Libecap, G. (1989) *Contracting for Property Rights*. Cambridge: Cambridge University Press.
Lindeboom, H.J. (1995) Protected Areas in the North Sea. *Helgoländer Meeresuntersuchungen* 49(1-4):591-602.
—— (2000) The Need for Closed Areas as Conservation Tools. In: M.J. Kaiser and S.J. de Groot (eds.). *Effects of Fishing on Non-Target Species and Habitats*. Oxford: Blackwell Science, pp. 290-302.
Lindeboom, H.J. and S.J. de Groot (eds.) (1998) *Impact-II. The Effects of Different Types of Fisheries on North Sea and Irish Sea Benthic Ecosystems*. Texel: Netherlands Institute for Sea Research.
LiPuma, E. (1992) Social Identity and the European Community. *Maritime Anthropological Studies* 5(2):46-73.
Löfgren, O. (1972) Resource Management and Family Firms. In: R. Andersen and C. Wadel (eds.), *North Atlantic Fishermen*. St. John's: ISER, pp. 82 103.
—— (1977) Maritime Hunters in Industrial Society. In: O. Löfgren, *Fångstmän i industri samhället*. Lund: LiberLäromedel, pp. 226 235.
—— (1989) The Reluctant Competitors. *Maritime Anthropological Studies* 2(1):34-58.

Low, B. et al. (1999) Human-Ecosystem Interactions. *Ecological Economics* 31 (2):227-242.
Lummis, T. (1983) Luck. In: P. Thompson, T. Wailey and T. Lummis, *Living the fishing*. London: Routledge and Kegan Paul, pp. 182-202.
—— (1985) *Occupation and Society*. Cambridge: Cambridge University Press.
Mace, P.M. (1993) Will Private Owners Practice Prudent Resource Management? *Fisheries* 18(9):29-31.
Macpherson, C.B. (1975) Capitalism and the Changing Concept of Property. In: E. Kamenska and R.S. Neale (eds.), *Feudalism, Capitalism and Beyond*. London: Edward Arnold, pp. 104 124.
Magnússon, F. (1990) *The Hidden Class*. Aarhus: Aarhus University Press.
Malinowski, B. (1955) *Magic, Science and Religion*. Garden City: Doubleday.
Matthews, D.R. (1993) *Controlling Common Property*. Toronto: University of Toronto Press.
McCay, B.J. (1978) Systems Ecology, People Ecology and the Anthropology of Fishing Communities. *Human Ecology* 6:397-422.
—— (1995) Social and Ecological Implications of ITQs. *Ocean and Coastal Management* 28(1-3):3-22.
—— (1998) *Oyster Wars and the Public Trust*. Tucson: University of Arizona Press.
—— (2000) Sea Changes in Fisheries Policy. In: E.P. Durrenberger and T.D. King (eds.), *State and Community in Fisheries Management*. Westport, CT: Bergin and Garvey, pp. 201-217.
—— (2001) Environmental Anthropology at Sea. In: C.L. Crumley (ed.), *New Directions in Anthropology and Environment*. Walnut Creek: Altamira Press, pp. 254-272.
—— (2002) Emergence of Institutions for the Commons. In: E. Ostrom et al. (eds.), *The Drama of the Commons*. Washington, DC: National Academy Press, pp. 361-402.
McCay, B.J. and J.M. Acheson (1987a) Human Ecology of the Commons. In: B.J. McCay and J.M. Acheson (eds.), *The Question of the Commons*. Tucson: The University of Arizona Press, pp. 1-34.
—— (eds.) (1987b) *The Question of the Commons*. Tucson: The University of Arizona Press.
McCay, B.J. and C.F. Creed (1989) Dividing Up the Commons. In: J.S. Thomas, L. Maril and E.P. Durrenberger (eds.), Marine Resource Utilization. Mobile. *University of South Alabama College of Arts and Sciences Publication* Vol. 1, pp. 11-23.
McCay, B.J. et al. (1995) Individual Transferable Quotas (ITQs) in Canadian and US Fisheries. *Ocean and Coastal Management* 28(1-3):85-115.
McEvoy, A.F. (1986) *The Fisherman's Problem*. Cambridge: Cambridge University Press.
—— (1988) Toward an Interactive Theory of Nature and Culture. In: D. Worster (ed.), *The Ends of the Earth*. Cambridge: Cambridge University Press, pp. 211 229.
McGoodwin, J.R. (1990) *Crisis in the World's Fisheries*. Stanford: Stanford University Press.
—— (2001) *Understanding the Cultures of Fishing Communities*. Rome: FAO.
McGuire, T.R. (1991) Science and the Destruction of a Shrimp Fleet. *Maritime Anthropological Studies* 4(1):32-55.
McKean, M.A. (1992) Success on the Commons. *Journal of Theoretical Politics* 4 (3):247-281.

McNaghten, P. and J. Urry (1998) *Contested Natures*. London: Sage.
Meltzoff, S.K. (1995) Marisquadores of the Shellfish Revolution. *Journal of Political Ecology* 2:20-38.
Menzies, C.R. (2003) Fishing, Families, and the Survival of Artisanal Boat-Ownership in the Bigouden Region of France. *Maritime Studies* 2(1):73-90.
Moran, E.F. (1990) Levels of Analysis and Analytical Level Shifting. In: E.F. Moran (ed.), *The Ecosystem Approach in Anthropology*. Ann Arbor: The University of Michigan Press, pp. 279-308.
Moustgaard, P. (1984) The Fishing Community, the Gear and the Environment. In: B. Gunda (ed.), *The Fishing Culture of the World*. Budapest: Akadémiai Kiadó, pp. 337-357.
Munk-Madsen, E. (2000) Wife the Deckhand, Husband the Skipper. *Women's Studies International Forum* 23(3):333-342.
Nadel-Klein, J. (2003) *Fishing for Heritage*. Oxford: Berg.
Nadel-Klein, J. and D.L. Davis (eds.) (1988) *To Work and to Weep*. St. John's: ISER.
Neis, B. et al. (eds.) (2005) *Changing Tides*. Halifax: Fernwood Publishing.
Norr, J.L. and K.L. Norr (1978) Work Organization in Modern Fishing. *Human Organization* 37(2):163 171.
Okely, J. (1992) Anthropology and Autobiography. In: J. Okely and H. Callaway (eds.), *Anthropology and Autobiography*. London: Routledge, pp. 1-28.
Olson, M. (1971[1965]) *The Logic of Collective Action*. Cambridge: Harvard University Press.
Orbach, M.K. (1977) *Hunters, Seaman, and Entrepreneurs*. Berkeley: University of California Press.
—— (1980) Fishery Cooperatives on the Chesapeake Bay. *Anthropological Quarterly* 53:48 55.
Orlove, B.S. (1980) Ecological Anthropology. *Annual Review of Anthropology* 1980:235-273.
Ortner, S.B. (1984) Theory in Anthropology since the Sixties. *Comparative Studies in Society and History* 26(1):126-166.
—— (1996) *Making Gender*. Boston, Mass.: Beacon Press.
Ostrom, E. (1990) *Governing the Commons*. Cambridge: Cambridge University Press.
—— (1992) The Rudiments of a Theory of the Origin, Survival, and Performance of Common-Property Institutions. In: D.W. Bromley (ed.), *Making the Commons Work*. San Francisco: ICS Press, pp. 293-318.
—— (1999) Coping With Tragedies of the Commons. *Annual Review of Political Science* 2:493-535.
Pálsson, G. (1989) The Art of Fishing. *Maritime Anthropological Studies* 2(1):1 20.
—— (1991) *Coastal Economies, Cultural Accounts*. Manchester: Manchester University Press.
—— (1994) Enskilment at Sea. *Man* N.S. 29(4):901-927.
—— (1996) Human-Environmental Relations. In: P. Descola and G. Pálsson (eds.), *Nature and Society*. London: Routledge, pp. 63-81.
Pálsson, G. and E.P. Durrenberger (1982) To Dream of Fish. *Journal of Anthropological Research* 38(2):227-242.
—— (1990) Systems of Production and Social Discourse. *American Anthropologist* 92(1):130-141.
Pálsson, G. and A. Helgason (1995) Figuring Fish and Measuring Men. *Ocean and Coastal Management* 28(1-3):117-146.

—— (1998) Schooling and Skipperhood. *American Anthropologist* 100(4):908-923.
Paludanus, R. (1776) *Oudheid- en natuurkundige verhandelingen.* Leyden: P. van der Eyk en D. Vygh.
Pascoe, S., J.L. Andersen and J.W. de Wilde (2001) The Impact of Management Regulation on the Technical Efficiency of Vessels in the Dutch Beam Trawl Fishery. *European Review of Agricultural Economics* 28(2):187-206.
Pascual Fernández, J. (1999) The Share System in Fishing. *Europaea* 5(2):65-87.
Pastoors M.A., A.D. Rijnsdorp and F.A. van Beek (2000) Effects of a Partially Closed Area in the North Sea ('Plaice Box') on Stock Development of Plaice. *ICES Journal of Marine Science* 57(4):1014-1022.
Peace, A. (1991) Fishing, Drinking and the Construction of Identity in Rural Ireland. *Maritime Anthropological Studies* 4(2):3-16.
—— (1996) When the Salmon Comes. *Journal of Anthropological Research* 52(1):85-106.
Pettersen, L.T. (1996) Crisis Management and Household Strategies in Lofoten. *Sociologia Ruralis* 36(2):236-248.
Pickett, S.T.A. and R.S. Ostfeld (1995) The Shifting Paradigm in Ecology. In: R.L. Knight and S.F. Bates (eds.), *A New Century for Natural Resources Management.* Cambridge, MA: MIT Press, pp. 261-278.
Piet, G.J. et al. (2000) A Quantitative Evaluation of the Impact of Beam Trawling on Benthic Fauna in the Southern North Sea. *ICES Journal of Marine Science* 57 (5):1332–1339.
Pinkerton, E. (ed.) (1989) *Co-operative Management of Local Fisheries.* Vancouver: University of British Columbia Press.
Plomp, C. (1940) Urk en de Zuiderzeewerken. *Mens en Maatschappij* 16(6):372-392.
—— (1989 [1940]) *Urk.* Zutphen: Walburg Pers.
Poggie Jr, J.J. (1980) Small Scale Fishermen's Psychocultural Characteristics and Cooperative Formation. *Anthropological Quarterly* 53(1):20 28.
—— (1992) Intracultural and Intrasocial Variability as a Tool for Policy Making in Fisheries Development and Management. In: J.J. Poggie Jr. et al. (eds.), *Anthropological Research.* Albany, NY: State University of New York Press, pp. 49-64.
Pollnac, R.B. (1991) Social and Cultural Characteristics in Small-Scale Fishery Development. In: M.M. Cernea (ed.), *Putting People First.* 2nd edn. Oxford: Oxford University Press, pp. 259-299.
Pollnac, R.B. and J.J. Poggie (1991) Psychocultural Adaptation and Development Policy for Small-Scale Fishermen's Cooperatives in Ecuador. *Human Organization* 50(1):43-49.
Prattis, J.I. (1987) Organizational Change and Adaptation. *American Anthropologist* 89(3):567-580.
Productschap Vis (2001) *Eindrapport bemanningsonderzoek.* Rijswijk: Productschap Vis.
—— (2004) *Fishing on a Square Inch.* Rijswijk: Productschap Vis.
Rappaport, R.A. (1979) *Ecology, Meaning, and Religion.* Berkeley: North Atlantic Books.
Rijneveld, R., W. Smit and J.W. de Wilde (1973) *Ekonomische perpektieven van de kottervisserij in Nederland.* Den Haag: LEI.
Rijneveld, R. et al. (1981) De vrije visserij in een kader van beperkingen. *Visserij* 34 (3):115-131.

Rijnsdorp, A.D., N. Daan and W. Dekker (2006) Partial Fishing Mortality per Fishing Trip. *ICES Journal of Marine Science* 63(3):556-566.
Rijnsdorp, A.D. and P.I. van Leeuwen (1996) Changes in Growth of North Sea Plaice since 1950 in Relation to Density, Eutrophication, Beam-Trawl Effort, and Temperature. *ICES Journal of Marine Science* 53(6):1199-1213.
Rijnsdorp, A.D. et al. (1998) Micro-Scale Distribution of Beam Trawl Effort in the Southern North Sea Between 1993 and 1996 in Relation to the Trawling Frequency of the Sea Bed and the Impact on Benthic Organisms. *ICES Journal of Marine Science* 55(3):403-419.
—— (2000a) Competitive Interactions among Beam Trawlers Exploiting Local Patches of Flatfish in the North Sea. *ICES Journal of Marine Science* 57(4):894-902.
—— (2000b) Effects of Fishing Power and Competitive Interactions among Vessels on the Effort Allocation on the Trip Level of the Dutch Beam Trawl Fleet. *ICES Journal of Marine Science* 57(4):927-937.
—— (2007) Sustainable Use of Flatfish Resources. *Journal of Sea Research* 57(1):114-125.
Rodgers, S.C. (1991) *Shaping Modern Times in Rural France.* Princeton: Princeton University Press.
Rose, C.M. (2002) Common Property, Regulatory Property, and Environmental Protection. In: E. Ostrom et al. (eds.), *The Drama of the Commons.* Washington, DC: National Academic Press, pp. 233-258.
Rossiter, T. and S. Stead (2003) Days at Sea. *Marine Policy* 27(3):281-288.
Runge, C.F. (1986) Common Property and Collective Action in Economic Development. *World Development* 14(5):623-635.
Ruttan, L.M. (2006) Sociocultural Heterogeneity and the Commons. *Current Anthropology* 47(5):843-853.
Rutz, H.J. (1977) Individual Decisions and Functional Systems. *American Ethnologist* 4(1):156 174.
Ryang, S. (1997) Native Anthropology and Other Problems. *Dialectical Anthropology* 22(1):23-49.
Saal, C.D. (1948) Godsdienstigheid en levensstijl in Noord Holland. *Sociologisch Bulletin* 2(1):2 17.
Salz, P. (ed.) (1997) Fisheries Management Policies Towards 2010. The Hague. *NRLO Report* 97/29.
Santopietro, G.D. and L. Shabman (1992) Can Privatization Be Inefficient? *Journal of Economic Issues* 26(2):408-419.
Schaap, D. (1990) *Vissen voor je brood.* Naarden: Strengholt.
Schaper, A. (1962) *De IJsselmeervisserij.* Utrecht: Kemink en Zoon.
Schlager, E. and E. Ostrom (1992) Property-Rights Regimes and Natural Resources. *Land Economics* 68(3):249 262.
Schreiber, D.K. (2001) Co-Management Without Involvement. *Fisheries* 2(4):376-384.
Schreur, S. (1953) *De ontwikkeling van de visserij van Texel.* Haarlem: Economisch-Technologische Dienst voor Noord-Holland.
—— (1964) Veranderingen op Texel. *Noord-Holland* 9(6):214-224.
—— (1966) De visserij van Texel. *Waddenbulletin* 1(2):6-9.
Scoones, I. (1999) New Ecology and the Social Sciences. *Annual Review of Anthropology* 28:479-507.

Scott, A. (2000) Moving Through the Narrows. In: R. Shotton (ed.), *Use of Property Rights in Fisheries Management*. Rome: FAO.
Scott, J.C. (1998) *Seeing Like a State*. New Haven: Yale University Press.
Shotton, R. (ed.) (2000) *Use of Property Rights in Fisheries Management*. Rome: FAO.
Sinclair, P.R. (1985) *From Traps to Draggers*. St. John's: ISER.
—— (1996) Sustainable Development in Fisheries Dependent Regions? *Sociologia Ruralis* 36(2):224-235.
Singleton, S. (2000) Co-operation or Capture? *Environmental Politics* 9(2):1-21.
Smit, W. (2001) Dutch Demersal North Sea Fisheries. In: R. Shotton (ed.), *Case Studies on the Allocation of Transferable Quota Rights in Fisheries*. Rome: FAO, pp. 15-23.
Smit, J.G.P. and C. Taal (2007) *Sociaaleconomische indicatoren van de Nederlandse vissector*. Den Haag, LEI.
Smith, M.E. (1977) Comments on the Heuristic Utility of Maritime Anthropology. *The Maritime Anthropologist* 1:2-5, 8.
—— (1990) Chaos in Fisheries Management. *Maritime Anthropological Studies* 3 (2):1-13.
Smith, T.A. (1997) Nonlinear Dynamics and the Micro-Macro Bridge. In: R.A. Eve et al. (eds.), *Chaos, Complexity, and Sociology*. Thousand Oaks: Sage, pp. 52-63.
Steins, N.A. (1999) *All Hands on Deck*. Wageningen. Ph.D. Thesis Wageningen Universiteit.
Stuurgroep Biesheuvel (1992) *Beheerst vissen*. Rijswijk: Visserijcentrum.
Stuurgroep Nijpels (2005) *Eindrapport Stuurgroep Nijpels*. Rijswijk: Productschap Vis.
Subcommissie Visquoteringsregelingen (1987) *Visquoteringsregelingen. Rapport van de Subcommissie Visquoteringsregelingen*. Den Haag, Tweede Kamer, vergaderjaar 1986-1987.
Symes, D. (1997a) The European Community's Common Fisheries Policy. *Ocean and Coastal Management* 35(2-3):137-155.
—— (1997b) Fisheries Management. *Fisheries Research* 32(2):107-114.
—— (1999) Sleeping in Separate Beds. In: P. Salz (ed.), *Proceedings of the 10th EAFE Conference*. The Hague: LEI, pp. 137-147.
Symes, D. and Crean, K. (1995) Privatisation of the Commons. *Geoforum* 26(2):175-185.
Symes, D., N. Steins and J.L. Alegret (2003) Experiences with Fisheries Co-Management in Europe. In: D.C. Wilson and J.R. Nielsen (eds.), *The Fisheries Co-Management Experience*. Dordrecht: Kluwer, pp. 119-133.
Taal, C. et al. (2008) *Visserij in cijfers 2008*. Den Haag: LEI.
Task Force (2006) *Vissen met tegenwind*. Rijswijk: Productschap Vis.
Taylor, L.J. (1981) 'Man the Fisher'. *American Ethnologist* 8(4):774-788.
—— (1983) *Dutchmen on the Bay*. Philadelphia: University of Pennsylvania Press.
Thiessen, V., A. Davis and S. Jentoft (1992) The Veiled Crew. *Human Organization* 51(4):342-352.
Thomas, J.S., G.D. Johnson and C.A. Riordan (1995) Independence and Collective Action. *Human Organization* 54(2):143-152.
Thomas, W.I. and D.S. Thomas (1928) *The Child in America*. New York: Knopf.
Thompson, E.P. (1971) The Moral Economy of the English Crowd in the Eighteenth Century. *Past and Present* 50:76-136.
Thompson, P. (1985) Women in the Fishing. *Comparative Studies in Society and History* 27(1):3-32.

Thompson, P. et al. (1983) *Living the Fishing*. London: Routledge and Kegan Paul.
Thorlindsson, T. (1988) The Skipper Effect in the Icelandic Herring Fishery. *Human Organization* 47(3):199-212.
Townsend, R. and J.A. Wilson (1987) An Economic View of the Tragedy of the Commons. In: B.J. McCay and J.M. Acheson (eds.), *The Question of the Commons*. Tucson: University of Arizona Press, pp. 311-326.
Tunstall, J. (1962) *The Fishermen*. London: McGibbon and Kee.
Valatin, G. (2000a) Quota Trading Systems in EU Fisheries. *Review of European Community and International Environmental Law* 9(3):296-306.
—— (2001) Solving the 'Tragedy' of the Common Fisheries Policy. Paper presented at the 8th Annual Conference of the European Association of Fisheries Economists, Salerno, Italy, 18-20 April 2001.
Valdés Pizzini, M. (1990) Fishermen Associations in Puerto Rico. *Human Organization* 49:164 173.
van der Burg, T. (2000) Neo-Classical Economics, Institutional Economics and Improved Fisheries Management. *Marine Policy* 24(1):45-51.
van der Kooi, G. (2005) *De Wynberch des Heren*. Hilversum: Verloren.
van der Kroon, O. (1994) *Ministerie in crisis*. Amsterdam: Uitgeverij L.J. Veen.
van der Schans, J.W. (2001) *Governance of Marine Resources*. Delft: Eburon.
van der Vlis, J.A. (1977) *tLant van Texsel*. Den Burg: Langeveld & De Rooy.
van Geffen, J. et al. (2002) *Op weg naar evenwicht*. Ede/Wageningen: Expertisecentrum LNV.
van Ginkel, R. (1989) 'Plunderers' into Planters. In: J. Boissevain and J. Verrips (eds.), *Dutch Dilemmas*. Assen: Van Gorcum, pp. 89-105.
—— (1990) Farming the Edge of the Sea. *Maritime Anthropological Studies* 3(2):49-67.
—— (1991) The Sea of Bitterness. *Man* N.S. 26(4):691 707.
—— (1993) *Tussen Scylla en Charybdis*. Amsterdam: Het Spinhuis.
—— (1994) Writing Culture from Within. *Etnofoor* 7(1):5-23.
—— (1995a) *'Groen zwart, Texels in het hart'*. Amsterdam: Het Spinhuis.
—— (1995b) 'Texelian at Heart'. *Ethnos* 60(3/4):265-286.
—— (1996a) The Abundant Sea and Her Fates. *Comparative Studies in Society and History* 38(2):218-242.
—— (1996b) Cooperating Competitors. *Anthropological Quarterly* 69(1):51-65.
—— (1997) *Zostera marina* in Dispute. In: D. Symes (ed.), *Property Rights and Regulatory Systems in Fisheries*. Oxford: Blackwell Science, pp. 230-243.
—— (1998) The Repatriation of Anthropology. *Anthropology & Medicine* 5(3):251-267.
—— (1999a) The Dynamics of Fisheries. In: D. Symes (ed.), *Europe's Southern Waters* Oxford: Blackwell Science, pp. 19-32.
—— (1999b) Contextualizing Marine Resource Use. *Journal of Environmental Policy and Planning* 1(3):223-233.
—— (2001) Inshore Fishermen. In: D. Symes and J. Phillipson (eds.), *Inshore Fisheries Management*. Dordrecht: Kluwer, pp. 177-193.
—— (2007a) *Coastal Cultures*. Apeldoorn / Antwerpen: Spinhuis Publishers.
—— (2007b) Celebrating Localism. In: P.J. Margry and H. Roodenburg (eds.), *Between Otherness and Authenticity*. Aldershot: Ashgate, pp. 37-57.
van Hoof, L. (2005) Sharing Responsibilities in Fisheries Management. Proceedings of the 17th EAFE Conference, Thessaloniki, 21-23 March 2005.

van Vliet, M. (1998a) From Economic Fish Stock Management to Integrated Management. In: D. Symes (ed.), *Northern Waters*. Oxford: Blackwell Science, pp. 216-226.
—— (1998b) Fishing as a Collective Enterprise. In: D. Symes (ed.), *Property Rights and Regulatory Systems in Fisheries*. Oxford: Blackwell Science, pp. 67-79.
—— (1999) The Tragedy of the Commons Model and the Revisionist Criticism. In: P. Salz (ed.), *Proceedings of the 10th EAFE Conference*. The Hague: LEI, pp. 157-188.
van Wijk, M.O. and J.W. de Wilde (2004) *Modellen voor de platvisvisserij op de Noordzee*. Den Haag: LEI.
Vayda, A.P. (1983) Progressive Contextualization. *Human Ecology* 11(3):265-282.
—— (1986) Holism and Individualism in Ecological Anthropology. *Reviews in Anthropology* 13(4):295-313.
Venema, N. (2001) An Overview of the Netherlands Open Sea and Coastal Fisheries. *Fisheries Centre Research Reports* 9(3):141-161.
Verhey, E. and G. van Westerloo (1977) Het geheim van Urk. *Vrij Nederland Bijvoegsel* 45 (12 November 1977):2-39.
Verrips, J. (1989) Dutch Bargees and the Problem of Solidarity. In: J. Boissevain and J. Verrips (eds.), *Dutch Dilemmas*. Assen: Van Gorcum, pp. 106-121.
Verslag zeevisserijen (1858-1911) *Verslag omtrent den staat der zeevisserijen*. Den Haag: Van Weelden en Mingelen.
Vervaele, J.A.E., D. Ruimschotel and R.J.G.M. Widdershoven (1990) *Rechtshandhaving bij visquotering*. Utrecht: Nederlands Instituut voor Sociaal en Economisch Recht.
Vestergaard, E. (1996) Voices From the Fishermen. *MARE Working Paper* No. 6. Aarhus, Aarhus University.
Vestergaard, T. (1990) The Fishermen and the Nation. *Maritime Anthropological Studies* 3(2):14-34.
—— (1992) Migration and the Occupational Identity of Danish Fishermen. In: L.R. Fischer et al. (eds.), *The North Sea*. Stavanger: Stavanger Maritime Museum, pp. 161-180.
—— (1996) Social Adaptations to a Fluctuating Resource. In: K. Crean and D. Symes (eds.), *Fisheries Management in Crisis*. Oxford: Blackwell Science, pp. 87-91.
Visser, M.A.J. (1946-47) De sociale positie van de vissersbevolking. *Het Gemeenebest* 7:311-316.
—— (1950) *Enkhuizen*. Enkhuizen: Vereniging 'Oud Enkhuizen'.
Vissersbond (1991) *Help... De visserij verzuipt!* Emmeloord: Nederlandse Vissersbond.
—— (1994) *De Nederlandse Vissersbond 1934-1994*. Emmeloord: Nederlandse Vissersbond.
Vlist, R. van der (1970) *Verschillen in groepsprestaties in de Nederlandse zeevisserij*. Groningen: Wolters Noordhoff.
Wade, R. (1987) The Management of Common Property Resources. *Cambridge Journal of Economics* 11(2):95-106.
Weber, M. (1969 [1920]) *Die protestantische Ethik und der Geist des Kapitalismus*. Hrsg. J. Winckelmann. München: Siebenstern.
Wiber, M.G. (2000) Fishing Rights as an Example of the Economic Rhetoric of Privatization. *Canadian Review of Sociology and Anthropology* 37(7):267-288.
Wilson, J.A. (1990) Fishing for Knowledge. *Land Economics* 66(1):12-29.

Wolf, E. (1982) *Europe and the People Without History*. Berkeley: University of California Press.
Zulaika, J. (1981) *Terranova*. St. John's ISER.

Index